The Language of Plants

The Language of Plants

SCIENCE, PHILOSOPHY, LITERATURE

*Monica Gagliano, John C. Ryan,
and Patrícia Vieira, Editors*

 University of Minnesota Press
Minneapolis
London

An earlier version of chapter 4 was published as "Breaking the Silence—Language and the Making of Meaning in Plants," *Ecopsychology* 7, no. 3 (September 2015): 145–52. DOI: 10.1089/eco.2015.0023. An earlier version of chapter 11 was published as "Phytographia: Literature as Plant Writing," *Environmental Philosophy* 12, no. 2 (2015): 205–20.

Poetry in chapter 14 was previously published as Elisabeth Bletsoe, "Stinking Iris" and "The Leafy Speaker," in *Pharmacopoeia and Early Selected Works* (Exeter: Shearsman Books, 2010). Copyright Elisabeth Bletsoe, 1999, 2010. Reprinted with permission.

Poetry in chapter 14 was previously published as "The White Iris," "The Jacob's Ladder," "The White Rose," "Matins," and "Scilla" in Louise Glück, *The Wild Iris*. Published in the United States by Harper Collins Publishers, 1992, and in the United Kingdom by Carcanet Press Ltd., 1996. Copyright 1992 by Louise Glück. Reprinted by permission of Harper Collins Publishers and Carcanet Press Ltd.

Published by the University of Minnesota Press
111 Third Avenue South, Suite 290
Minneapolis, MN 55401-2520
http://www.upress.umn.edu

Printed in the United States of America on acid-free paper

The University of Minnesota is an equal-opportunity educator and employer.

24 23   10 9 8 7 6 5

Library of Congress Cataloging-in-Publication Data
Names: Gagliano, Monica, 1976– editor. | Ryan, John (John Charles) (Poet),
    editor. | Vieira, Patrícia I., 1977– editor.
Title: The language of plants : science, philosophy, literature / Monica
    Gagliano, John C. Ryan, and Patrícia Vieira, editors.
Description: Minneapolis : University of Minnesota Press, 2017. |
    Includes bibliographical references and index.
Identifiers: LCCN 2016037101 | ISBN 978-1-5179-0184-4 (hc) |
    ISBN 978-1-5179-0185-1 (pb)
Subjects: LCSH: Plant cellular signal transduction. | Plant cell interaction.
    | Plant ecophysiology. | Chemical ecology.
Classification: LCC QK725 .L275 2017 | DDC 571.7/42—dc23
LC record available at https://lccn.loc.gov/2016037101

Contents

Part III. Literature

Introduction

Monica Gagliano, John C. Ryan, and Patrícia Vieira

PLANTS AND US

Plants are perhaps the most fundamental form of life, providing suste-
nance, and thus enabling the existence of all animals, including us
humans. Their evolutionary transition from Paleozoic aquatic beginnings to a vegetative life out of water is undoubtedly one of the farthest-
reaching events in the history of the earth. It was the silent yet relentless
colonization of terrestrial environments by the earliest land plants that
transformed the global landscape and radically altered the geochemical
cycles of the planet. This resulted in lowered concentrations of atmospheric carbon dioxide and thus set the scene for the emergence of terrestrial animals about 350 million years ago. Over the subsequent circa
200 million years, as Mesozoic forests of ferns, conifers, and cycads flourished and flowering plants made their appearance, so the first reptiles,
and then mammals and birds emerged. The first *Homo* species did not
arise until about 2.8 million years ago and our modern human ances-
tors were nowhere to be seen until approximately 0.05 million years ago.
If millions of years could be measured in meters, the history of plants
would equate to a 500-meter-long walk, while ours would be no more
than a few centimeters—a notably brief exhalation in the history of our
planet, yet one our species has branded with a sense of utmost preemi-
nence in search for its own meaning.

While humans have often pondered their place in the midst of the
myriad life forms that inhabit the earth, plants have not always been at
the forefront of their concerns. Science has traditionally privileged the
study of animals, in an attempt to pinpoint what distinguishes us from
our closest relatives in the evolutionary scale. Although both Darwinian and post-Darwinian scientific views emphasized the continuities that
bind all life together with its rich diversity—in Darwin's own theory of
descent with modification, according to which the process of natural

selection explains life in terms of both unity by descent and diversity by modification[1]—, the study of animals has remained the central scientific concern. A quick survey of the scientific literature of the last five years reveals that, on average, only one paper is published on plants for every two published on animals.[2] This tendency is evident among the leading multidisciplinary scientific journals, such as *Nature* and *Science,* and it is also found among the top journals that cover the specific area of biology, where only 35 percent of papers are published on plants. As in other cases of "taxonomic chauvinism," the level of personal interest in different kinds of organisms, and particularly a preference for vertebrates such as mammals and birds, seems to be the source of the partiality among scientists.[3] Theorists have referred to the emphasis on animals as "zoocentrism" and have argued that it reinforces the privileging of animals as sentient, intelligent, and mobile, at the same time as it contributes to the marginalization of plants as relatively passive life forms.[4] No matter what we call it, the bias is real and has resulted in an impoverished appreciation of plants and their role in the natural world.[5]

Science has recently noticed that our "plant blindness"[6] and concomitant predilection for animals over plants might be ancestrally derived, as our visual attention toward human and nonhuman animals is inherently higher than when we see plants.[7] This extremely prevalent condition, whereby we struggle even to notice plants as being alive, clearly remains a significant source of negligence toward the botanical world and, more generally, the environment. While our inattention to plants might be due to a perceptual bias embedded in our physiology, it is also true that botany has been a part of medicine for most of ancient history and throughout the medieval era. During this time, the major focus of interest was not the *being* of plants but rather their *usefulness* to humans as food and medicine. A more integrated ecological thinking of plants was only to emerge about a century ago, when, in 1895, Eugenius Warming published *Plantesamfund,* the first book on plant ecology, where he introduced new ideas on plant communities, their adaptations, and environmental influences.[8] The history of the relationship between humans and plants suggests, then, that the causal root of plant blindness goes beyond physiological underpinnings and harks back, in Matthew Hall's words, to a deeply ingrained "cultural-philosophical attitude."[9]

Similarly to science, philosophy has historically focused on the study of human beings, foregrounding what distinguishes us from our closest evolutionary relatives. The human has traditionally been understood, as Aristotle famously put it, as a special, "political" animal gifted with *logos*—reason, speech, language.[10] The living beings deemed to be more akin to humanity were regarded as inherently superior to the more dissimilar ones. Such gradation was codified in the so-called *scala naturae*, literally, the "ladder of nature," also known as the "great chain of being." This view of the world that coalesced during Greek Neoplatonism and remained highly influential throughout the medieval and early modern periods in Europe posited a hierarchical universe where inanimate beings occupied the lowest level of the scale, followed by plants, animals, humans, angels, and, finally, God, who stood at the top of the pyramidal construction as the image of perfection. Even though continuity was one of the key principles of the chain, with all beings related to one another as just another gradation of divine power, those entities placed further down the scale were thought to be inferior to the ones higher up. Plants, ranking just above inanimate things, were therefore considered to be less perfect than animals and humans, a lower form of life that could not compete in complexity with the higher ones.

The notion that plants are imperfect and ontologically lacking the characteristics that render animals superior, including movement, intentionality, or the ability to communicate, was to remain a philosophical tenet long after the Renaissance. If some vegetal beings, and particularly their flowers, have long been recognized as a source of aesthetic pleasure, most plants continued to be relegated to the margins of philosophy as less worthy of attention than animals. For instance, in his *Philosophy of Nature* (1842), G. W. F. Hegel, who decisively influenced contemporary thought, argued that plants are but a step to be dialectically superseded by animals in the fulfillment of Spirit in nature. According to Hegel, plants are unable to preserve "inwardly the unity of the self," that is, to return to themselves, and, thus, do not evince the subjectivity and inwardness that animals display *in nuce* and that is then fully developed in human self-consciousness.[11]

The favoring of animals, and especially mammals, over plants continues in contemporary philosophy. While works dissecting the complex relation between humans and the more intelligent animals abound, from the groundbreaking *Animal Liberation* (1976) by Peter Singer, to the more recent *When Species Meet* (2008) by Donna Haraway or Cynthia Willett's *Interspecies Ethics* (2014),[12] philosophical approaches to vegetal life are much less common.[13] This situation is reflected, in turn, in ethical and political attitudes. Animal protection movements have existed in the West at least since the nineteenth century, and some form of animal rights is enshrined in the legal system of numerous countries. Conversely, many people would still regard the notion of plant rights as whimsical.[14] The scientific, philosophical, and socio-political prejudices against plants call for new approaches to challenge implicit attitudes and automatic stereotypes,[15] thus inviting fresh perspectives on plant, as well as human, modes of existence.

When compared to science and philosophy, literature has been more attentive to the plant world, although not without its appropriations of, and conventions and biases toward, the vegetal. Generally speaking, works of poetry and prose in the Western tradition tend to represent plants as part of the landscape or as the backdrop for human and, on occasion, animal dramas, as evident in some fairy tales and fables.[16] For many writers, plants become, at most, the correlatives of human emotions, eliciting feelings of pleasure and displeasure, triggering memories, and reflecting human states of mind, including inner turmoil or spiritual meditation. The mysterious intricacies of vegetal lives, obscure to the human subject, are largely cast aside and relegated to narrative blind spots in a wide array of literary works.[17]

An example of this literary treatment of plants can be found in the work of American poet and painter Washington Allston, who, before T. S. Eliot popularized the term *objective correlative*, revealingly employed a vegetal metaphor to posit this very idea. In his posthumous *Lectures on Art, and Poems* (1850), Allston regards air, earth, heat, and water as the elemental constituents of the cabbage and cauliflower, two representatives of "the lower forms of organic life." The "peculiar form" of the cabbage and cauliflower (those qualities that make each recognizable as *not* the other and as different from animals) originates as a concept in the human mind and then materializes in the world in correspondence to the elements. Through this vegetally inspired *logos*, Allston then goes on to conceptualize the objective correlative in terms that seem to reduce the phenomenal world to a series of signifiers of human pathos: "the presence of some outward object [the plant], predetermined to correspond to the preexisting idea of its living power, is essential to the evolution of its proper end,—the pleasurable emotion [of the human]."[18] Of course, such a reductive, utilitarian, and anthropocentric view of plants is not unique to American Romantics, such as Allston. The predominant literary discourse on plants renders their lives symbolic or figurative, as organic referents for animal meaning. Even Gilles Deleuze and Félix Guattari's philosophical championing of the rhizome, which has been highly influential in literary and cultural studies,[19] appropriates the material and relational existence of *actual* rhizomes, reifying them as ideational instruments of a supposedly interconnected form of human being. In contrast, a phytocentric—or, preferably, biocentric—form of literary criticism would seriously regard the lives of plants in relation to humankind in terms that would look beyond the purely symbolic or "correlative" dimensions of the vegetal.

ENCOUNTERS WITH PLANTS

The prevalent scientific, philosophical, and literary biases toward vegetal life that have resulted in a particularly zoocentric mode of thinking have not gone undisputed. At a time when they were still mainly viewed as passive automatons, Charles Darwin's grandfather, the naturalist Erasmus Darwin, for example, argued that plants are animate, living beings and attributed to them sensation, movement, and a certain degree of mental activity, emphasizing the continuity between humankind and plant existence.[20] Charles Darwin himself was a distinguished botanical experimentalist, who departed from the classical view of plants as insensate to reveal their sensitivity and power of movement, akin to that observed in some animals.[21] While his name is famously associated with the animals (and particularly the finches) of the Galapagos Islands, and he is better known for his studies of animal and human evolution and behavior, Darwin's botanical work played a pivotal role in the development of his ideas on evolution, natural selection, and the interconnectedness of life.[22] More recently, the understanding of plants as active and communicative organisms exhibiting as wide a range of behaviors as observed in animals has reemerged,[23] giving rise to new approaches to plant biology research and even new research fields, such as plant neuro-biology.[24] The proponents of this neurobiological perspective are plant physiologists who argue that plants can no longer be portrayed as passive organisms merely subjected to environmental circumstances. Instead, they view plants as information-processing organisms with complex communication strategies.[25]

Still, attempts to offer a biocentric and unified view of life have, more often than not, highlighted the implicit nature of our zoocentric attitude and the inadequacy of language in articulating a prejudice-free human relationship to plant life. For instance, by adopting the metaphors of Linnaeus's taxonomic schemata, particularly its sexual tropes, and rendering flora with animal-like qualities in his poetry, Erasmus Darwin suggested that plants are more akin to humans than one might ordinarily assume. Thus, he attempted to make the science of botany more interesting and accessible to his readers by zoologizing plants.[26] Similarly, in the monograph *The Power of Movement in Plants,* Charles Darwin, together with his son Francis, used a neurological metaphor to acknowledge the sensitivity of plant roots when he proposed that the tip of roots acts like the brain of some animals,[27] even though plants possess neither actual brains nor nerves. And, recently, the idea of the root tip of plants functioning as a "brain-like" organ, together with the so-called "root-brain hypothesis"[28] and a general "phyto-cerebrated" view of plants, has been revived in plant neurobiology.[29] Irrespective of whether this neurological metaphor is correct[30] or, more generally, the modern application of neuroscience terminology and concepts to plants is appropriate,[31] the discussion on plant behavior has again reverted to correspondences to the animal world. In other words, in our use of language, plants are still expected to exhibit animal-*like* qualities in order to be acknowledged as sensitive living organisms, rather than being appreciated in their own right and on their own terms.

As shown by the scientific literature survey mentioned previously, this implicit zoocentric attitude is strong in biological research, which is generally focused on understanding the "structure" of things in terms of immediate mechanical, physical, physiological, and biochemical/molecular factors (i.e., investigating the *"how* does it work?" question, or proximate causation).[32] Interestingly, however, such zoocentric bias disappears in research that moves away from the mechanistic and proximate description of organisms and is interested, instead, in understanding their ecological lives.[33] This might simply be because ecologically and environmentally focused research aims to understand the "function" of organisms within their context—the interactions of individuals with their surrounding physical environment and the wider community of organisms living within it. As a result, the scope of these research areas is naturally broader, more inclusive, and process oriented (i.e., investigating the "*why* it occurs?" question, or ultimate causation). Although we believe that both proximate and ultimate questions are interconnected and lead to a better understanding of living systems, it might be that a more ecological approach to plants can assist us in integrating a biocentric view of living forms, changing how we think about plants and, possibly, helping us understand how plants "think" about themselves.

The reflections on the vegetal world that have thrived on the margins of Western science seem to have gained new momentum in recent decades, attempting to understand plants in their own terms and offering a more integrated, ecological approach to plant life. Following decades of single-minded focus on molecular biology to understand mechanisms of plant growth under controlled laboratory conditions, combined with intense efforts toward plant genetic engineering,[34] research in plant science has recently enjoyed a renaissance that has involved a real celebration of "plantness." It has revived and built upon the original observations of plant behavior by the great botanists of the eighteenth, nineteenth, and twentieth centuries, offering a new synthesis of plant sensitivities in works such as Richard Karban's *Plant Sensing and Communication,* Anthony Trewavas's *Plant Behaviour and Intelligence,* Dov Koller's *The Restless Plant,* and František Baluška's *Plant-Environment Interactions: From Sensory Plant Biology to Active Plant Behavior.*[35] These recent studies have shifted the paradigm of plant science from a mechanicist approach to vegetable life, reduced to its constituent parts, to a perspective attributing greater agency to plants in dynamic relation to their environments. As such, these studies help to disassemble the rigid, long-standing hierarchies separating botanical and zoological forms of life.

In philosophy, as in science, several authors have attributed to plants a key role in their systems of thought. Neoplatonic philosopher Plotinus, for instance, believed that everything in existence is part of the One, which is shaped like a tree. All beings issue forth from this universal tree as its branches, leaves, fruits, and so on, or, to put it differently, every life form is a particular, materialized thought of the One. Plants represent, in this constellation, the growing thinking of the One, while animals are its sensory and humans its rational thinking. Still, even though Plotinus foregrounded plants, which gave the One its shape, he never abandoned a hierarchical view of life forms, whereby vegetal beings, more entangled with matter than animals, remained at the bottom of the ladder. His metaphorical tree of the One, which provided the structure for his philosophical edifice, stood in sharp contrast to the place of actual plants in his thought.

Centuries after Plotinus, during the European Enlightenment, Jean-Jacques Rousseau would turn to plants and their study by botany as a welcome distraction from his political writings. For Rousseau, what began as a pastime soon turned into a "passion," and he confessed to being "obsessed with this [botany's] madness."[36] Decrying those who adopt an instrumental approach to vegetal beings, seeing in them only a source of nourishment or medicine, he regarded his botanizing as a way to purify his soul from the corruption of human society: "Brightly colored flowers, the varied flora of the meadows, cool shade, streams, woods, and greenery, come and purify my imagination."[37] The charm and "effortless pleasure" afforded by botany depended precisely on the fact that it did not dismember, dissect, or chemically analyze plants but, rather, observed them in their natural environment, an activity that did not exclude aesthetic contemplation.[38]

In tandem with recent developments in plant science, contemporary philosophical works have tried to recover the reflections on plants by authors such as Plotinus and Rousseau, as well as those originating in non-Western systems of thought, and has taken these into new directions. In his *Plants as Persons: A Philosophical Botany* (2011), for instance, Matthew Hall goes back to the Hindu, Jain, Buddhist, and animist traditions in an effort to extend the category of personhood to plants.[39] The work of Michael Marder, in turn, reveals how the deconstruction of Western metaphysics can be carried out with reference to the processes and phenomena of vegetal life. On this basis, it reconsiders the concept of nature, as well as ethical and political relations, that appear in a different light when contemplated from the perspectives of plant ontology and epistemology. Learning from plants entails thinking outside the totalizing categories of Western metaphysics and, therefore, paves the way for a more open-ended, less instrumental, approach not only to social relations but also to the environment. Following in the footsteps of Marder's reflections, Jeffrey Nealon's *Plant Theory* (2016) tries to redress the "the strange and consistent elision of plants within the voluminous work on life within contemporary theory and philosophy," to which the studies of Hall and Marder are exceptions.[40] Interested in issues of bio-power and vegetable life, Nealon recognizes the growing body of scientific research into plant behavior and touches upon some of the topics developed in this volume, noting, for instance, that plants exhibit "a certain kind of language—they share information concerning soil conditions and the presence of predators."[41]

Alongside philosophers, there were also botanically minded writers who notably engaged with the lives of plants in direct, sensory, and embodied—rather than purely metaphorical and symbolic—ways. English poet John Clare resisted "Linnaeus's dark system," developing a personal, emotive relationship to the vegetal world based in folk knowledge, rapport with individual plants, and awareness of the environmental specificities of places.[42] Clare was exposed to genus–species Linnaean classification, based sexually and visually on the number of flower parts, through Scottish horticulturalist James Lee's *An Introduction to Botany* (1760). However, the Enlightenment's project of classifying life into hierarchies had no charm for Clare, who believed the system stifled a strong affective connection to local wild plant personae.[43]

Paralleling Clare's attraction to plants, the works of Henry David Thoreau express an ongoing fascination for scientific thinking interwoven with skepticism toward many of its epistemological tenets, at a time of rapid industrialization and scientific development in the United States. His belief that "the mystery of the life of plants is kindred with that of our own lives" (stated in his journal in 1852)[44] inspired his development of radically interdisciplinary approaches to the vegetal world that would lead to his identification of forest ecology principles *avant la lettre* in *Faith in a Seed* (published posthumously in 1993).[45] Already in the twentieth century, through a mode of conversation involving the direct address of plants, Australian poet Judith Wright drew attention to the nameless "Swamp Plant" and the loss of the poetry of flora—"Only science, then, has noticed you / not poetry / . . . / no words but dog-Latin / have tagged you."[46]

The writings of Clare, Thoreau, and Wright exemplify a kind of botanically curious and inflected literary discourse that is attuned to the lives of plants—one that resists the figuration of the vegetal for human constructs. Their works also provide the foundation for resisting formalism in literary criticism of plants, in which a botanical text would be analyzed as a relatively self-sufficient artifact in the absence of its broader cultural, ecological, and ethical contexts or implications.[47] We propose that a textual work with plant-based themes—let us generally call this *vegetal textuality*—always comes into existence in dynamic relation to actual, living flora. Therefore, literary discourse, even at its most metaphorical, is necessarily beholden to the material being of the vegetal.

Investigations on plant behavior in botany, novel philosophical approaches to plants, and ecocriticism's rise to prominence in the fields of literature and cultural studies have prompted a reconceptualization of the relation between plants and humans. Science has reopened the door to a cognitive approach to plants,[48] lifting tacit self-imposed restrictions not only on questions of plant behavior but also on matters of embodiment, agency, and consciousness, which together represent an exciting new frontier of exploration. Research on the philosophical import of plants has revealed the indebtedness not only of aesthetics but also of ontology, epistemology, ethics, and even political philosophy to conceptions of vegetal being. Often relegated to the margins of thought, plants have nevertheless been the foundation of the philosophical notion of the human and of life itself. Recent literary studies of plants are beginning to focus on the representation of the intelligence, behavior, and subjectivity of the vegetal world in works of poetry and prose, fiction, and nonfiction.[49] Importantly, these new texts release the vegetal from a background position in literary discourse and underscore the vital role of plant narration, voice, presence, and sensoriality.

Critical plant studies (also known as human–plant studies or, simply, plant studies) has emerged as a broad framework for reevaluating plants, their representations, and human–plant interactions, much as human–animal studies developed in the 1990s in response to the animal ethics debates coalescing during the two previous decades.[50] Principles of ethics, aesthetics, poetics, agency, cognition, intentionality, communication, and language inform this new field of study, which encompasses the work of scientists, philosophers, linguists, poets, literary scholars, geographers, cultural theorists, and others. This is the broad context in which we situate *The Language of Plants,* one that relies on the trandisciplinary interdigitation of research areas, in the spirit of biologist Edward O. Wilson's notion (adopted from nineteenth-century philosopher William Whewell) of consilience as the "jumping together" of knowledge.[51]

Speaking of Plant Language

Broadly understood as communication, expression, and articulation, language has been a key concern in the endeavor to reconsider plant life, since many of our notions about the vegetal world hinge on what kinds of language(s) we ascribe to it. In light of the diverse contents of this book, we posit two modes of the language of plants: *extrinsic* and *intrinsic* language. *Extrinsic language* refers to the ways in which scientists, theorists, writers, artists, and others express what is "peculiar" (to echo Allston, referenced earlier in this introduction) about plant being. This includes the scientific language about plants (especially the taxonomic terminology that shapes how we speak and think about the botanical world), the philosophical language deployed to articulate the particularities of plant ontology, and the representation of vegetality in literary works. The various ways in which we refer to plants (chapters 7 and 8), the "language of flowers" (chapter 10), plants as narrators (chapter 13), the use of direct address to plants (as having subjectivity, that is, as a "you"; chapter 14), and the diction that expresses to a reader the complex botanical world all may be considered extrinsic modes of plant language. Humans devise and levy this language, largely independent of the vegetal's own powers, in response to plants. If used without care or ethical concern, extrinsic language is imposed upon plants as a means of dissecting, ordering, or consigning them to the background. Alternatively, in a positive sense, extrinsic language modes provide us with insight into the mysteries of plant lives and inspire societies to figure out what a distinctly vegetal form of thinking, being, and acting would be like.[52]

In contrast, the *intrinsic* language of plants encompasses the modes of communication and articulation used by vegetal species to negotiate ecologically with their biotic and abiotic environments. Some examples include the language of biochemistry—plant hormones, electrical signaling, pressure cues, and so on (chapter 1), as well as the multisensorial expressions of plants—their visual articulations (chapter 3), their olfactory bouquets (chapter 2), or their aural enunciations, revealed in the emergent field of plant bioacoustics.[53] Intrinsic language also includes the ecological interactions between plants and animals, soil microorganisms, and the environment, where "language," inclusively conceived, mediates these exchanges (chapter 4). Hence, we view language not as the mechanical result of an individuated living subject (plant or otherwise) but as an ecology produced by organisms in an interdependent and multispecies interrelation. Furthermore, *extrinsic* and *intrinsic* languages of plants are not watertight concepts but often interact, influence, and dialogue with one another (chapter 5). Rather than demarcating an absolute division, these two notions signal the privileging of a given perspective: that of humans, in the case of *extrinsic,* and that of plants, in *intrinsic* language. While extrinsic language strives to describe flora's existence, which includes its articulation in speech, the intrinsic language of plants that humans are able to grasp remains beholden to our hermeneutic efforts (chapters 6 and 9). Still, this impossibility to reach the plant per se should not spell out our capitulation in the effort to understand it. This book is based upon the premise that we should continue trying to listen to what plants tell us in their own modes of expression.

The concept of the language of plants is neither a flight of fancy nor a figure of speech, symbol, metaphor, or allegory. Its precursors are theories that decouple language from a linguistic or verbal root and instead conceptualize it as an inherent attribute of all living and nonliving phenomena. An important precursor is the medieval notion, codified in Jakob Böhme's *The Signature of All Things* (1621), that all entities bear a mark of God's design. These "signatures" form a nonverbal language to be interpreted by human beings. In the essay "On Language as Such and on the Language of Man," written in 1916, German philosopher and cultural critic Walter Benjamin takes Böhme's idea further by positing a language of things, human language being just a more complex example of a generalized phenomenon. Benjamin does not employ the term *language* metaphorically or anthropomorphically. He suggests that everything makes use of expression, which constitutes each being's particular language.[54] If the language of plants is nonverbal, then, we must turn to their specific forms of articulation to gain even the most rudimentary glimpse of their modes of being as distinct from our own.

The field of biosemiotics has contributed extensively to an inclusive conception of language that transcends its rigid alignment with verbal utterance. Particularly drawing on the work of American philosopher Charles Sanders Peirce, German biologist Jakob von Uexküll, and Danish biologist Jesper Hoffmeyer, contemporary biosemiotics generally conceives of language as an evolutionary response that humans share, albeit in different manifestations, with other forms of life.[55] Peirce famously claimed that the world is "perfused with signs, if it is not composed exclusively of signs."[56] Following in Peirce's footsteps, the biosemioticians of today have likewise argued that language is "pervasive in all life."[57] As semiosis (a system of meaningful signs), language is more than the audible communication carried out by humans; it encompasses the complexities of intersubjective and interspecies dialogue, involving nature (including plants) and humanity.

For ecocritical biosemiotics, language is about "world making," or *poiesis*—how beings bring forth their lifeworlds (or *Umwelten*) in dynamic conjunction with the lifeworlds of other entities. Within biosemiotics, some scholars attend to vegetal semiotics, recognizing plants as autonomous beings with physiological and semiotic processes that differ from those of humans and other animals.[58] In 1981, Martin Krampen inaugurated the field of phytosemiotics in an article exploring the importance of sign processes within and between plants.[59] Krampen and other contemporary phytosemiotic theorists build upon the work of Uexküll, who in 1922 noted the "comfortable calm" of plants (an allusion to their perceived immobility and particular temporality), behind which there is a bustle of poietic activity hidden from unaided human perception. Uexküll was fascinated with how "an uninterrupted stream of liquids enters by the roots, rising along the stem and branches out in all directions to the leaves where it evaporates again in a well-controlled fashion."[60] Thus, phytosemiotics considers the languages of plants as expressions of their physiologies bearing semiotic resonance.

The question of the language of beings (not only human) has also been central to ecocriticism, defined as "the study of the relationship between literature and the physical environment, usually considered from out of the current global environmental crisis and its revisionist challenge to given modes of thought and practice."[61] Along with its concerted exploration of the environmental dimensions of human language, one of ecocriticism's aims is to consider language beyond us and to reflect upon the polyvocality of the world. In an attempt to transgress a conception of language based upon human subjectivity, ecocritical theory foregrounds sensuous, bodily, and material forms of expression that most animate and inanimate beings possess.[62] According to ecocritic Leonard M. Scigaj, a "living language" recognizes "our shared biocentric relationships with all orders of sentient beings."[63] The inclusive concept of language forwarded by some ecocritics and espoused in this volume embraces the diverse expressions *about* and *of* plants. It implies our attunement to a speaking without words, a listening without hearing, that humankind must learn to cultivate for the sake of the future we wish to share with each other and with other beings.

The language of plants has implications for ethics, politics, and sustainability—themes particularly explored in Part II of this collection. Rather than a mechanism of hierarchical separation, as it has been historically constructed, language, cast in a fresh light, allows new narratives to emerge through the complex interdependencies between plants and humans. We should acknowledge that the link between ecological ethics and language is not new. For instance, in 1977 the philosopher George Sessions, drawing from Baruch Spinoza and Robinson Jeffers, asserted the importance of language to environmental ethics in arguing against metaphysical dualism (the ontological division between nature and humanity) and for metaphysical monism (the essential interconnectedness between organisms).[64] More recently, scholars have begun to examine the relationship between environmental sustainability and human language, especially the impact of metaphors, tropes, and other figures of speech on our everyday practices.[65] Still, these studies lack a recognition of what we described previously in this introduction as *intrinsic* language—the modes of expressiveness proper to plants—and how this influences the *extrinsic* language we deploy to represent vegetal beings (or the natural world more generally) in cultural productions.

In a scientific context, language imparts a greater sense of unity to the kingdoms or, conversely, perpetuates a mode of thinking that renders plants as the mere objects of human use and manipulation. Similarly, in terms of philosophy and literature, language fosters greater cohesion between ideas, texts (broadly defined), and living beings or, alternatively, becomes a means of projecting our imagination, affect, and ratiocination onto plants, thus discounting or marginalizing their inherent capacities. The essays collected in this volume commence a conversation about the convergences and divergences between plant language and the human language of plants, as well as about the implications of these exchanges for our attitudes toward vegetal life.

SCIENTIFIC, PHILOSOPHICAL, AND LITERARY LANGUAGES OF PLANTS

The chapters in the book highlight that more nuanced conceptions of the language of plants can feed into one another and that this interplay can further contribute to our evolving understanding of vegetal life. For this reason, the volume strives not only to describe the various modes of plant communication and expression but also to question what it means to talk about the language of plants. Is *language* the right word to use in the context of the vegetal world? Should we distinguish between the language of plants and other forms of human and animal expression? Who/what uses languages in the case of plant life? These questions, in turn, point toward the ethical and political dimensions of the topic. In which ways does language underpin the moral consideration of the plant world? Does language foster concern for flora and resistance to the exploitation of plants as the inert materials of human consumption (for example, as sources of food, fiber, fuel, and medicine)? Will reflections on the language of plants lead us to think about them anew?

This volume offers three angles for approaching the cluster of questions arising from a reflection on the language of plants: science, philosophy, and literature. A few words are in order to explain this tripartite division that retraces a late modern, hierarchical understanding of these fields of study, according to which science is the foundation of knowledge, philosophy provides a reflection upon and justification for the operations of science, and literature, as well as the other arts, embellishes the other, more serious human practices. Needless to say, it was not our intention to subscribe to and reinforce stereotypes about disciplinary roles and boundaries. In fact, it would have been perfectly conceivable to have started the collection with the literature section, followed by philosophy and then science, a structure inspired by a prescientific worldview, in which literature often encompassed all forms of knowledge. Similarly, we could have begun with philosophy, followed by either science or literature, adopting an early modern perspective, which regarded philosophy as the basis for all other forms of inquiry. The current structure pays tribute to recent scientific developments on plants that have challenged long-standing beliefs about flora. Still, the disciplinary division espoused here should be understood dynamically, rather than hierarchically. Readers can start with any section and proceed to the other two at their will, or meander through various chapters, moving between the three parts of the volume, depending on their interests. We are fully aware that science, philosophy, and literature each has its own conventions and biases, reflected even in details such as the differences in citation styles in Part I ("Science") and Parts II and III ("Philosophy" and "Literature") of the book. It is our hope that, in dialogue, these disciplinary limitations might be overcome and that the transdisciplinary exchange set in motion in *The Language of Plants* will challenge the underlying assumptions, methodologies, and frameworks of these fields of study, so as to contribute to a better appreciation of language and botanical life.

In Part I, Richard Karban's "The Language of Plant Communication (and How It Compares to Animal Communication)" (chapter 1) takes the reader on an audacious journey into the vegetal world, where plants pry on their neighbors to acquire information about future risks of competition and potential attacks by pathogens and herbivores; where plants make business transactions with those microorganisms that allow them to forage more efficiently and with the animals that assist their sex lives and transport their progeny to places where they are likely to thrive. By exploring the sophisticated ways in which plants sense and respond to environmental cues, Karban shows that, in their interactions with their environment, plants exhibit many of the characteristics, which, according to linguists, define language, even if our understanding of the precise lexicon they have at their command is still limited.

Focusing their discussion on plant chemistry as one of the primary languages through which plants interact with their surroundings and communicate with friends, foes, and among themselves, Robert Raguso and André Kessler further develop the linguistic analogy for plant chemical signaling in chapter 2, "Speaking in Chemical Tongues: Decoding the Language of Plant Volatiles." Raguso and Kessler explore the role plant volatiles play in the four main sender–receiver interactions that characterize the evolution of signaling in animals. They examine whether different kinds of interactions require different lexicons and how the information content of these volatile signals might be impacted by a plant's environment. The chapter takes a fresh look at the emerging patterns in the rapidly growing field of plant volatile communication, examining the idea that plants are active agents in their own lives and ensure their success through an ancient chemical language common across kingdoms.

In "Unraveling the 'Radiometric Signals' from Green Leaves" (chapter 3), Christian Nansen opens a window into the visual world of green leaves. Like other parts of plants, green leaves emit volatiles and visual signals that are, at least partially, associated with complex internal physiological processes taking place inside the leaf. Insects and other herbivores with sensitive vision in the right sections of the light spectrum can look at a green leaf and assess the quality of the plant as a food source or oviposition site. In his chapter, Nansen explores how humans may also "see" these signals by using advanced imaging technologies to gain knowledge about a plant's health, as well as stress condition, thus offering insights on plant language from a human perspective.

In the final chapter of Part I, Monica Gagliano's "Breaking the Silence: Green Mudras and the Faculty of Language in Plants" (chapter 4) calls into question the notion that language is a form of behavior that makes humans unique. She examines linguistic processes from an ecological perspective and, then, from a wider biological viewpoint, one that enables us to explore language as a meaning-making activity at the core of every form of life, including plants. By providing an overview of recent empirically grounded advances in our understanding of the language of nonhumans, and particularly plants, she proposes that the nonhuman world is not lacking language, as we might be led to believe. Ultimately, she invites the emergence of a new, truly interdisciplinary dialogue, where language and its power are refocused toward conceptualizing a more integrated perception of the world.

Part II of the book opens with Michael Marder's "To Hear Plants Speak" (chapter 5), where he treats the question of the language of plants as a problem of translation into the more or less familiar frameworks of human discourse. He discusses four possible modalities of such translation: (1) the symbolic "language of flowers"; (2) the "talking trees" that pervade various cultures, from the sacred grove of Dodona in ancient Greece, to the vegetalized beings in literature; (3) biochemical communication among plants, as well as between plants and animals, such as insects, quantified in plant sciences; and (4) the plants' participation in the language of things, with its spatial nexuses and articulations. He argues that, in order to hear plants speak, we must leave plenty of room for the untranslatable (and, hence, the unspeakable) in these practices of translation.

Luce Irigaray's "What the Vegetal World Says to Us" (chapter 6) continues to discuss the difficulties inherent in, and the ethics of, translating the language of plants into human forms of expression. Irigaray urges us to learn from plants and their modes of being in the world, including a focus on becoming, rather than being, a relativization of preassigned roles and a nonappropriative relation to one's surroundings. Expressing themselves through a language without words, plants inhabit a world where there is no separation between signifier and signified, between subject and object. The chapter concludes by advocating for a communion of beings that express their essence in different but equally valid forms of existence.

In "The Intelligence of Plants and the Problem of Language: A Wittgensteinian Approach" (chapter 7), Nancy Baker turns to the work of philosopher Ludwig Wittgenstein to assess whether mental concepts such as "intelligence," "consciousness," "knowledge," "learning," and so on, should be used to describe the behavior of plants. Wittgenstein shows not only that the application of our concepts is context dependent, but also that the criteria for their application change depending on the developmental level of the behavior in question. Baker brings these insights to bear on the debate about the appropriate language to be used when referring to plant activity and argues for the existence of a continuum between human, animal, and plant forms of behavior.

Karen Houle's "A Tree by Any Other Name: Language Use and Linguistic Responsibility" (chapter 8) also highlights the centrality of language in defining our relation to the vegetal world. She points out that there are many different meanings of *justice,* a word we often use without reflecting upon its various connotations. In everyday parlance, in institutional protocol, but also in scientific discourse, a single notion of justice dominates, *distributive justice,* which conceives of justice in terms of measurable and fungible states, goods or qualities, and amounts of possession or dispossession thereof. Moving away from this "conceptual monoculture," Houle turns to *procedural* and *linguistic* justice. Applying these other notions of what is "just" to our approaches to the environment leads to a more balanced relationship between human beings and plants.

The last chapter in Part II, Timothy Morton's "What Vegetables Are Saying about Themselves" (chapter 9), argues that plants are similar to algorithms, since algorithms are not knowledgeable about numbers and simply execute computations. In this sense, a trope is also an algorithm—a twist of language that emerges as meaning by simply following a recipe. For Morton, what is disturbing about rhetoric, algorithms, and plants is that they exhibit a zero degree of intelligence, or not; we cannot know in advance. Morton goes back to German philosophers Arthur Schopenhauer and Friedrich Nietzsche to bolster his argument that a plant is the zero degree of personhood, a situation that lies at the core of our difficult relationship with the vegetal world.

In Part III, Isabel Kranz's "The Language of Flowers in Popular Culture and Botany" (chapter 10) provides an account of the flower-based ideographic sign system that arose in France during the classical age— the so-called *langage des fleurs,* according to which flower combinations could encode secret messages. Parallel to the rise of the language of flowers, the emergence of scientific botany placed emphasis on floral signs as key indicators of the order of the botanical kingdom. In the systematization of plants and the structuring of scientific knowledge through flowers, an undercurrent emerged based on the notion that plants convey a surplus of meaning. In her chapter, Kranz examines the two versions of flower languages, arguing that they are interrelated. By linking the foundations of modern botany to a popular tradition, imbued with Orientalist notions and bourgeois ideas of love, the chapter illustrates how floral systems share similar attitudes toward language, order, and knowledge.

Patrícia Vieira's "*Phytographia*: Literature as Plant Writing" (chapter 11) develops the notion of plant writing, or *phytographia,* the roots of which go back to the early modern concept of *signatura rerum,* described, among others, by Jakob Böhme, as well as, more recently, to Walter Benjamin's idea of a "language of things" and to Jacques Derrida's notions of "arche-writing" and "the trace." According to Vieira, *phytographia* designates the encounter between the plants' inscription in the world and the traces of that imprint left in literary works, mediated by the artistic perspective of the author. The final section of the chapter turns to the so-called "jungle novels," set in the Amazonian rainforest, as one of the possible instantiations of *phytographia.*

Joni Adamson and Catriona Sandilands's "Insinuations: Thinking Plant Politics with *The Day of the Triffids*" (chapter 12) examines the fictional triffids of John Wyndham's 1951 novel. Previously treated by critics as allegorical or incidental to the narrative, the vegetality of the triffids embodies their capacity—and that more generally of plants—to "insinuate" themselves into human milieus. Adamson and Sandilands interpret the novel as a symptom of exploitative relations with the plant world and of the glossing over of the agential capabilities of flora, fictional or otherwise. From an examination of the mobility and agency of the triffids, their chapter turns to the real-life insinuations of the tumbleweed and the dog-strangling vine in the U.S. Southwest, where plant mobility instigates a particularly weed-averse biopolitics.

Applying the theory of material ecocriticism, Erin James's "What the Plant Says: Plant Narrators and the Ecosocial Imaginary" (chapter 13) develops the concepts of "plant narrator" and "the material language of plants" through a reading of four texts featuring talkative flora: Stephen Wright's *Meditations in Green,* Ursula Le Guin's story "Direction of the Road," Orhan Pamuk's novel *My Name Is Red,* and Aldo Leopold's *A Sand County Almanac.* Through her analysis, James considers how plants describe their experiences of the world and how authors narrate their observations of plants. She argues that the speaking, ventriloquizing plant is only one expression of the plant-as-narrator form. Heteroglossia and polyvocality link vegetal agency to the phenomenon of narration, in which plants have a say, albeit a nonverbal one. The material narration of plants—exemplified by the oak rings in Leopold's arboreal biography—is a mode of signification proper to the vegetal. James's discussion prompts us to reimagine how plants might tell their own spatially articulated stories, implying that listening also involves remaining sensorially open and in contact with plants.

Part III and the book conclude with John C. Ryan's "In the Key of Green? The Silent Voices of Plants in Poetry" (chapter 14). Ryan offers a critical perspective on the concept of voice in nature, asserting that, to date, ecocriticism has only superficially addressed the possibility of nature speaking. Through a framework merging communication theory and critical plant studies, Ryan proffers a model of plant voice that resists the ventriloquization of plants through technological mediation. The voice of plants (their auditory utterances, which bioacoustic research increasingly affirms) and the giving of voice to plants (by us, in art, literature, society, and culture) involve an interplay between human and vegetal voicing. Rather than dwelling on the prospect that plants vocalize their intentions and desires, however, the chapter develops the idea of nonverbal, ecological, and corporeal voice as the manifestation of vegetal presence and human recognition of it through our capacities for taste, smell, touch, and proprioception. The poetry of Louise Glück and Elisabeth Bletsoe, focused on vegetal subjects, provides a basis for thinking about voice as the spatial and material articulation of plants.

Through the complementary perspectives of science, philosophy, and literature, this volume highlights the potential of language itself, as well as of language in the vegetal world. As editors, we hope *The Language of Plants* elicits more questions than it resolves and thus stimulates further directions for multidisciplinary approaches to critical plant studies. The mundane observation that nonverbal beings cannot vocalize should not exclude them from the domain of language. Rather, it is the critical examination of the discourse of language—including its bias toward human expression and its presumption of verbal capacities—that can inspire us to see language, plants, and ourselves in a new light. The volume's focus on the relationship between human and vegetal languages foregrounds the diverse forms of expression of the botanical realm and deepens the appreciation of our interdependencies.

Notes

1. Charles Darwin, *On the Origin of Species* (London: John Murray, 1859). Specifically, see chapter 6 for an in-depth discussion of this theory.

2. This assertion is based on a Web of Science search of all papers published between 2010 and 2015 on the topic *plant* (895,430 records) and *animal* (1,514,084 records). For a closer examination of these results, the top three journals within four main categories were selected based on the ISI Journal Citation Reports. Specifically, *Nature, Science,* and *Proceedings of the National Academy of Sciences* were selected for the "multidisciplinary sciences" category (which includes resources of a very broad or general character in the sciences; it covers the spectrum of major scientific disciplines such as physics, chemistry, mathematics, and biology); *Biological Reviews, PLOS Biology,* and *BMC Biology* for the "biology" category (which includes resources having a broad or interdisciplinary approach to biology; in addition, it includes materials that cover a specific area of biology, such as theoretical biology, mathematical biology, thermal biology, cryobiology, and biological rhythm research); *Trends in Ecology and Evolution, Ecology Letters,* and *Annual Review of Ecology and Systematics* for the "ecology" category (which covers resources concerning many areas relating to the study of the interrelationship of organisms and their environments, including ecological economics, ecological engineering, ecotoxicology, ecological modeling, evolutionary ecology, biogeography, chemical ecology, marine ecology, wildlife research, microbial ecology, molecular ecology, and population ecology; this category also includes general ecology resources and ones devoted to particular ecological systems); and *Nature Climate Change, Annual Review of Environment and Resources,* and *Global Environmental Change* for the "environmental studies" category (which covers resources that are multidisciplinary in nature; these include environmental policy, regional science, planning and law, management of natural resources, energy policy, and environmental psychology). The average ratio for the number of papers published on animals to those published on plants varies within category: multidisciplinary sciences (2.3), biology (1.9), ecology (1.0), and environmental studies (0.9). There is a bias favoring animal studies within the multidisciplinary sciences and biology, while there is no bias in the specific area of ecology and environmental studies. Bibliographic data was extracted on July 16, 2015.

3. Xavier Bonnet, Richard Shine, and Olivier Lourdais, "Taxonomic Chauvinism," *Trends in Ecology and Evolution* 17, no. 1 (2002): 1–3.

4. For a discussion of zoocentrism, see Leena Vilkka, *The Intrinsic Value of Nature* (Amsterdam: Rodopi, 1997), chapter 4. On the relationship between

language and plants, see John Ryan, *Green Sense: The Aesthetics of Plants, Place, and Language* (Oxford: TrueHeart Press, 2012), 217–58.

5. See Francis Hallé, *In Praise of Plants* (Portland, Ore.: Timber Press, 2002), chapter 1.

6. James H. Wandersee and Elisabeth E. Schussler, "Preventing Plant Blindness," *American Biology Teacher* 61, no. 2 (1999): 82–86, doi:10.2307/4450624. According to Wandersee and Schussler, the term *plant blindness* refers to "the inability to see or notice plants in one's own environment, leading to the inability to recognize the importance of plants in the biosphere and in human affairs." Given the fundamental role plants play in maintaining life on earth, understanding how we can move beyond plant blindness and recognize plant conservation as one of humanity's most urgent issues is imperative for a sustainable world. Since the term was first introduced, research efforts toward addressing the social and educational biases that cause plant blindness and exploring learning experiences that can create shifts in perception away from plant blindness toward seeing the importance of plants have indeed increased.

7. Benjamin Balas and Jennifer Momsen, "Attention 'Blinks' Differently for Plants and Animals," *CBE: Life Sciences Education* 13, no. 3 (2014): 437–43.

8. Fourteen years later, Warming's book was made available to all English and American botanists in its expanded and revised English translation. Eugenius Warming and Martin Vahl, *Oecology of Plants: An Introduction to the Study of Plant-Communities* (Oxford: Clarendon Press, 1909). See also Henry C. Cowles, "Review: Ecology of Plants," *Botanical Gazette* 48, no. 2 (1909): 149–52.

9. Experimental support of plant blindness as a perceptual or cognitive phenomenon is currently available but limited. Recent critiques of plant blindness have pointed out that the concept takes zoocentric attitudes to be a human "default condition," thus normal and inevitable, and suggested that the challenge to a zoocentric focus is a cultural-philosophical one. See Matthew Hall, *Plants as Persons: A Philosophical Botany* (New York: State University of New York Press, 2011), 6.

10. Aristotle, *The Politics* (London: Heinemann; New York: G. P. Putnam's Sons, 1932), I, 1253a.

11. G. W. F. Hegel, *Philosophy of Nature* (Oxford: Oxford University Press, 2004), §350.

12. Peter Singer, *Animal Liberation: A New Ethics for Our Treatment of Animals* (London: Cape, 1976); Donna J. Haraway, *When Species Meet* (Minneapolis: University of Minnesota Press, 2008); Cynthia Willett, *Interspecies Ethics* (New York: Columbia University Press, 2014).

13. Some exceptions are referenced below.

14. For a discussion of plant rights, see Michael Marder, "Should Plants Have Rights?" *Philosopher's Magazine* 62 (2013): 56–60, doi:10.5840/tpm20136293. See also Alessandro Pelizzon and Monica Gagliano, "The Sentience of Plants: Animal Rights and Rights of Nature Intersecting?," *Australian Animal Protection Law Journal* 11 (2015): 5–13.

15. Implicit bias, automatic stereotypes, and prejudice were long assumed to be fixed behavioral responses, the influence of which was inescapable as it occurred without conscious control. Several studies have now demonstrated that these attitudes are malleable and responsive to a wide range of strategic, social, and contextual influences. Interestingly, the focus of attention and the amount of time spent learning about a person's unique attributes are strongly influential on the automatic operation of stereotypical attitudes. For some examples, see Irene Blair, "The Malleability of Automatic Stereotypes and Prejudice," *Personality and Social Psychology Review* 6, no. 3 (2002): 242–61; and Leslie Roos et al., "Can Singular Examples Change Implicit Attitudes in the Real-World?," *Frontiers in Psychology* 4 (2013): 594, doi:10.3389/fpsyg.2013.00594. On plants, see Jana Fančovičová and Pavol Prokop, "Plants Have a Chance: Outdoor Education Programmes Alter Students' Knowledge and Attitudes Towards Plants," *Environmental Education Research* 17, no. 4 (2011): 537–51; and Eva Nyberg and Dawn Sanders, "Drawing Attention to the 'Green Side of Life,'" *Journal of Biological Education* 48, no. 3 (2014): 142–53, doi:10.1080/00219266.2013.849282.

16. Consider the English fairy tale "Jack and the Beanstalk," first appearing in print in 1807. Although the beanstalk connects heaven and earth and evokes the Yggdrasil, or world tree, mythologies, it serves a utilitarian role in the story.

17. Some recent work in ecocritical plant studies redresses this trend by emphasizing the role of plants in literature. See Randy Laist, ed., *Plants and Literature: Essays in Critical Plant Studies* (Amsterdam: Rodopi/Brill, 2013).

18. Washington Allston, *Lectures on Art, and Poems,* ed. Richard Henry Dana Jr. (New York: Baker and Scribner, 1850), 16.

19. See, for example, Neil Campbell, *The Rhizomatic West: Representing the American West in a Transnational, Global, Media Age* (Lincoln: University of Nebraska Press, 2008).

20. See Janet E. Browne, "Botany for Gentlemen: Erasmus Darwin and 'The Loves of the Plants,'" *Isis* 80, no. 4 (1989): 593–621.

21. See Charles Darwin, *The Power of Movement in Plants* (London: John Murray, 1880); and Charles Darwin, *The Movements and Habits of Climbing Plants* (London: John Murray, 1882).

22. As pointed out by David Kohn and his coauthors, "indeed, when he first landed in the Galapagos, Darwin obviously thought the plants were more interesting than the birds, so he took due care with labeling." David Kohn et al.,

"What Henslow Taught Darwin," *Nature* 436 (2005): 643–45, doi:10.1038/436643a. After the publication of his seminal book *On the Origin of the Species* in 1859, Darwin spent the following twenty years studying botany and plant physiology and published several botanical books.

23. See Anthony Trewavas, *Plant Behaviour and Intelligence* (Oxford: Oxford University Press, 2014); and Richard Karban, *Plant Sensing and Communication* (Chicago: University of Chicago Press, 2015).

24. See František Baluška et al., "Introduction," in *Communication in Plants: Neuronal Aspects of Plant Life,* ed. František Baluška, Stefano Mancuso, and Dieter Volkmann (Berlin: Springer-Verlag, 2006).

25. Eric D. Brenner et al., "Plant Neurobiology: An Integrated View of Plant Signaling," *Trends in Plant Science* 11, no. 8 (2006): 413–19.

26. Erasmus Darwin, *The Botanic Garden: The Loves of Plants* (London: J. Johnson, 1789).

27. See Darwin, *Power of Movement,* 572–73.

28. Peter Barlow, "Charles Darwin and the Plant Root Apex: Closing a Gap in Living Systems Theory as Applied to Plants," in Baluška, Mancuso, and Volkmann, *Communication in Plants,* 37–51.

29. See František Baluška et al., "The 'Root-Brain' Hypothesis of Charles and Francis Darwin: Revival after more than 125 Years," *Plant Signaling and Behavior* 4, no. 12 (2009): 1121–27.

30. See the discussion by Anthony Trewavas, "Response to Alpi *et al.*: Plant Neurobiology—All Metaphors Have Value," *Trends in Plant Science* 12, no. 6 (2007): 231–33.

31. See the discussion by Amedeo Alpi et al., "Plant Neurobiology: No Brain, No Gain?," *Trends in Plant Science* 12, no. 4 (2007): 135–36.

32. The nature of causation in biology was defined by Ernst Mayr more than fifty years ago. He distinguished between *proximate* causes as the immediate, mechanical influences on a specific trait and *ultimate* causes as those explaining the historical evolution of an organism. The proximate–ultimate dichotomy has shaped the perspective of most contemporary biologists, as well as philosophers of science, and it is still widely accepted today. See Ernst Mayr, "Cause and Effect in Biology," *Science, n.s.,* 134, no. 3489 (1961): 1501–6. See also alternative perspectives on the topic by Kevin N. Laland et al., "Cause and Effect in Biology Revisited: Is Mayr's Proximate-Ultimate Dichotomy Still Useful?," *Science* 334, no. 6062 (2011): 1512–16, doi:10.1126/science.1210879.

33. No zoocentric bias was found among the top journals that cover the specific area of ecology and environmental studies, where on average 51 percent and 55 percent of papers, respectively, are published on plants (i.e., for every paper published on animals, one is published on plants).

34. This time was defined by the so-called "Plant Revolution"; see Philip Abelson and Pamela Hines, "The Plant Revolution," *Science* 285, no. 5426 (1999): 367–68, doi:10.1126/science.285.5426.367. See also Monica Gagliano and Michael Marder, "What Plant Revolution Would You Opt For?," *Los Angeles Review of Books Blog*, November 2014, http://philosoplant.lareviewofbooks.org/?p=82.

35. Karban, *Plant Sensing and Communication*; Trewavas, *Plant Behaviour and Intelligence*; Dov Koller, *The Restless Plant,* ed. Elizabeth Van Volkenburgh (Cambridge, Mass.: Harvard University Press, 2011); František Baluška, ed., *Plant-Environment Interactions: From Sensory Plant Biology to Active Plant Behavior* (Berlin: Springer-Verlag, 2009).

36. Jean-Jacques Rousseau, *Reveries of the Solitary Walker* (Oxford: Oxford University Press, 2011), 51, 69.

37. Ibid., 77.

38. Ibid., 78.

39. Hall, *Plants as Persons*.

40. Jeffrey Nealon, *Plant Theory: Biopower and Vegetable Life* (Stanford, Calif.: Stanford University Press, 2016), 11.

41. Ibid., 12.

42. Molly Mahood, *The Poet as Botanist* (Cambridge, UK: Cambridge University Press, 2008), 112–46.

43. Jonathan Bate, *John Clare: A Biography* (London: Picador, 2004).

44. Henry David Thoreau, *Material Faith: Thoreau on Science,* ed. Laura Dassow Walls (Boston: Houghton Mifflin, 1999), 89–90.

45. Henry David Thoreau, *Faith in a Seed: The Dispersion of Seeds and Other Late Natural History Writings,* ed. Bradley P. Dean (Washington, D.C.: Island Press, 1993).

46. Judith Wright, *Collected Poems, 1942–1985* (Sydney: Angus & Robertson, 1994), 367.

47. Timothy Clark, *The Cambridge Introduction to Literature and the Environment* (Cambridge, UK: Cambridge University Press, 2011), 47.

48. Monica Gagliano, "In a Green Frame of Mind: Perspectives on the Behavioural Ecology and Cognitive Nature of Plants," *AoB Plants* 7 (2015): plu075, doi:10.1093/aobpla/plu075.

49. See Laist, *Plants and Literature*.

50. See, for example, Kenneth Shapiro and Margo DeMello, "The State of Human-Animal Studies," *Society & Animals* 18 (2010): 2–17.

51. Edward O. Wilson, *Consilience: The Unity of Knowledge* (New York: Alfred A. Knopf, 1998).

52. Michael Marder, *Plant-Thinking: A Philosophy of Vegetal Life* (New York: Columbia University Press, 2013), 10.

53. Monica Gagliano, "The Flowering of Plant Bioacoustics: How and Why," *Behavioral Ecology* 24 (2013): 800–801, doi:10.1093/beheco/art021.

54. Wolfgang Bock, "Walter Benjamin's Criticism of Language and Literature," in *A Companion to the Works of Walter Benjamin,* ed. Rolf J. Goebel (Rochester, N.Y.: Camden House, 2009), 23–45 [25].

55. Wendy Wheeler and Hugh Dunkerley, "Introduction," in "Earthographies: Ecocriticism and Culture," ed. Wendy Wheeler and Hugh Dunkerley, special issue, *New Formations* 64 (2008): 7–14 [8].

56. Charles Sanders Peirce, quoted in Joseph Brent, *Charles Sanders Peirce: A Life* (Bloomington: Indiana University Press, 1998), 346.

57. Wheeler and Dunkerley, "Introduction," 8.

58. Kalevi Kull, "An Introduction to Phytosemiotics: Semiotic Botany and Vegetative Sign Systems," *Sign Systems Studies* 28 (2000): 326–50.

59. Martin Krampen, "Phytosemiotics," *Semiotica* 36, nos. 3–4 (1981): 187–209.

60. Jakob von Uexküll, quoted in Martin Krampen, "Phytosemiotics," in *Essential Readings in Biosemiotics: Anthology and Commentary,* ed. Donald Favareau (Dordrecht: Springer, 2010), 266.

61. Clark, *Cambridge Introduction,* xiii.

62. Ibid., 46.

63. Leonard M. Scigaj, *Sustainable Poetry: Four American Ecopoets* (Lexington: University Press of Kentucky, 1999), 109.

64. George Sessions, "Spinoza and Jeffers on Man in Nature," *Inquiry: An Interdisciplinary Journal of Philosophy* 20 (1977): 481–528; for further discussion of language and environmental philosophy, see Max Oelschlaeger, ed., *Postmodern Environmental Ethics* (Albany: State University of New York Press, 1995); and David R. Keller, ed., *Environmental Ethics: The Big Questions* (Chichester, UK: John Wiley & Sons, 2010).

65. For example, see Brendon Larson, *Metaphors for Environmental Sustainability: Redefining Our Relationship with Nature* (New Haven, Conn.: Yale University Press, 2011).

I. Science

The Language of Plant Communication (and How It Compares to Animal Communication)

Richard Karban

PLANTS HAVE BEEN VIEWED IN POPULAR CULTURE AS ORGANISMS that are unable to sense their environments and are unresponsive as a result. In fact, this view is incorrect; plants perceive their own state and many aspects of their surroundings and adjust numerous traits depending upon these conditions. Many "plant behaviors" increase the fitness of the individuals that display them relative to individuals displaying other alternative behaviors. In addition to sensing, plants communicate among tissues on the same individual to coordinate their responses to the environment. They eavesdrop on neighboring individuals to acquire information about future risks of competition, disease, and herbivores. They communicate with those microbes that allow them to forage more effectively and with animals that facilitate mating and move their seeds to locations where they are likely to thrive. This chapter will first explore the sophisticated ways in which plants sense and respond to environmental cues. Increasingly, we are finding that the cues used by plants exhibit many of the characteristics that linguists require of language. Our understanding of the precise lexicon of plant communication is less complete than our grasp of its varied functions. I develop these themes in much greater detail in the book *Plant Sensing and Communication* (Karban 2015).

Plants lack central nervous systems, so the mechanisms involved in sensing and communicating are very different from those of animals. However, plants have been faced with many of the same challenges as animals and they have evolved different systems that provide many of the same functions. Plant tissues and organs are much less specialized than those of animals (White 1984; Herrera 2009). Plants are made up of repeated units, many of which are capable of sensing stimuli, acquiring resources, becoming reproductive organs, and any number of other functions that are performed solely by specialized organs in animals. Unlike animals, plants can lose some of these redundant units without dying.

PLANT PERCEPTION

Perception of Light

Plants can sense a wide variety of cues that provide reliable information; these may be produced by other organisms or by their abiotic environments. Plants and animals detect light when receptor molecules absorb electromagnetic radiation; absorbed energy causes changes in receptors that trigger responses in photosynthetically active plant tissues or in the nerve cells of animals. Photosynthetic pigments, such as chlorophyll in leaves, selectively absorb red wavelengths when leaves produce carbohydrates from carbon dioxide (CO_2) and water (Wolken 1995; Ballaré 1999). Green and far-red wavelengths are reflected and not absorbed by leaves. Pigments within individual cells detect the ratio of red to far-red light, which regulates the expression of many plant genes. In full sun, the ratio of red to far-red light is 1.15 but it is reduced to 0.05 under dense vegetation (Hutchings and de Kroon 2004). Plants that experience low red to far-red light change physiologically to become more efficient at trapping light (Pearcy et al. 1994) or change morphologically, growing toward the light to avoid being shaded (Hutchings and de Kroon 2004). By responding to light quality, plants detect and, in many cases, avoid competing neighbors. Plants differentiate shade that is cast by a photosynthetically active competitor with that cast by an inanimate object and preferentially grow away from the competitor (Novoplansky et al. 1990; Novoplansky 1991). Plants are also able to anticipate future light conditions. Seedlings respond to the far-red light that is scattered by surrounding foliage before actually being shaded and gain information about the relative positions of neighbors at a very early stage in canopy development (Ballaré et al. 1990). Plants also have receptors that respond to blue light. These enable them to assess the quantity of light in addition to its quality (Vandenbussche et al. 2005). Responses to the light environments produced by competitors allow plants to grow away from shade and to maximize their fitness in environments with either low or high levels of competition (Dudley and Schmitt 1996).

Individual plant cells sense their local light environment, enabling plants to respond at very fine spatial scales, elongating buds and shoots that are in the sun and shedding organs that are in the shade. Animals have specialized organs that contain retinal pigments (a form of vitamin A) that trap incoming radiation, similar to the phytochrome pigments of plants (Wolken 1995). Photoreceptor cells of animals tend to be densely packed in a layer called the retina that connects to nerve cells, allowing vision. Animals that are sensitive to more than one wavelength have the ability to distinguish "color"; humans have receptors that are sensitive to blue, green, and yellow light. Light receptors of animals are located in specialized organs, eyes. Unlike plants, animals can only detect gradients of light when these specialized organs are stimulated. Animals must move their eyes into appropriate positions in order to intercept relevant visual information. In contrast, plants have receptors located throughout each individual.

Chemical Perception

Plants require chemical resources (CO_2, water, mineral nutrients) to grow, and they locate these resources using chemical receptors. Plants also detect other organisms, both harmful and beneficial, by responding to chemical cues. As was the case for light receptors, chemical receptors are found on cells throughout an individual plant. Receptors are often located on the plasma membrane that surrounds the cell; one end of the receptor is outside and can perceive a particular chemical and the other end is inside and triggers a cascade of reactions once that chemical has been detected. Plants also sense many important chemicals as the result of feedbacks rather than dedicated chemical receptors. For example, plants acquire CO_2 from the atmosphere through pores, called stomata, located on the surfaces of leaves and other organs. Feedback based on the internal and external concentrations of CO_2, water, and other factors controls the opening and closing of stomata (Assmann 1993).

Plants respond to patchily distributed soil nutrients by selectively proliferating roots in richer microenvironments (Weaver 1926; Drew et al. 1973). Roots are repeated units that can respond locally to fine-scale heterogeneity in resources. Differential growth in response to the relative abundance of nutrients allows plants to "forage" effectively, placing more roots in rich patches and leaving poor ones. Roots detect and grow away from inanimate objects (Falik et al. 2005), and they respond to competing roots when these are encountered (Mahall and Callaway 1991; Gersani et al. 2001). The ability to strategically place rootlets in rich patches has been associated with increased fitness measures, although the relationship is not as clear as in the case of foraging for light (Fitter et al. 2002; Hodge 2004).

Selective root growth and shedding is a relatively slow process, and plants also rapidly adjust their uptake of soil nutrients in response to local conditions. Roots selectively exude chemicals into the soil, making nutrients more available (Metlen et al. 2009). They also form associations with mycorrhizal fungi, which are more efficient at acquiring resources (Smith and Read 2008; see "Communication with Animals and Microbes" section below). Plants release chemical cues into the soil that help mycorrhizae colonize them (Giovannetti et al. 1996; Akiyama et al. 2005). Plants also sense how many nutrients their mycorrhizal associates are providing and reward or punish them accordingly (Kiers et al. 2011; Fellbaum et al. 2012).

Plants assess their own internal balance and allocate toward the limiting function. For example, individuals limited by water or soil nutrients invest more heavily in roots, while those limited by light or carbohydrates invest more heavily in shoots and leaves (Bloom et al. 1985). They balance information acquired at different spatial and temporal scales (de Kroon et al. 2009). Only by reacting to cues at very fine scales can responses be precise enough to track environmental heterogeneity. However, feedback from other plant parts provides needed information about the individual's overall condition.

Plants detect herbivores and pathogens primarily through chemical cues that are associated with these attackers. Once herbivores or pathogens have been detected, plants change their defensive phenotypes such that they become more resistant to attackers (induced resistance) or more tolerant of attack. Plants will only benefit from these responses if cues reliably predict future risk. For example, plants respond by producing antibacterial compounds when they detect flagellin (Zipfel and Felix 2005; Bent and Mackey 2007). This chemical is indispensable and highly conserved among bacteria so that it is a reliable cue indicating their presence. Many plants respond to the chemicals present in insect eggs or the glue that holds eggs to leaf surfaces (Hilker and Meiners 2010). The presence of eggs is likely to be highly predictive that feeding larvae are coming soon (Beyaert et al. 2012). Previous attack is the most obvious cue of future events, although plants may respond to any cue they can detect that reliably predicts impending conditions (Karban et al. 1999).

Early examples of induced resistance against both pathogens and herbivores involved responses that provided systemic protection throughout the entire plant (Chester 1933; Green and Ryan 1972; Haukioja and Hakala 1975; Kuć 1995). More recently, we have learned that many induced responses are localized. Most plants are sectored to some extent so that nutrients, hormones, and defensive chemicals are not readily exchanged among spatially distant tissues (Orians and Jones 2001). Vascular communication among distant plant organs often requires active transpiration and growth, and these processes are expensive or impossible at times (Waisel et al. 1972; Schenk et al. 2008). Communication among plant tissues using volatile cues may overcome these important constraints and may be more rapid than vascular communication. Volatile communication also has several disadvantages relative to vascular signals. Volatile cues degrade over relatively short distances (Karban et al. 2014). When volatile signals are released into the air, they become public information, while vascular signals remain private (Gershenzon 2007). Volatile communication is dependent on patterns of airflow, and information may be lost or used by competing plants, receptive herbivores, or other plant parasites. Communication systems may become more private if they rely on cues that are genetically constrained. For example, communication between sagebrush plants was more effective if sender and receiver were genetically identical (Karban and Shiojiri 2009), genetically related (Karban et al. 2013), or shared the same volatile profile (Karban et al. 2014).

Volatile chemicals emitted by plants that have been attacked by herbivores and pathogens may also communicate with predators and parasites of those attackers. Plants provide food and shelter to carnivorous insects that remove herbivores. They also provide reliable information that presumably enables the predators and parasites to find prey more efficiently (Dicke and Sabelis 1988; Turlings et al. 1990). Many of the volatiles emitted by damaged plants are not found in the emissions of undamaged or artificially damaged plants (Dicke and Van Poecke 2002). Volatiles emitted by damaged plants are more attractive to predators and parasitoids than volatiles from the herbivores or their frass (Sabelis et al. 1984; Turlings et al. 1991; Turlings and Wackers 2004). Plants appear to remember past attacks so that individuals damaged more than once increase emissions of attractive volatiles following subsequent attacks (Ton et al. 2007). In some instances, the volatiles emitted by plants contained useful and specific information for predators and parasites of the herbivores. For example, parasitoids recognized the volatiles emitted in response to their caterpillar host, *Heliothis virescens*, in preference to those emitted in response to a nonhost, *Helicoverpa zea* (De Moraes et al. 1998). However, not all responses of carnivores to plant volatiles showed fine-tuning, and some predators and parasitoids were attracted to cues of hosts that they were unable to exploit (Agrawal and Colfer 2000; Thaler et al. 2002). Some parasitoids and predators learn to associate particular plant volatiles with specific prey (Allison and Hare 2009). This has enabled these carnivores to hunt more successfully and increase their fitness (Dukas and Duan 2000). It is likely that emitting volatile cues that attract predators of herbivores benefits plants, although this has not been demonstrated conclusively (Allison and Hare 2009; Kessler and Heil 2011).

Perception of Touch

Plants respond to touch in a variety of circumstances. These include roots that avoid obstacles (Darwin 1880), leaves and leaflets of "sensitive plants" that rapidly fold up following contact (Braam 2005), carnivorous plants that catch insects when these prey spring trigger hairs (Darwin 1893; Forterre et al. 2005), stems that become mechanically sturdier following repeated touch (Biddington 1986), vines that attach to nearby supports (Braam 2005), and reproductive organs that respond to visitors, increasing the likelihood of outcrossing (Simons 1992). We do not yet understand the detailed mechanisms that allow plants to detect touch, although the diverse phenomena described above all appear to involve touch receptors located at the boundary between the cell wall and the plasma membrane (Telewski 2006; Chehab et al. 2009). Both electrical signals and the plant hormone jasmonic acid have been found to be involved with conveying touch signals within plants (Volkov et al. 2010; Chehab et al. 2012).

Plants sense gravity, allowing shoots to grow up and roots down. This is accomplished by detecting mechanical pressure at the interface between the cell wall and plasma membrane (Telewski 2006). Cells that sense gravity contain dense particles that place more mechanical pressure on the bottom of the cell than on the top (Morita 2010). Some leaves have also been reported to detect the footsteps of insects, and a similar pressure mechanism is believed to be involved, although these insects may also trigger leaf hairs (Peiffer et al. 2009).

Perception of Temperature

Plants that are exposed to cold undergo many changes, although some of these may represent responses to damage rather than sensing temperature. Temperature alters the plasma membrane, changing the structure and function of this important interface (Chinnusamy et al. 2010). Plants that are exposed gradually to cold temperatures alter their plasma membranes and the genes that they express, making them more tolerant of chilling (Thomashow 1999; Lee et al. 2005).

A few plants are able to thermoregulate by producing more heat as external temperatures fall, particularly for their inflorescences (Nagy et al. 1972; Knutson 1974). Several species are able to maintain a relatively constant temperature that is independent of ambient air (Seymour 2004). These plants are able to sense temperature and regulate respiratory rates in response to varying conditions. The mechanisms and receptors involved with sensing temperature have not yet been described.

Perception of Electricity and Sound

Both plant and animal cells use electrical signals to transmit information and many of the responses described in this chapter involve electrical signaling within and between cells. Electrical signals generate action potentials, in which the electrical potential of a cell membrane rapidly rises and falls (Beilby 2007). Action potentials stimulate nerves and muscles in animals. In plants they are involved in perception of light, touch, wounding, water status, and temperature, among other things, although many plant biologists have been skeptical of their roles in the past. Recent work demonstrated that responses to herbivores involved changes in potentials, and injecting plants with electrical currents duplicated these responses (Mousavi et al. 2013). Plants apparently detect electrical signals with glutamate-like receptors that are functionally similar to the glutamate receptors that mediate neural communication, memory formation, and learning in vertebrate nervous systems.

There has been speculation for centuries that plants produce and respond to music (Tompkins and Bird 1973). Sound is produced when molecules collide, for example, when the tension in a plant's water transport system is abruptly released. However, there is no convincing evidence that this sound production is beneficial rather than an unavoidable consequence of damaging water stress (Kikuta et al. 1997; Gagliano 2013).

Under some circumstances plants sense and respond to sound. Exposure of seeds and seedlings to ultrasound (pressure waves with a frequency outside the range of human hearing) causes changes in plant chemistry, germination rates, and root metabolism and growth (Telewski 2006; Rokhina et al. 2009; Gagliano et al. 2012). Recent evidence suggests that plants respond to sound produced by herbivore feeding to increase their production of defensive chemicals (Appel and Cocroft 2014).

What Is Language?

Linguists and animal behaviorists have had a difficult time agreeing on what constitutes language. Most scholars define language as a set of abstract symbols that represent objects, actions, or feelings that can be used to communicate among individuals. Plants respond to cues that are directly caused by, or can be associated with, their environment. They respond differently depending upon the cue received, allowing them to change allocation patterns and other behaviors that increase their fitness in specific situations. Many linguists stipulate that true language involves symbols (cues) that are not causally linked to the objects or actions being described, in which case bees and most other nonhuman animals would not exhibit language. Indeed most animal cues are only reliable symbols when they are caused by a particular state, and therefore reliably indicating its existence. This is also the case for plants that will only be favored by selection to respond to cues that provide reliable information. These reliable cues are directly linked to their cause in most instances.

In addition to the symbolic representation used in language, humans and some other animals are able to differentiate between self and nonself (Gallup 1970; Plotnik et al. 2006). Self-awareness allows individuals of these intelligent species to recognize themselves in mirrors and to behave differently toward themselves than they would toward other individuals. Interestingly, plants also appear to have the ability to distinguish and to respond differently to cues from self and nonself. Plants distinguish their own roots from those of other individuals and avoid competing with themselves by growing shorter and fewer roots or by growing away from other "self" roots (Falik et al. 2003; Holzapfel and Alpert 2003; Gruntman and Novoplansky 2004). Roots that were not physically connected were not recognized as self in these cases. Sagebrush responded more effectively to cues from genetically identical plant tissues compared to nonself tissues to become more resistant to herbivory (Karban and Shiojiri 2009). Some plants also distinguish kin from unrelated strangers. Individuals of *Cakile edentula* that encountered roots of kin allocated fewer resources to fine roots than individuals that encountered roots of strangers (Dudley and File 2007). Sagebrush responded more effectively to cues from more closely related neighbors to induce resistance against damage by herbivores (Karban et al. 2013). In this case, volatile cues emitted by damaged plants were perceived by sagebrush, although the precise chemical nature of the cues and mechanisms involved in perception remain unknown. This ability to distinguish kin from strangers is important since it makes cooperation and other "social" behaviors more likely to evolve (Hamilton 1964).

THE LEXICON OF PLANTS

Since plants perceive so many different kinds of visual, chemical, tactile, and electrical stimuli, they, like animals, rely on multiple senses. Some of these senses are primarily used for perceiving information about a plant's own internal state, others are primarily used for perceiving information about the abiotic environment or other interacting species. The discussion that follows will focus on the cues that plants use to communicate with other individuals.

Animal behaviorists often distinguish between cues and signals. Cues contain information about the abiotic environment or other organisms; organisms that emit cues are not assumed to increase their fitness by doing so. The cue may be released inadvertently or as the by-product of some other useful process. In contrast, signals are emitted by organisms and have been shaped by natural selection such that they increase the sender's average fitness (Otte 1974; Bradbury and Vehrencamp 1998; Maynard Smith and Harper 2003). In practice, distinguishing cues from signals is difficult, although the distinction is conceptually useful. Although plants benefit by responding to stimuli in many instances, most of the examples involving plant communication must be considered as cues rather than signals since there is rarely evidence that the individual emitting the cue benefits. Signals are more common in communication between tissues within an individual.

In many instances, cues provide receivers with specific information. However, most plant responses that are well understood have been found to be elicited by any of multiple cues with somewhat redundant functions. Below I discuss some of the lexicon of cues used by plants to communicate with neighboring individuals about competition for light and about risk of attack by herbivores and pathogens. These two topics have received considerable attention from researchers in recent years, although they probably represent only a small fraction of the information that plants can sense and communicate.

Competition for Light

Plants in many environments are limited by light and may benefit by signaling their presence to competitors and by growing away from, or over the top of, neighbors that they perceive. As discussed above, plants are sensitive to the fluxes of blue, red, and far-red wavelengths, as well as the total of all photosynthetically active wavelengths, and any of these cues are sufficient to stimulate shade avoidance responses (Ballaré 2009). Plants are also sensitive to the ratio of wavelengths; photosynthetic plant pigments preferentially absorb red wavelengths so that low ratios of red wavelengths indicate the presence of competitors (Smith 2000). Mutants that failed to respond to light cues by elongating their stems were outcompeted by responsive wild type neighbors (Schmitt et al. 1995).

In addition to light cues, plants also use concentrations of ethylene to detect competitors (Pierik et al. 2006; Kegge and Pierik 2010). Ethylene builds up in dense stands of vegetation and stimulates plant growth, stem elongation, and other shade avoidance responses. Ethylene provides information about the overall competitive environment but little specific information about the identity of the neighbor, or its strength as a competitor, because ethylene is produced by essentially all plants (Kegge and Pierik 2010). Perception of ethylene may allow individuals to assess their own position relative to that of their neighbors; insensitive tobacco plants displayed delayed detection of neighbors and were outcompeted by wild type plants that could respond to ethylene (Pierik et al. 2003).

Risk of Attack by Herbivores and Pathogens

Plants are at risk from many different herbivores and microbes, so attempting to defend simultaneously against all of these threats is difficult and costly when those defenses are not needed. At a fine scale, plants first recognize risk of attack by experiencing a change in potential of the plasma membrane that surrounds the cells (Maffei et al. 2007). This membrane depolarization is triggered by many different stimuli including mechanical damage, hydrogen peroxide (H_2O_2) introduced by feeding herbivores, oral secretions from herbivores, and cell wall fragments from invading pathogens, although the signature of the depolarization is specific to the stimulus (Zebelo and Maffei 2012). In essence, the plant converts the cues it receives from herbivores and pathogens into an electrical code that it recognizes. This localized depolarization is followed by changes in the action potential of plasma membranes that can travel rapidly among different cells (Maffei et al. 2007; Mousavi et al. 2013). Other internal chemical signaling systems have also been described, including fluxes of calcium ions (Ca^{2+}), reactive oxygen species such as hydrogen peroxide, mitogen-activated protein kinases (MAPKs), and the plant hormones salicylic acid and jasmonic acid; these show varying levels of information content and specificity that depend upon the attacker and the plant.

These signaling mechanisms allow plants to coordinate defenses against attacking herbivores and pathogens at the scale of single cells and systemically among distant cells within an individual. Plants also emit and respond to volatile cues when distant tissues or even neighbors have been attacked. Tobacco plants that were infested with tobacco mosaic virus emitted relatively large quantities of volatile methyl salicylate that coordinates systemic resistance (Shulaev et al. 1997; Park et al. 2007). Other individuals that were exposed to methyl salicylate produced antimicrobial proteins and lesions that prevented the spread of viruses (Shulaev et al. 1997; Vlot et al. 2008). Volatile methyl jasmonate or other related chemicals may play a similar role in acting as an inter-plant signal that orchestrates responses to herbivore damage. Clipped sagebrush branches emitted high concentrations of methyl jasmonate in lab experiments and these stimulated production of defensive compounds in tomato plants that shared air space (Farmer and Ryan 1990). Native tobacco plants and other sagebrush individuals that occurred naturally in the field near experimentally clipped sagebrush became more resistant to herbivores (Karban et al. 2000; 2006). Volatile methyl jasmonate is taken up into leaves and rapidly converted to biologically active jasmonic acid (Tamogami et al. 2008). Wild tobacco plants with silenced jasmonic acid pathways in the field were more susceptible to both adapted insects and generalist herbivores that normally do not feed on this host (Kessler et al. 2004).

Plants produce other volatile chemicals after damage by herbivores, pathogens, and other stresses that certainly act as internal signals and may also convey information among individuals. These include ethylene, which was discussed above, and green leaf volatiles that humans perceive as the smell of freshly cut grass. The precise role of green leaf volatiles in plant communication is not understood although they are emitted in greater concentrations following attack; they also move between plants and stimulate the production of many defensive plant responses (Arimura et al. 2001).

Communication with Animals and Microbes

In addition to plant–plant communication, plants exchange information with animals and microbes that can strongly influence the success of the participants. Many of the cues that plants release into the environment, especially visual and chemical cues, are public information that can be used by any organism capable of perceiving it. For example, herbivores and plant parasites often locate their hosts using these cues (Visser 1986). Some plant volatiles are perceived by only a small subset of organisms, providing the possibility for "private channels" although this notion remains controversial (Raguso 2008). Even with public messages, there is the possibility for considerable specific information. For example, some of the volatiles induced by herbivory are specific to the particular herbivore and the particular host plant, and predators and parasites of these herbivores can discriminate among these slight variations (Takabayashi et al. 1995; De Moraes et al. 1998).

Plants may be able to shift interactions with other organisms in their favor by controlling the exchange of information. As mentioned above, plants emit volatiles that are attractive to the predators and parasites of their herbivores; these carnivores may provide an indirect defense for the plant (Dicke et al. 1990; Turlings et al. 1990). Many plants rely on external agents for mating and dispersing their seeds to favorable locations. Since many visitors to flowers and fruits consume these organs rather than providing pollination or dispersal services, plants attempt to preferentially attract or facilitate that subset of visitors that are most beneficial. This may be accomplished by offering rewards that are difficult to access or use. Unusual floral patterns such as nectar guides are visible to only a limited set of visitors (Sprengel 1793, 1996). Rewardless flowers that mimic female insects attract an even more specialized group of visitors (Renner 2006). Flowers often reduce their attractiveness to visitors once they have been successfully pollinated (Willmer 2011). Similarly fruits adjust the cues they advertise to animals as their seeds mature and reach a state where they can germinate; ripe fruit is typically more nutritious, less chemically defended, softer, more brightly colored, and more aromatic than unripe fruit (Cipollini and Levey 1997; Schaefer et al. 2003). These changes in the signals that fruit provide tend to be "honest" since both the plant and consumers can benefit from the transfer of reliable information.

Many interactions play out over very small scales but have huge consequences for plant fitness. Important communication may occur over microscopic distances. For example, a chemical conversation between pollen and receptive stigmas favors outcrossed pollen over self pollen (Wheeler et al. 2010). Plants are able to recognize the microbes that reach their surfaces and actively attract or filter potential colonists (Lundberg et al. 2012; Junker and Tholl 2013). Some are pathogens to be blocked or destroyed but other microbes are essential to plant success. For example, some plants form symbiotic relationships with nitrogen-fixing bacteria, housing the bacteria and supplying them with carbohydrates and other nutrients in exchange for biologically available nitrogen. Especially in situations where nitrogen is limiting, the partners seek each other out by engaging in an intricate dialogue of sophisticated chemical signals that culminates in the plants producing specialized structures for the bacteria, called nodules (Oldroyd 2013). Plants can detect how much nitrogen particular colonies of bacteria are providing and reward only those that are the most cooperative (Kiers et al. 2003).

As discussed above, associations with mycorrhizal fungi allow plants to more effectively absorb nutrients and water from the soil; plants and mycorrhizal fungi engage in a back-and-forth signaling that results in plants facilitating partners that provide the most benefit (Kiers et al. 2011). Many mycorrhizae are generalists and associate with multiple plant individuals, forming underground networks. Plants may communicate with one another using information transferred by these networks. For example, uninfested tomato plants that were linked by mycorrhizal networks responded when connected neighbors were attacked by pathogenic fungi before being attacked themselves (Song et al. 2010). Communication through these networks may be possible over greater distances than could occur using other forms of communication. Bean plants that were linked by mycorrhizae to neighbors infested with aphids emitted volatiles that repelled aphids and attracted their parasitoids (Babikova et al. 2013). In this case, the bean plants prepared for aphid attack despite no direct contact with these insects. The cues that move within mycorrhizal networks are not well known at this time.

The Communication Machinery Limits the Message

Among linguists and animal behaviorists, there is general consensus that language constrains the information that can be communicated. Noam Chomsky (1965) argued that the biologically determined structure of the human brain constrained our use of language and that all of the world's languages share a uniquely human universal grammar. This notion was modified by Marshall McLuhan (1962, 1964), who argued that the technology of communication affects the information content of the message. Researchers studying animal communication also believe that the receiver's basic biology restricts the cues it can perceive and the nature of the "language" that it uses (Otte 1974; Bradbury and Vehrencamp 1998). Precursors to cues and signals must have already existed but have been modified for their eventual roles in communication.

Preexisting variation as well as physical and biological constraints have surely shaped communication systems involving plants. For example, vascular communication is feasible among leaves that share vascular connections or "plumbing" (Orians and Jones 2001), volatile communication is feasible for leaves lacking these connections but only over relatively short distances (Heil and Adame-Alvarez 2010), and mycorrhizal networks may allow communication among individuals over longer distances (Babikova et al. 2013). Only some organisms can perceive particular cues, and this "receiver bias" shapes the communication channel. For example, nectar guides visible in ultraviolet will be helpful to some insects but invisible to many vertebrate visitors, while red flowers will attract vertebrates but not many insect visitors.

In summary, considerable evidence has accumulated invalidating the old notion that plants are unable to perceive environmental cues and respond in ways that increase their fitness. In fact, plants respond to cues from distant tissues from the same individual, from other individuals, and even from different species. While plants have much less ability to communicate than animals, the more developed disciplines of animal behavior and linguistics have much to offer the nascent field of plant communication. Although we know little about the "language" of plants, it appears to be extremely complex, using many different modalities (visual, electrical, chemical cues) and capable of conveying sophisticated information about both past and future events.

References

Agrawal, A. A., and R. G. Colfer. 2000. "Consequences of Thrips-Infested Plants for Attraction of Conspecifics and Parasitoids." *Ecological Entomology* 25:493–96.

Akiyama, K., K. Matsuzaki, and H. Hayashi. 2005. "Plant Sesquiterpenes Induce Hyphal Branching in Arbuscular Mycorrhizal Fungi." *Nature* 435:824–27.

Allison, J. D., and J. D. Hare. 2009. "Learned and Naive Natural Enemy Responses and the Interpretation of Volatile Organic Compounds as Cues or Signals." *New Phytologist* 184:768–82.

Appel H. M., and R. B. Cocroft. 2014. "Plants Respond to Leaf Vibrations Caused by Insect Herbivore Chewing." *Oecologia* 175:1257–66.

Arimura, G., R. Ozawa, J. Horiuchi, T. Nishioka, and J. Takabayashi. 2001. "Plant–Plant Interactions Mediated by Volatiles Emitted from Plants Infested by Spider Mites." *Biochemical Systematics and Ecology* 29:1049–61.

Assmann, S. 1993. "Signal Transduction in Guard Cells." *Annual Review of Cell Biology* 9:345–75.

Babikova, Z., L. Gilbert, T. J. A. Bruce, M. Birkett, J. C. Caulfield, C. Woodcock, J. A. Pickett, and D. Johnson. 2013. "Underground Signals Carried through Common Mycelial Networks Warn Neighboring Plants of Aphid Attack." *Ecology Letters* 16:835–43.

Ballaré, C. L. 1999. "Keeping Up with the Neighbours: Phytochrome Sensing and Other Signalling Mechanisms." *Trends in Plant Science* 4:97–102.

———. 2009. "Illuminated Behaviour: Phytochrome as a Key Regulator of Light Foraging and Plant Anti-herbivore Defence." *Plant, Cell & Environment* 32:713–25.

Ballaré, C. L., A. L. Scopel, and R. A. Sánchez. 1990. "Far-Red Radiation Reflected from Adjacent Leaves: An Early Signal of Competition in Plant Canopies." *Science* 247:329–32.

Beilby, M. J. 2007. "Action Potential in Charophytes." In *International Review of Cytology,* edited by K. W. Jeon, 43–82. Amsterdam: Elsevier.

Bent, A. F., and D. Mackey. 2007. "Elicitors, Effectors, and R Genes: The New Paradigm and a Lifetime Supply of Questions." *Annual Review of Plant Pathology* 45:399–436.

Beyaert, I., D. Köpke, J. Stiller, A. Hammerbacher, K. Yoneya, A. Schmidt, J. Gershenzon, and M. Hilker. 2012. "Can Insect Egg Deposition 'Warn' a Plant of Future Feeding Damage by Herbivorous Larvae?" *Proceedings of the Royal Society B* 279:101–8.

Biddington, N. L. 1986. "The Effects of Mechanically Induced Stress in Plants—a Review." *Plant Growth Regulation* 4:103–23.

Bloom, A. J., F. S. Chapin, and H. A. Mooney. 1985. "Resource Limitation in Plants—an Economic Analogy." *Annual Review of Ecology and Systematics* 16:363–92.

Braam, J. 2005. "In Touch: Plant Responses to Mechanical Stimuli." *New Phytologist* 165:373–89.

Bradbury, J. W., and S. L. Vehrencamp. 1998. *Principles of Animal Communication.* Sunderland, Mass.: Sinauer.

Chehab, E. W., E. Eich, and J. Braam. 2009. "Thigmomorphogenesis: A Complex Plant Response to Mechano-Stimulation." *Journal of Experimental Botany* 60:43–56.

Chehab, E. W., C. Yao, Z. Henderson, S. Kim, J. Braam. 2012. "Arabidopsis Touch-Induced Morphogenesis Is Jasmonate Mediated and Protects against Pests." *Current Biology* 22:701–6.

Chester, K. S. 1933. "The Problem of Acquired Physiological Immunity in Plants." *Quarterly Review of Biology* 8:129–54, 275–324.

Chinnusamy, V., J.-K. Zhu, and R. Sunkar. 2010. "Gene Regulation during Cold Stress Acclimation in Plants." In *Plant Stress Tolerance: Methods and Protocols, Methods in Molecular Biology 639*, edited by R. Sunkar, 39–55. New York: Humana Press.

Chomsky, N. 1965. *Aspects of the Theory of Syntax*. Cambridge, Mass.: MIT Press.

Cipollini, M. L., and D. J. Levey. 1997. "Why Are Some Fruits Toxic? Glycoalkaloids in *Solanum* and Fruit Choice by Vertebrates." *Ecology* 78:782–98.

Darwin, C. 1880. *The Power of Movement in Plants*. London: John Murray.

———. 1893. *Insectivorous Plants*. London: John Murray.

de Kroon, H., E. J. W. Visser, H. Huber, L. Mommer, and M. J. Hutchings. 2009. "A Modular Concept of Plant Foraging Behaviour: The Interplay between Local Responses and Systemic Control." *Plant, Cell & Environment* 32: 704–12.

De Moraes, C. M., W. J. Lewis, P. W. Paré, H. T. Alborn, and J. H. Tumlinson. 1998. "Herbivore-Infested Plants Selectively Attract Parasitoids." *Nature* 393:570–74.

Dicke, M., and M. W. Sabelis. 1988. "How Plants Obtain Predatory Mites as Bodyguards." *Netherlands Journal of Zoology* 38:148–65.

Dicke, M., M. W. Sabelis, J. Takabayashi, J. Bruin, and M. A. Posthumus. 1990. "Plant Strategies of Manipulating Predator–Prey Interactions through Allelochemicals: Prospects for Application in Pest Control." *Journal of Chemical Ecology* 16:3091–3117.

Dicke, M., and R. M. P. Van Poecke. 2002. "Signalling Plant–Insect Interactions: Signal Transduction in Direct and Indirect Plant Defence." In *Plant Signal Transduction*, edited by D. Scheel and C. Wasternack, 289–316. Oxford: Oxford University Press.

Drew, M. C., L. R. Saker, and T. W. Ashley. 1973. "Nutrient Supply and the Growth of the Seminal Root System in Barley: I. The Effect of Nitrate Concentration on the Growth of Axes and Laterals." *Journal of Experimental Botany* 24:1189–1202.

Dudley, S. A., and A. L. File. 2007. "Kin Recognition in an Annual Plant." *Biology Letters* 3:435–38.

Dudley, S. A., and J. Schmitt. 1996. "Testing the Adaptive Plasticity Hypothesis: Density-Dependent Selection on Manipulated Stem Length in *Impatiens capensis*." *American Naturalist* 147:445–65.

Dukas, R., and J. J. Duan. 2000. "Potential Fitness Consequences of Associative Learning in a Parasitoid Wasp." *Behavioral Ecology* 11:536–43.

Falik, O., P. Reides, M. Gersani, and A. Novoplansky. 2003. "Self/Non-self Discrimination in Roots." *Journal of Ecology* 91:525–31.

———. 2005. "Root Navigation by Self Inhibition." *Plant, Cell & Environment* 28:562–69.

Farmer, E. E., and C. A. Ryan. 1990. "Interplant Communication: Airborne Methyl Jasmonate Induces Synthesis of Proteinase Inhibitors." *Proceedings of the National Academy of Sciences* 87:7713–16.

Fellbaum, C. R., E. W. Gachomo, Y. Beesetty, S. Choudhari, G. D. Strahan, P. E. Pfeffer, E. T. Kiers, and H. Bucking. 2012. "Carbon Availability Triggers Fungal Nitrogen Uptake and Transport in Arbuscular Mycorrhizal Symbiosis." *Proceedings of the National Academy of Sciences* 109:2666–71.

Fitter, A., L. Williamson, B. Linkohr, and O. Leyser. 2002. "Root System Architecture Determines Fitness in an *Arabidopsis* Mutant in Competition for Immobile Phosphate Ions but Not for Nitrate Ions." *Proceedings of the Royal Society B* 269:2017–22.

Forterre, Y., J. M. Skotheim, J. Dumais, and L. Mahadevan. 2005. "How the Venus Flytrap Snaps." *Nature* 433:421–25.

Gagliano, M. 2013. "Green Symphonies: A Call for Studies on Acoustic Communication in Plants." *Behavioral Ecology* 24:789–96.

Gagliano, M., S. Mancuso, and D. Robert. 2012. "Towards Understanding Plant Bioacoustics." *Trends in Plant Science* 17:323–25.

Gallup, G. G. 1970. "Chimpanzees: Self-Recognition." *Science* 167:86–87.

Gersani, M., J. S. Brown, E. E. O'Brien, G. M. Maina, and Z. Abramsky. 2001. "Tragedy of the Commons as a Result of Root Competition." *Journal of Ecology* 89:660–69.

Gershenzon, J. 2007. "Plant Volatiles Carry Both Public and Private Messages." *Proceedings of the National Academy of Sciences* 104:5257–58.

Giovannetti, M., C. Sbrana, A. Silvia, and L. Avio. 1996. "Analysis of Factors Involved in Fungal Recognition Response to Host-Derived Signals by Arbuscular Mycorrhizal Fungi." *New Phytologist* 133:65–71.

Green, T. R., and C. A. Ryan. 1972. "Wound-Induced Proteinase Inhibitor in Plant Leaves: A Possible Defense Mechanism Against Insects." *Science* 175:776–77.

Gruntman, M., and A. Novoplansky. 2004. "Physiologically Mediated Self/Non-self Discrimination in Roots." *Proceedings of the National Academy of Sciences* 101:3863–67.

Hamilton, W. D. 1964. "The Genetical Evolution of Social Behaviour." *Journal of Theoretical Biology* 7:1–52.

Haukioja, E., and T. Hakala. 1975. "Herbivore Cycles and Periodic Outbreaks: Formulation of a General Hypothesis." *Report from the Kevo Subarctic Research Station* 12:1–9.

Heil, M., and R. M. Adame-Alvarez. 2010. "Short Signalling Distances Make Plant Communication a Soliloquy." *Biology Letters* 6:843–45.

Herrera, C. M. 2009. *Multiplicity in Unity.* Chicago: University of Chicago Press.

Hilker, M., and T. Meiners. 2010. "How Do Plants 'Notice' Attack by Herbivorous Arthropods?" *Biological Reviews* 85:267–80.

Hodge, A. 2004. "The Plastic Plant: Root Responses to Heterogeneous Supplies of Nutrients." *New Phytologist* 162:9–24.

Holzapfel, C., and P. Alpert. 2003. "Root Cooperation in a Clonal Plant: Connected Strawberries Segregate Roots." *Oecologia* 134:72–77.

Hutchings, M. J., and H. de Kroon. 1994. "Foraging in Plants: The Role of Morphological Plasticity in Resource Acquisition." *Advances in Ecological Research* 25:159–238.

Junker, R. R., and D. Tholl. 2013. "Volatile Organic Compound Mediated Interactions at the Plant–Microbe Interface." *Journal of Chemical Ecology* 39:810–25.

Karban, R. 2015. *Plant Sensing and Communication.* Chicago: University of Chicago Press.

Karban, R., A. A. Agrawal, J. S. Thaler, and L. S. Adler. 1999. "Induced Plant Responses and Information Content about Risk of Herbivory." *Trends in Ecology and Evolution* 14:443–47.

Karban, R., I. T. Baldwin, K. J. Baxter, G. Laue, and G. W. Felton. 2000. "Communication between Plants: Induced Resistance in Wild Tobacco Plants following Clipping of Neighboring Sagebrush." *Oecologia* 125:66–71.

Karban, R., and K. Shiojiri. 2009. "Self-Recognition Affects Plant Communication and Defense." *Ecology Letters* 12:502–6.

Karban, R., K. Shiojiri, M. Huntzinger, and A. C. McCall. 2006. "Damage-Induced Resistance in Sagebrush: Volatiles Are Key to Intra- and Interplant Communication." *Ecology* 87:922–30.

Karban, R., K. Shiojiri, S. Ishizaki, W. C. Wetzel, and R. Y. Evans. 2013. "Kin Recognition Affects Plant Communication and Defence." *Proceedings of the Royal Society B* 280:20123062. doi:10.1098/rspb.2012.3062.

Karban, R., L. H. Yang, and K. F. Edwards. 2014. "Volatile Communication between Plants That Affects Herbivory: A Meta-Analysis." *Ecology Letters* 17:44–52.

Kegge, W., and R. Pierik. 2010. "Biogenic Volatile Organic Compounds and Plant Competition." *Trends in Plant Science* 15:126–32.

Kessler, A., R. Halitschke, and I. T. Baldwin. 2004. "Silencing the Jasmonate Cascade: Induced Plant Defenses and Insect Populations." *Science* 305:665–68.

Kessler, A., and M. Heil. 2011. "The Multiple Faces of Indirect Defences and Their Agents of Natural Selection." *Functional Ecology* 25:348–57.

Kiers, E. T., M. Duhamel, Y. Beesetty, J. A. Mensah, O. Franken, E. Verbruggen, C. R. Fellbaum, et al. 2011. "Reciprocal Rewards Stabilize Cooperation in the Mycorrhizal Symbiosis." *Science* 333:880–82.

Kiers, E. T., R. A. Rousseau, S. A. West, and R. F. Denison. 2003. "Host Sanctions and the Legume–Rhizobium Mutualism." *Nature* 425:78–81.

Kikuta, S. B., M. A. Lo Gullo, A. Nardini, H. Richter, and S. Salleo. 1997. "Ultrasound Acoustic Emissions from Dehydrating Leaves of Deciduous and Evergreen Trees." *Plant, Cell & Environment* 20:1381–90.

Knutson, R. M. 1974. "Heat Production and Temperature Regulation in Eastern Skunk Cabbage." *Science* 186:746–47.

Kuć, J. 1995. "Induced Systemic Resistance—An Overview." In *Induced Resistance to Disease in Plants,* edited by R. Hammerschmidt and J. Kuć, 169–75. Dordrecht: Kluwer.

Lee, B. H., D. A. Henderson, and J.-K. Zhu. 2005. "The *Arabidopsis* Cold-Responsive Transcriptome and Its Regulation by ICE1." *Plant Cell* 17: 3155–75.

Lundberg, D. S., S. L. Lebeis, S. H. Paredes, S. Yourstone, J. Gehring, S. Malfatti, J. Tremblay, et al. 2012. "Defining the Core *Arabidopsis thaliana* Root Microbiome." *Nature* 488:86–90.

Maffei, M., A. Mithofer, and W. Boland. 2007. "Before Gene Expression: Early Events in Plant–Insect Interaction." *Trends in Plant Science* 12:310–16.

Mahall, B. E., and R. M. Callaway. 1991. "Root Communication among Desert Shrubs." *Proceedings of the National Academy of Sciences* 88:874–76.

Maynard Smith, J., and D. Harper. 2003. *Animal Signals.* Oxford: Oxford University Press.

McLuhan, M. 1962. *The Gutenberg Galaxy: The Making of Typographic Man.* Toronto: University of Toronto Press.

———. 1964. *Understanding Media: The Extensions of Man.* New York: McGraw-Hill.

Metlen, K. L., E. T. Aschehoug, and R. M. Callaway. 2009. "Plant Behavioural Ecology: Dynamic Plasticity in Secondary Metabolites." *Plant, Cell & Environment* 32:641–53.

Morita, M. T. 2010. "Directional Gravity Sensing in Gravitropism." *Annual Review of Plant Biology* 61:705–20.

Mousavi, S. A. R., A. Chauvin, F. Pascaud, S. Kellenberger, and E. E. Farmer. 2013. "Glutamate Receptor-Like Genes Mediate Leaf-to-Leaf Wound Signaling." *Nature* 500:422–26.

Nagy, K. A., D. K. Odell, and R. S. Seymour. 1972. "Temperature Regulation by Inflorescence of Philodendron." *Science* 178:1195–97.

Novoplansky, A. 1991. "Developmental Responses of *Portulaca* Seedlings to Conflicting Spectral Signals." *Oecologia* 88:138–40.

Novoplansky, A., D. Cohen, and T. Sachs. 1990. "How *Portulaca* Seedlings Avoid Their Neighbors." *Oecologia* 82:490–93.

Oldroyd, G. E. D. 2013. "Speak, Friend, and Enter: Signalling Systems That Promote Beneficial Symbiotic Associations in Plants." *Nature Reviews Microbiology* 11:252–63.

Orians, C. M., and C. G. Jones. 2001. "Plants as Mosaics: A Functional Model for Predicting Patterns of Within-Plant Resource Heterogeneity to Consumers Based on Vascular Architecture and Local Environmental Variability." *Oikos* 94:493–504.

Otte, D. 1974. "Effects and Functions in the Evolution of Signalling Systems." *Annual Review of Ecology and Systematics* 4:385–417.

Park, S.-W., E. Kaimoyo, D. Kumar, S. Mosher, and D. F. Klessig. 2007. "Methyl Salicylate Is a Critical Mobile Signal for Plant Systemic Acquired Resistance." *Science* 318:113–16.

Pearcy, R. W., R. L. Chazdon, L. J. Gross, and K. A. Mott. 1994. "Photosynthetic Utilization of Sunflecks: A Temporally Patchy Resource on a Time Scale of Seconds to Minutes." In *Exploitation of Environmental Heterogeneity by Plants,* edited by M. W. Caldwell and R. W. Pearcy, 175–208. San Diego: Academic Press.

Peiffer, M., J. F. Tooker, D. S. Luthe, and G. W. Felton. 2009. "Plants on Early Alert: Glandular Trichomes as Sensors for Insect Herbivores." *New Phytologist* 184:644–56.

Pierik, R., D. Tholena, H. Poorter, E. J. W. Visser, and L. A. C. J. Voesenek. 2006. "The Janus Face of Ethylene: Growth Inhibition and Stimulation." *Trends in Plant Science* 11:176–83.

Pierik, R., E. J. W. Visser, H. de Kroon, and L. A. C. J. Voesenek. 2003. "Ethylene Is Required in Tobacco to Successfully Compete with Proximate Neighbours." *Plant, Cell & Environment* 26:1229–34.

Plotnik, J. M., F. B. M. de Waal, and D. Reiss. 2006. "Self-Recognition in an Asian Elephant." *Proceedings of the National Academy of Sciences* 103:17053–57.

Raguso, R. A. 2008. "Wake Up and Smell the Roses: The Ecology and Evolution of Floral Scent." *Annual Review of Ecology and Systematics* 39: 549–69.

Renner, S. S. 2006. "Rewardless Flowers in the Angiosperms and the Role of Insect Cognition in Their Evolution." In *Plant–Pollinator Interactions: From*

Specialization to Generalization, edited by N. M. Waser and J. Ollerton, 123–44. Chicago: University of Chicago Press.

Rokhina, E. V., P. Lens, and J. Virkutyte. 2009. "Low-Frequency Ultrasound in Biotechnology: State of the Art." *Trends in Biotechnology* 27:298–306.

Sabelis, M. W., B. P. Afman, and P. J. Slim. 1984. "Location of Distant Spider Mite Colonies by *Phytoseiulus persimilis*: Localization of a Kairomone." In *Acarology VI,* vol. 1, edited by D. A. Griffiths and C. E. Bowman, 431–40. New York: Halsted Press.

Schaefer, H. M., V. Schmidt, and H. Winkler. 2003. "Testing the Defence Trade-Off Hypothesis: How Contents of Nutrients and Secondary Compounds Affect Fruit Removal." *Oikos* 102:318–28.

Schenk, H. J., S. Espino, C. M. Goedhart, M. Nordenstahl, H. I. Martínez Cabrera, and C. S. Jones. 2008. "Hydraulic Integration and Shrub Growth Form Linked across Continental Aridity Gradients." *Proceedings of the National Academy of Sciences* 105:11248–53.

Schmitt, J., A. C. McCormac, and H. Smith. 1995. "A Test of the Adaptive Plasticity Hypothesis Using Transgenic and Mutant Plants Disabled in Phytochrome-Mediated Elongation Responses to Neighbors." *American Naturalist* 146:937–53.

Seymour, R. S. 2004. "Dynamics and Precision of Thermoregulatory Responses of Eastern Skunk Cabbage *Symplocarpus foetidus*." *Plant, Cell & Environment* 27:1014–22.

Shulaev, V., P. Silverman, and I. Raskin. 1997. "Airborne Signalling by Methyl Salicylate in Plant Pathogen Resistance." *Nature* 385:718–21.

Simons, P. 1992. *The Action Plant: Movement and Nervous Behaviour in Plants.* Oxford: Blackwell.

Smith, H. 2000. "Phytochromes and Light Signal Perception by Plants—an Emerging Synthesis." *Nature* 407:585–91.

Smith, S. E., and D. J. Read. 2008. *Mycorrhizal Symbiosis.* 3rd ed. New York: Academic Press.

Song, Y. Y., R. S. Zeng, J. F. Xu, J. Li, X. Shen, and W. G. Yihdego. 2010. "Interplant Communication of Tomato Plants through Underground Common Mycorrhizal Networks." *PLOS One* 5:e13324.

Sprengel, C. K. 1793. *Das Entdeckte Geheimniss der Natur im Bau und in der Befruchtung der Blumen.* Berlin: Vieweg.

———. 1996. "Discovery of the Secret of Nature in the Structure and Fertilization of Flowers." In *Floral Biology: Studies on Floral Evolution in Animal-Pollinated Plants,* edited by D. G. Lloyd and S. C. H. Barrett, 3–43. New York: Chapman and Hall.

Takabayashi, J., S. Takahashi, M. Dicke, and M. A. Posthumus. 1995. "Developmental Stage of Herbivore *Pseudaletia separata* Affects Herbivore-Induced Synomone by Corn Plants." *Journal of Chemical Ecology* 21:273–87.

Tamogami, S., R. Ralkwal, and G. K. Agrawal. 2008. "Interplant Communication: Airborne Methyl Jasmonate Is Essentially Converted into JA and JA-Ile Activating Jasmonate Signaling Pathway and VOCs Emission." *Biochemical and Biophysical Research Communications* 376:723–27.

Telewski, F. W. 2006. "A Unified Hypothesis of Mechanoperception in Plants." *American Journal of Botany* 93:1466–76.

Thaler, J. S., M. A. Farag, P. W. Paré, and M. Dicke. 2002. "Jasmonate-Deficient Plants Have Reduced Direct and Indirect Defences Against Herbivores." *Ecology Letters* 5:764–74.

Thomashow, M. F. 1999. "Plant Cold Acclimation: Freezing Tolerance Genes and Regulatory Mechanisms." *Annual Review of Plant Physiology and Plant Molecular Biology* 50:571–99.

Tompkins, P., and C. Bird. 1973. *The Secret Lives of Plants.* New York: Harper & Row.

Ton, J., M. D'Alessandro, V. Jourdie, G. Jakab, D. Karlen, M. Held, B. Mauch-Mani, and T. C. J. Turlings. 2007. "Priming by Airborne Signals Boosts Direct and Indirect Resistance in Maize." *Plant Journal* 49:16–26.

Turlings, T. C. J., J. H. Tumlinson, F. J. Eller, and W. J. Lewis. 1991. "Larval-Damaged Plants: Sources of Volatile Synomones That Guide the Parasitoid *Cotesia marginiventris* to the Micro-habitat of Its Hosts." *Entomologia Experimentalis et Applicata* 58:75–82.

Turlings, T. C. J., J. H. Tumlinson, and W. J. Lewis. 1990. "Exploitation of Herbivore-Induced Plant Odors by Host-Seeking Parasitic Wasps." *Science* 250:1251–53.

Turlings, T. C. J., and F. Wackers. 2004. "Recruitment of Predators and Parasitoids by Herbivore-Injured Plants." In *Advances in Insect Chemical Ecology,* edited by R. T. Cardé and J. G. Millar, 21–75. Cambridge: Cambridge University Press.

Vandenbussche, F., R. Pierik, F. F. Millenaar, L. A. C. J. Voesenek, and V. D. Straeten. 2005. "Reaching Out of the Shade." *Current Opinion in Plant Biology* 8:462–68.

Visser, J. H. 1986. "Host Odor Perception in Phytophagous Insects." *Annual Review of Entomology* 31:121–44.

Vlot, A. C., P.-P. Liu, R. K. Cameron, S.-W. Park, Y. Yang, D. Kumar, F. Zhou, et al. 2008. "Identification of Likely Orthologs of Tobacco Salicylic Acid-Binding Protein 2 and Their Role in Systemic Acquired Resistance in *Arabidopsis thaliana*." *Plant Journal* 56:445–56.

Volkov, A. G., J. C. Foster, T. A. Ashby, R. K. Walker, J. A. Johnson, and V. S. Markin. 2010. "*Mimosa pudica*: Electrical and Mechanical Stimulation of Plant Movements." *Plant, Cell & Environment* 33:163–73.

Waisel, Y., N. Liphschitz, and Z. Kuller. 1972. "Patterns of Water Movement in Trees and Shrubs." *Ecology* 53:520–23.

Weaver, J. E. 1926. *Root Development in Field Crops*. New York: McGraw-Hill.

Wheeler, M. J., S. Vatovec, and V. E. Franklin-Tong. 2010. "The Pollen S-Determinant in *Papaver*: Comparisons with Known Plant Receptors and Protein Ligand Partners." *Journal of Experimental Botany* 61:2015–25.

White, J. 1984. "Plant Metamerism." In *Perspectives in Plant Population Ecology*, edited by R. Dirzo and J. Sarukhan, 15–47. Sunderland, Mass.: Sinauer.

Willmer, P. 2011. *Pollination and Floral Ecology*. Princeton, N.J.: Princeton University Press.

Wolken, J. J. 1995. *Light Detectors, Photoreceptors, and Imaging Systems in Nature*. Oxford: Oxford University Press.

Zebelo, S. A., and M. Maffei. 2012. "Signal Transduction in Plant–Insect Interactions: From Membrane Potential Variations to Metabolomics." In *Plant Electrophysiology*, edited by A. G. Volkov, 143–72. Berlin: Springer-Verlag.

Zipfel, C., and G. Felix. 2005. "Plants and Animals: A Different Taste for Microbes?" *Current Opinion in Plant Biology* 8:353–60.

TWO

Speaking in Chemical Tongues
Decoding the Language of Plant Volatiles

Robert A. Raguso and André Kessler

PLANT CHEMICALS MEDIATE A FULL SPECTRUM OF PLANT–ENVIRONMENT interactions. For example volatile organic compounds (VOCs) emitted by plants drive antagonistic interactions with herbivores and pathogens, mutualistic interactions with pollinators, mycorrhizal fungi, and rhizobia, and often play crucial roles when an organism deceives or exploits another in the scramble for costly resources. Thus, plant chemistry provides one of the primary "languages" through which plants interact with their environment (Karban, this book), and VOCs emitted from flowers, leaves, and roots epitomize the ubiquity and multifunctionality of chemical communication. Several thousand plant VOCs have been identified, including short chain (C6–C8) membrane lipid–derived alcohols, aldehydes, and esters (the so-called "green leaf volatiles," or GLVs), large classes of terpenoids—the ten-carbon (monoterpene) and fifteen-carbon (sesquiterpene) compounds derived from the greenhouse gas isoprene, aromatic compounds with a benzene-ring skeleton (e.g., benzoic acid derivatives and phenylpropanoids), and diverse sets of compounds containing nitrogen (N) atoms or sulfur (S) atoms (Dudareva et al. 2004). The challenge of interpreting how volatile chemicals function lies in our ability to formulate and test critical hypotheses underlying the transfer of chemical information associated with diverse plant–environment interactions.

In this chapter, we will use certain linguistic analogies to explore how plants communicate with friends, foes, and themselves through the release of VOCs. First, we ask whether different kinds of interactions require different "lexicons," by exploring the chemical identity and diversity of plant volatiles associated with antagonism, mutualism, deception, and eavesdropping, the four factorial sender–receiver interactions that characterize the evolution of animal signaling. We ask whether and how often "neologisms" arise in plant volatile communication, as passwords or secret handshakes that mitigate the risks of eavesdropping by harmful third parties. Second, we examine how biotic and abiotic aspects of a plant's environment determine important contextual limits that shape the information content of volatile signals. We discuss how the "syntax" of plant volatile communication—which compounds are produced when, and in what physiological and ecological contexts—can shift the meaning and outcome of interactions with other organisms. The study of plant VOCs is a rapidly growing field and is reviewed frequently (Raguso 2008; Heil and Karban 2010; Farré-Armengol et al. 2013). Here we take a fresh look at the emerging patterns in this field, united by the ideas that plants take active agency in their own survival and reproduction and do so through an ancient chemical language shared with bacteria, fungi, and animals.

CHEMICAL INFORMATION ACROSS THE INTERACTION MATRIX

Antagonistic Interactions Mediated by VOCs

Organisms interact in diverse ways, and plant VOCs can mediate antagonistic as well as mutualistic interactions. How might the "syntax" of volatile blends maximize benefits for the sender organism, and how should such information be perceived in order to benefit the receiver organism? Should these guiding principles differ between antagonistic and mutualistic interactions, when the outcomes of such interactions are not set in stone, but instead are often conditional upon the presence of other community members, the availability of resources, or the timing of a particular interaction?

Volatile mediation of antagonistic plant–herbivore interactions depends on the association of VOC emissions with traits that impact herbivore performance (resistance) and, in consequence, plant performance (defense). Such conditions are satisfied when VOCs are toxic and directly affect herbivore performance, or when VOC production is correlated with chemical or physical traits that reduce herbivore performance. Plants that are toxic should preemptively warn and repel herbivores (chemical aposematism), minimizing damage to plant tissues.

Conversely, poorly defended plants should be chemically "silent" or cryptic. Thus, the abundance and composition of volatile information should be positively correlated with plant resistance. Accordingly, female *Manduca sexta* (Kessler and Baldwin 2001, 2004) and *Heliothis virescens* moths (De Moraes et al. 2001) respond to herbivore-induced leaf VOC emissions by laying fewer eggs on damaged tobacco plants. Presumably, this behavior has been driven by reduced survivorship of larvae that ingest damage-induced defense compounds (e.g., nicotine) in leaves or suffer increased mortality due to predators attracted to such plants. Below we discuss the composition of VOC blends commonly found to mediate plant–herbivore interactions and the factors influencing their functionality.

The importance of herbivore-induced VOC emissions has been studied primarily in economically important systems, especially bark beetle attack on coniferous trees. Bark beetle aggregation on hosts serves not only to locate mates but also to overcome both constitutive and induced defenses of the host plant, including oleoresins (a physical defense) and the insecticidal monoterpenes, diterpene acids, and phenolic compounds included therein (Raffa 2001). VOC-mediated interactions between bark beetles and conifers cover a full spectrum of outcomes (Seybold et al. 2006), with low monoterpene concentrations triggering beetle attack, intermediate concentrations (e.g., of α-pinene) synergizing large beetle aggregations, and high concentrations being repellent or toxic (Raffa and Smalley 1995; Wallin and Raffa 2000). Similar dosage effects typify plant–pollinator interactions in the Australian cycad *Macrozamia lucida,* for which low concentrations of β-myrcene attracts pollen-eating thrips, but high concentrations of β-myrcene and increased cone temperature repels the thrips, as is needed to effect pollination (Terry et al. 2014). Thus, the same volatile can mediate a range of interactions through the spatial, temporal, and environmental contexts in which it is presented.

Context-dependent effects between host VOCs and insect pheromones (compounds that convey information between members of the same species) are frequently observed. Reddy and Guerrero (2004) listed plant VOCs that synergize the attraction of insect herbivores to their aggregation and sex pheromones. Often the active compounds were

GLVs emitted by wounded leaves, along with other products of cell degradation (benzaldehyde) or fermentation (ethyl acetate). As with monoterpenes in conifers, these compounds are directly associated with plant tissue damage and are reliable cues reflecting the plants' metabolic state, and insect herbivores are particularly sensitive to them. For example, *Lygus* bugs show similar sensitivity in the physiological responses of their antennal olfactory receptors (so-called "electroantennogram" assays) to host-produced GLVs (hexyl butyrate, (*E*)-2-hexenyl butyrate, (*E*)-2-hexenol, and (*E*)-2-hexenal) and a monoterpene (geraniol), as they do to insect-produced pheromones (Chinta et al. 1994). It is possible that the small roster of VOCs known to mediate plant–herbivore interactions is an artifact of researcher bias or analytical methods. However, the repeated identification of the same group of ubiquitous plant VOCs compels us to ask how specific ratios of common compounds can encode different kinds of information (Bruce et al. 2005).

Another important theme is the apparent context dependency of chemical signaling in antagonistic interactions. Herbivores (and their predators) routinely differentiate between VOCs emitted from herbivore-damaged plants versus those with no damage, mechanical damage, or damage from different herbivore species. Also, different sexes (Arab et al. 2007) or developmental stages may respond differently to the same VOC bouquet, likely due to stage- or sex-specific requirements for host utilization. Unlike the *Manduca* and *Heliothis* adult females described above, the larvae of *Spodoptera frugiperda* moths preferentially orient toward herbivore-induced VOCs over those from undamaged plants in Y-tube choice assays (Carroll et al. 2006), despite the higher food quality of the undamaged plants. Thus, our predicted correlations between plant VOC signaling strength and food quality will depend conditionally upon the identity of the receiver organism, its biological imperatives and capabilities, and the kind of selective pressure it might bring to bear on the plant. Specialized herbivores that have circumvented the plant's defenses (Renwick 2002) can be attracted to the chemistry of damaged plants, rather than repelled by it. When such species are the primary selective agents, we would expect plants to reduce or eliminate the provision of volatile information.

Additional selective agents on VOC emission include predators and parasitoids, pathogens, other plants, and abiotic factors, which might exert positive or negative selection on the correlation between VOCs and resistance expression. An example addressing both constitutive and induced expression of traits is the perennial herb *Solanum carolinense,* in which inbreeding reduces the expression of resistance-mediating secondary metabolites and consequently reduces resistance to major herbivores, which exacerbates inbreeding depression (Campbell et al. 2013). Inbred genotypes emit less abundant and less diverse constitutive VOCs, which alters the ecological interactions between different trophic levels (plants, herbivores, and their predators) and may affect herbivore host choice (Kariyat et al. 2012; 2014). This constitutive pattern is mirrored by herbivore responses, as outbred plants induce stronger resistance and higher VOC emission than inbred genotypes (Kariyat et al. 2012). How can we differentiate between herbivore selection on VOC signals and the potential for linked expression of resistance and information-mediating compounds? In populations in which herbivores are major agents of natural selection, resistance should be more tightly correlated with VOC emission than in populations with low herbivory. This prediction also hints at a viable experimental approach to test for the evolutionary effects on VOC signaling with herbivores as agents of natural selection.

Mutualistic Interactions

Plants engage in mutualistic interactions at all scales, including belowground nutritional partnerships (mycorrhizae, rhizobia), aboveground defensive partnerships (endophytes, bodyguards, carnivores), and animal-mediated reproductive partnerships (pollination, seed dispersal). Some of these mutualisms are facultative—they occur opportunistically but may involve other partners under other circumstances. However, other mutualisms are obligate—typified by high mutual codependence among partners. A key concept in obligate mutualism is the idea that cooperative interactions are vulnerable to individuals that benefit from the interaction while withholding services. Two mechanisms for holding such cheaters in check are partner choice (specific recognition signals to initiate mutualistic interactions) and sanctions (reprisals against the defecting party). An example of the former is the unique sequence of chemical signals exchanged between legume roots (luteolin, a flavonoid) and rhizobial bacteria (nod factor, a lipochitooligosaccharide), which trigger the formation of nitrogen-fixing nodules through rhizobial colonization of root hairs (Long 1996). These distinctive chemical signals represent a "secret handshake" or private channel of selective screening of honest partners from a community of opportunists (Hossaert-McKey et al. 2010). Sanctions are less well documented, but take the form of severe reprisals such as selective abortion of yucca flowers in which yucca moth females have deposited too many eggs (Pellmyr and Huth 1994).

It is not yet clear how selection shapes which mutualistic interactions require chemically unique signals to enforce partner choice and which can persist using more generic channels of communication (Raguso 2008). The best-studied examples come from an extreme form of obligate mutualism in plant–pollinator interactions called "nursery pollination," due to the usage of floral resources (ovules or developing seeds, sometimes pollen) to nourish the pollinator's offspring, outside of the more typical nectar- or pollen-based floral market economy. Nursery pollination has evolved in over one thousand species in at least eight angiosperm families, roughly eight hundred of which are figs (Hossaert-McKey et al. 2010). The available data suggest that there are several ways to enforce partner choice in nursery pollination systems. In the majority of fig species studied thus far, receptive flowers produce species-specific VOC blends dominated by terpenoids, and fig wasp attraction has been shown to be guided either by precise ratios of enantiomers (alternative mirror-image forms of the same chemical structures: e.g., (R)-(+) linalool and (-) -β-pinene in *Ficus hispida*; Chen and Song 2008) or by less common aromatic compounds (4-methylanisol in *F. semicordata*; Chen et al. 2009). Fig volatiles also attract nonpollinating fig wasps, which parasitize the fig–wasp mutualism and complicate the evolution of this interaction (Proffit et al. 2009).

In contrast, North American *Yucca* (Agavaceae) plants produce strong floral scents characterized by a ubiquitous VOC ((E)-4,8-dimethyl-1,3,7-nonatriene = DMNT) and its oxidized derivatives (Svensson et al. 2005). Although *Yucca* VOC blends are attractive to *Tegeticula* yucca moths and the novel compounds stimulate the moths' antennae (Svensson et al. 2011), they may not represent "private channels" because *Yucca* flowers attract other species of flies, beetles, moths, and true bugs. Although floral scent is necessary for pollinator attraction in these systems, it is unclear how pollinator discrimination, genetic drift, phylogenetic and environmental factors differentially shape the evolution of scent chemistry. For example, might some of these species display geographic "dialects" of VOC composition, reflecting pollinator shifts or competition with related species in local communities? *Yucca* species with large distributional ranges lack geographic variation in scent chemistry, even in populations with different pollinators (Svensson et al. 2005), whereas fig species that range from the Indian subcontinent to South China (*Ficus racemosa* and *F. hispida*) show divergence in terpenoid scent components, potentially due to geographic isolation (Soler et al. 2011). Nursery pollination systems have served as excellent model systems for understanding the costs and benefits of mutualisms, and our understanding of how chemical signals mediate such interactions remains incomplete and largely descriptive.

Unlike the obligate pollination systems described above, most plant–pollinator interactions hinge on the exchange of nutritious floral products (nectar, pollen) for incidental pollinator services (effective pollen removal and deposition). These exchanges take place in "floral marketplaces" in which plants compete for pollinator attention and fidelity by advertising the presence of floral rewards through visual and olfactory display (Raguso 2004). Food-based plant–pollinator mutualisms show greater degrees of flexibility and generalization than nursery pollination systems, at least at the taxonomic level (e.g., species of bumblebees) if not at the functional level (pollinators with similar morphology, behavior, and effectiveness). VOCs play diverse roles in such systems, depending upon the mode of pollinator attraction and behavior (Raguso 2008), the extent to which learning impacts floral constancy (Schiestl and Wright 2009), and selective pressures by natural enemies ranging from microbes to ungulates (Theis et al. 2007; Junker and Tholl 2013).

Deception in the Air

Deception occurs when the sender of a signal increases its own fitness at the expense of a receiver (Vereecken and McNeil 2010). Deceptive signaling is a form of aggressive mimicry in which the fitness costs of ignoring a bogus signal (related to courtship, parental care, or social cohesion) are so high that the receiver must respond (Lehtonen and Whitehead 2014). Many examples of aggressive mimicry are driven by deceptive chemical signals (Vereecken and McNeil 2010). *Aggressive sexual mimicry*—the exploitation of a signal related to mating success—has evolved independently in several orchid genera, particularly in Mediterranean climates (Schiestl 2005; Ayasse et al. 2011). The primary fitness benefits of sexual deception are twofold: (1) high pollinator fidelity due to the specificity of the interaction (unlike food rewarding flowers), and (2) increased probability of outcrossing due to male dispersal after mating (Gaskett 2011). However, deceptive plants harm their pollinators through opportunity costs for males and mating disruption for females (Wong and Schiestl 2004). As with other examples of mimicry, the fitness of aggressive mimics often is constrained by negative frequency-dependent selection; that is, mimics are most effective when they are less abundant than their models (Gigord et al. 2001).

The breakthrough in understanding the chemistry of sexual mimicry came with the coupling of chemical analysis with electrophysiological and behavioral assays (Ayasse et al. 2011). These studies revealed selective antennal responses of male bee or wasp pollinators to a subset of VOCs, whose biological activity was confirmed behaviorally, first in the Mediterranean orchid genus *Ophrys* (Schiestl et al. 1999) and later in the Australian orchid genus *Chiloglottis* (Schiestl et al. 2003). Most *Ophrys* orchids studied thus far are pollinated by solitary bees that utilize specific ratios of a common set of unsaturated long-chain hydrocarbons as sex pheromones. This phenomenon mirrors the use of hydrocarbon "bar codes" (variable ratios of the same set of compounds) by ants, bees, and wasps to track species or social-group identity (Dani et al. 2001). In contrast, the Australian orchids studied thus far (*Chiloglottis* and *Drakaea*) are pollinated by parasitic thynnine wasps, whose wingless females use chemically unique VOCs (alkyl-substituted cyclohexane diketones in *Chiloglottis* and tetra-substituted pyrazines in *Drakaea*) as sex pheromones (Bohman et al. 2014). An intriguing question emerging on both continents is whether sexually deceptive orchid chemistry closely tracks that of the female insects that they mimic or, alternatively, the flowers produce modified blends that are more attractive to male insects than their conspecific mates (Vereecken and Schiestl 2008; Bohman et al. 2014). The latter alternative—the exploitation of preexisting perceptual bias—has placed floral mimicry into a much broader context and challenged biologists to reconsider how deception evolves (Schaefer and Ruxton 2009).

Generalized food deception—the use of common floral traits to falsely advertise the presence of nectar or pollen—has the distinction of being more common but less well understood than sexual deception. Roughly one-third of all orchid species lack nectar rewards, and most of these are food deceptive (Cozzolino and Widmer 2005). However, food-deceptive strategies have evolved in other angiosperm lineages, including dioecious species in which rewardless female flowers mimic rewarding male flowers of the same species (Dafni 1984; Jersáková et al. 2006). Food deception increases outcrossing by reducing geitonogamy (pollinator-assisted self-pollination) and pollen discounting (loss of exported pollen) through fewer visits to flowers of the same plant (Johnson and Nilsson 1999; Jersáková and Johnson 2006). Perhaps the largest surprise in recent studies of food-deceptive plants is that they rarely engage in strict floral mimicry, which would predict that their fitness decreases in the absence of a look-alike, food-rewarding model (Schiestl and Johnson 2013). In most cases, food-deceptive flowers utilize flower color polymorphism, high individual variation in scent composition, or other means (e.g., the absence of scent) to interfere with associative learning by the duped pollinators (Salzmann et al. 2007; Schatz et al. 2013).

Just as sexually deceptive flowers exploit male insects by co-opting their responses to reproductive signals of conspecific females, *brood-site-deceptive* flowers mimic the cues of dead or decaying substrates—feces, urine, carrion, rotting fruits, and fungi—used by female insects as oviposition sites—places to lay their eggs (Johnson and Jürgens 2010; Urru et al. 2011). Recent studies have identified the critical VOCs in brood-site-deceptive flowers, sometimes showing how visual, tactile, thermal, and respiratory (CO_2) components complement scent to manipulate pollinator behavior (Raguso 2004). The aromatic compounds indole, phenol, and cresol, as well as the aliphatic ketone 2-heptanone, typify flowers that mimic herbivore feces, whereas small, highly volatile sulfides characterize rotting meat, fruity esters and other yeast fermentation products (ethanol, acetoin, butanediol) signify rotting fruit or sap mimicry, and eight-carbon alcohols and ketones indicate fungal mimicry (Vereecken and McNeil 2010; Urru et al. 2011).

Our survey of deceptive communication in flowers has raised some interesting themes. Highly reliable, stereotyped chemical signals, such as sex pheromones and the signatures of microbial decay, represent potential "adaptive peaks" of phenotypic selection for deceptive plants, due to the probability of evoking a behavioral response (Ollerton and Raguso 2006). This is true not only for the floral examples provided above, but also for the chemical interplay between plant roots and soil microorganisms in the rhizosphere (Denison et al. 2003). Another theme is that the boundaries between different forms of deception often blur. Recent studies on the Eurasian orchid genus *Epipactis* provide some challenging examples of "bait and switch" deceptive pollination. Flowers of *E. helleborine* and *E. purpurata* attract vespid wasps with GLVs, but instead of finding insect larvae (their prey) on the plants, the wasps find floral nectar, which they drink while pollinating the flowers (Brodmann et al. 2008). A related species, *E. veratrifolia*, attracts syrphid flies as pollinators by reproducing the simple but distinctive monoterpene alarm pheromone of *Megoura* aphids, which the syrphids use as a larval food source (Stökl et al. 2011). This is brood-site mimicry, as the flies often oviposit on the flowers of *E. veratrifolia*, but they also are provisioned with small amounts of nectar, comparable to a honeydew meal taken when attracted to actual aphids. As the natural history and chemical ecology of more diverse pollinator groups (flies, wasps, beetles) continue to be explored, we will no doubt discover more unusual but equally logical examples of chemically mediated deceptive pollination.

Eavesdropping

Eavesdropping occurs when unintended bystanders utilize information exchanged in a communication network. Eavesdropping is widespread in animal behavior (Valone 2007) but has not been widely considered with respect to plant VOCs, for which most work has focused on plant–plant interactions (Arimura et al. 2010). Plant VOCs can prime or directly induce changes in a neighboring plant's metabolism (Baldwin et al. 2006). Often these changes include the increased expression of herbivore resistance–mediating traits and thus greater preemptive induction of resistance prior to a herbivore's arrival. Plant responses to VOCs emitted from damaged plant tissues have been interpreted as means by which plants overcome the slow progress of wound signal communication through the plants' vascular system (Orians 2005; Arimura et al. 2010), allowing for rapid induction of plant-wide resistance (Frost et al. 2007). Studies of lima bean (*Phaseolus lunata*; Heil and Silva Bueno 2007), poplar

(*Populus deltoids* × *nigra*; Frost et al. 2007) and sagebrush (*Artemisia tridentata*; Karban and Shiojiri 2009) support this hypothesis, finding that VOC-mediated information transfer within plants as well as between genetically related plant tissues results in stronger induction of resistance in the receiving tissue. The elevated responses of neighboring plants to VOCs from a damaged relative over those from unrelated neighbors suggest that VOC perception contributes to kin recognition and potentially to kin selection in plants (Karban and Shiojiri 2009). Nevertheless, the information encoded in the herbivore-induced VOCs, once released, is available to all organisms able to perceive it and thus is vulnerable to eavesdropping.

If within-plant signaling is a primary function of VOC-mediated plant–plant communication, any utilization of such information by an organism other than the sender or its close relatives constitutes eavesdropping. Eavesdroppers can be other plant genotypes and species that gain competitive benefits over the sender (Karban et al. 2003; Kessler et al. 2006), but also herbivores or even natural enemies of the herbivores that use herbivore-induced VOC emission to find their host (discussed above) or prey (Dicke and van Loon 2000). The potential fitness effect of eavesdropping on the sender depends on how the eavesdropper uses the information. When a neighboring plant exploits a neighbor's VOC cues to gain a competitive benefit, it is analogous to social eavesdropping that can affect plant population and community dynamics. Similarly, herbivores that tap into plant communication are interceptive eavesdroppers, as are their natural enemies, although the latter should not impose negative fitness effects on the sender plant. To the contrary, this should have positive consequences and is considered, as in the case of VOC-mediated indirect resistance, to be a major function of herbivore-induced VOC emission (Dicke and van Loon 2000; Kessler and Heil 2011).

What chemical features typify the VOC bouquets that mediate plant–plant communication? The observation that closely related plant genotypes respond more strongly to each other's herbivore-induced VOCs than to those of distantly related genotypes (Frost et al. 2007; Karban and Shiojiri 2009) supports the hypothesis of a private channel shared by close relatives. However, it is unclear how plants maintain private channels in plant–plant communication (as for mutualism, see above), as well as how plant cells perceive VOCs. Studies of specific compounds that are active in inducing responses in undamaged neighbors have usually identified ubiquitous VOCs emitted in response to damage, including GLVs and terpenoids (Arimura et al. 2000; Frost et al. 2008; Kessler et al. 2006). Recent studies have begun to reveal how common compounds can trigger high specificity in plant–plant communication. Perhaps the best illustration of this idea is the behavioral discrimination shown by parasitic dodder plants *(Cuscuta pentagona)* that grow toward monoterpenes from healthy tomato seedling hosts (α-pinene, β-myrcene) and away from wounded hosts GLV ((Z)-3-hexenyl acetate) in olfactometer assays (Runyon et al. 2006).

VOCs can induce changes in receiver plant metabolism, either through direct interactions with cell membranes and alteration of membrane potential, or through interactions with yet unknown receptors. Studies of *Arabidopsis* (Asai et al. 2009) and tomato (Zebelo et al. 2012) found variation in plasma membrane potential and alterations in cellular calcium ion concentration after exposure to common GLVs, mono- and sesquiterpenes. Differential responses to VOCs from related versus unrelated conspecific genotypes or other species could reflect a specific interaction of the VOC bouquet of the sender with the cell membranes of the receiver, without specific receptor molecules. Studies of VOC perception in sagebrush *(Artemisia tridentata)* identified two common, heritable monoterpene chemotypes (camphor or thujone) for which VOCs of a neighbor with the same chemotype induced stronger resistance than VOCs from the different chemotype, providing a mechanism for stronger responses to closely related genotypes (Karban, Wetzel, et al. 2014). There are two additional ways in which sender VOCs can affect herbivore resistance in the receiver. A recent study on tomato suggests that a GLV ((Z)-3-hexenol) is directly converted into a glycoside—a nonvolatile compound linked with sugar, (Z)-3-hexenyl viscianoside—upon entering the receiver plant, where it functions as a toxin (Sugimoto et al. 2014). Moreover, birch tree leaves *(Betula spp.)* directly adsorb VOCs emitted by neighboring *Rhododendron tomentosum* plants and subsequently reemit these compounds with repellent effects on herbivores (Himanen et al. 2010). It is not clear how common these mechanisms are, but they could affect the outcome and thus the evolution of VOC-mediated information transfer.

When discussing eavesdropping in plant–plant interactions, an additional factor is that metabolic changes and associated resistance may not be immediately induced upon VOC exposure. Instead, VOC-mediated information transfer may prime a plant to respond more effectively when a herbivore actually attacks (Engelberth et al. 2004; Kessler et al. 2006). These priming responses might allow any plant to respond to a neighbor's VOC emissions and utilize information about herbivory in their environment, which explains findings of interspecific VOC-mediated plant–plant interactions (Kessler et al. 2006). Through priming, plants need not invest in costly metabolic reconfigurations as they usually do in response to direct herbivory, but can prepare for incipient herbivory and so gain resistance and fitness advantages over competitive neighbors (Morrell and Kessler 2014). It is not yet clear how commonly direct induction versus priming occurs in natural systems and if the magnitude of response is always correlated with relatedness. While much needs to be done to identify private-channel communication between plants, some VOC-mediated interactions with other organisms, such as pollinators and predators/parasitoids can provide insights into the factors that need to be considered.

There is increasing evidence that floral VOCs are crucial for information exchange with pollinators (see above) and also can provide reliable cues for eavesdropping antagonists, such as nectar robbers and floral herbivores. In *Cucurbita pepo*, 1,4-dimethoxybenzene is the major floral attractant not only for specialized pollinator bees *(Peponapis pruinosa)* but also for specialized herbivorous beetles *(Acalymma vittatum)*. Experimental augmentation of this VOC increased beetle attraction and reduced both pollinator attraction and seed set, such that increased VOC emission affected plant fitness negatively in both direct and indirect ways (Theis and Adler 2011). Thus, the presence of mutualists and antagonists should select for intermediate VOC phenotypes or, more likely, population-level variation in scent reflecting the prevalence of the conflicting agents of selection. In flowers of *Polemonium viscosum*, the dominant compound (2-phenylethanol) mediates interactions with mutualists and antagonists alike in a concentration-dependent manner, and emissions of this VOC can vary two orders of magnitude within a population (Galen et al. 2011). Natural selection could balance the availability of information encoded in individual active compounds in plants through two mechanisms. First, variation in the amount of an active compound emitted into the environment and perceived by the interacting organisms can determine its information content: whether the compound is attractive or repellent. The two floral examples above fall into this category. Second, the chemical context in which the active compound is presented can affect its efficacy as an attractant or repellent (discussed below). These are equivalent to the two major classes of hypotheses originally proposed by Hebets and Papaj (2005) for interpreting signal evolution in animals. Beyond the alteration of common plant VOCs, the acquisition of novel compounds would provide another way to minimize eavesdropping. The brood-site- and sexually deceptive flowers discussed above fall into this category, as their floral VOCs represent innate or learned information that is crucial for the fitness of the "intended" receiver but is not typically associated with intact plant tissue.

Although the full extent of eavesdropping on plant communication remains unknown, recent network approaches to the study of VOC-mediated information transfer between plants and mutualistic parasitoids that attack herbivores revealed significant levels of eavesdropping on tritrophic interactions. The hyperparasitoid wasp *Lysipia nana* uses herbivore-induced VOC emissions from *Brassica oleracea* to find *Pieris rapae* caterpillars that are parasitized by their hosts, parasitoid wasps in the genus *Cotesia* (Poelman et al. 2012). The hyperparasitoid eavesdrops on the mutualistic VOC-mediated interaction between the plant and the parasitoid and so compromises the information transfer that is thought to be beneficial to the sender, as well as to the receiver. These findings suggest that if the probability of eavesdropping on an interaction has significant fitness consequences, selection should favor the tuning of information transfer to a private channel.

The Environmental Context

Abiotic Factors Affecting Information Content of Plant Volatiles

In order to be detected and perceived effectively, chemical ligands released into a fluid medium (water or air) need to travel a certain distance from the sender, maintain their structural integrity (the key to information content), and should contrast with the surrounding chemical milieu. Interactions between the flow velocity and viscosity of the fluid medium (as influenced by convection) impact how chemical plumes diffuse within a boundary layer or are broken into filaments or "pockets" by turbulence (Vickers 2000; Murlis et al. 2000). These phenomena help to determine the "active space" within which chemical information is detected by receiver organisms at appropriate spatial and temporal scales (Zimmer and Zimmer 2008). Such considerations have guided studies of chemical tracking by crabs, moths, and ants (Cardé and Willis 2008) but are equally relevant to studies of plant–plant communication and eavesdropping, as reviewed by Blande et al. (2010) and Karban (this book).

By focusing on airborne chemical signals and cues produced by terrestrial plants, we limit our discussion to VOCs whose chemical properties endow them with sufficient vapor pressure under typical atmospheric conditions. The known universe of plant VOCs, either alone or in conjunction with microbial, fungal, or animal partners, includes compounds of diverse biosynthetic origins, molecular structures, and functional groups. Abiotic factors such as soil nutrients, photoperiod and light quality, ambient temperature, and relative humidity impact signal production through altering photosynthetic gas exchange, VOC biosynthesis, and emission (Kolosova et al. 2001; Gouinguené and Turlings 2002). Industrial pollutants and ozone also affect the quality and quantity of volatile signals as they are emitted (Himanen et al. 2009) through VOC diffusion into leaf or floral boundary layers, as well as plume composition farther from the source tissues (Pinto et al. 2010). We discuss these phenomena below after taking a closer look at structural and evolutionary aspects of plant-emitted VOCs.

Despite the apparent structural diversity of VOCs, certain compounds are overrepresented in the plant headspace as currently understood in the literature. This is especially true for floral VOCs including limonene, (E)-β-ocimene, β-myrcene, linalool, α- and β-pinene and β-caryophyllene (terpenoids), benzaldehyde, methyl salicylate, benzyl alcohol and 2-phenylethanol (aromatics), and sulcatone (methyl-5-hepten-2-one, a carotenoid derivative). These compounds have been identified from 52–71 percent of all taxa investigated thus far, based on a survey of nearly one thousand plant species representing ninety families and thirty-eight orders (Knudsen et al. 2006). Why are these twelve VOCs so overrepresented, when more than 1,700 volatile structures have been identified from floral headspace? One (efficacy-related) possibility is that their physical and chemical properties (volatility, polarity, structural stability, redox potential) satisfy a number of criteria for optimal chemical communication within the range of ambient conditions in which plants interact with other organisms. Another (content-related) possibility is that these and similar compounds are somehow ideally suited for information perception in terrestrial organisms. Most of these compounds demonstrate multiple ecological functions depending upon dosage and context (Raguso 2008). Schiestl (2010) used a meta-analysis to address whether these and related compounds represent a kind of preexisting sensory channel for terrestrial insect–plant interactions, predating the evolution of flowering plants and pollination (Rodriguez and Levin 1976; Pellmyr and Thien 1986). Using statistical tests of phylogenetic commonness, Schiestl (2010) tracked sixty-three major VOCs known to be produced by both vascular plants and insects (in which they serve as pheromones or defense compounds) and found strong relationships between monoterpenes and herbivorous insects, and between aromatic compounds and pollinators, but no similar relationships involving GLVs or sesquiterpenes. Most of the common monoterpenes are ubiquitous essential oil components of gymnosperm lineages, suggesting that they are primitive characters whose ancestral function was likely to have been antiherbivore toxins or repellents (discussed above). Building on this theme, Junker and Blüthgen (2010) used field bioassays to demonstrate that common floral monoterpenes are attractive to insects that must feed from flowers but are repellent to insects that only do so opportunistically (and thus may not have evolved resistance). This pattern was not as pronounced for aromatic floral volatiles, which are more commonly associated with floral attraction of moth mutualists and antagonists (Theis and Adler 2011). However, even some aromatic compounds have been shown to deter floral enemies (as discussed above for ants and 2-phenylethanol; Galen et al. 2011). Gene silencing experiments showed that when isoeugenol and benzyl benzoate are experimentally removed from the scent of *Petunia* flowers, there is an increase of florivory by *Diabrotica* beetles and wood crickets in field settings (Kessler et al. 2013).

One theme emerging from these studies is that of an ancient chemical language shared among insects and plants, with the flexibility to assign new meanings to old chemical phrases used in novel contexts, ranging from complex networks of plant–pollinator interactions to the multitrophic interaction webs mediated by herbivore-induced plant volatiles (Holopainen 2004). Anthropogenic environmental change may present a different kind of novel context, in which increasing atmospheric temperature, carbon dioxide (CO_2) concentration, and pollution impact the production and transmission of VOCs from plant tissues (Rinnan et al. 2005; Himanen et al. 2009). The unsaturated VOC molecules emitted by damaged or herbivore-induced vegetation (e.g., monoterpenes, sesquiterpenes, and GLVs) are vulnerable to ozonolysis, resulting in derivative compounds and secondary organic aerosols (Pinto et al. 2010). Because predaceous spider mites and parasitoid wasps use these compounds to find herbivores on damaged plants, it was thought that increased atmospheric ozone (O_3) could impair plant indirect defenses (Vuorinen et al. 2004; Pinto, Blande, et al. 2007). In experimental systems including cabbage, *Plutella* diamondback moths, and *Cotesia* parasitoid wasps, high O_3 treatments significantly degraded some VOCs ((Z)-3-hexen-1-ol, limonene and other monoterpenes, DMNT and (*E,E*)-α-farnesene), but not others (1,8-cineole and benzyl cyanide; Pinto, Nerg, and Holopainen 2007). In binary choice tests, *Cotesia* wasps preferred VOC blends induced by *Plutella* moth larvae under ambient conditions versus (degraded) blends induced under high O_3 treatments, but preferred either of these VOC blends to those of undamaged cabbage plants (Pinto, Nerg, and Holopainen 2007). Thus the induced blend, however diminished, is sufficient to facilitate VOC-mediated communication between cabbage plants and the natural enemies of their herbivores. Future environments with enriched atmospheric O_3 may select for less reactive VOCs (e.g., benzyl cyanide) or even alternative chemical channels (e.g., nonvolatile root exudates) as signals (Pinto, Tiiva, et al. 2007; Pinto et al. 2010).

McFrederick et al. (2008) used a modeling approach to explore the potential for atmospheric degradation of the most abundant floral VOCs—linalool, β-myrcene, and (*E*)-β-ocimene—asking how ozonolysis of these compounds might impact the spatial and temporal structure of floral scent plumes. Their models suggest that oxidation would dramatically reduce the postemission lifetimes of these monoterpenes and β-caryophyllene and thus reduce the distance from which insect pollinators could be attracted. The authors further warn that experimental studies of VOC signal degradation should include other common atmospheric pollutants (OH and NO_3) and suggest that anthropogenic conditions might select for the most stable compounds (e.g., benzaldehyde) to increase in frequency as key floral signals (McFrederick et al. 2009). However, other atmospheric pollutants can have additional impacts on volatile communication. Wind tunnel studies demonstrate that the ability of *Manduca sexta* moths to track plumes of benzaldehyde is impaired in urban air, due to the presence of toluene as a pollutant that masks benzaldehyde perception and tracking by the moths' olfactory receptor neurons (Riffell et al. 2014).

Under present conditions, the distances over which VOCs promote plant–plant communication already are modest (Karban, Yang, and Edwards 2014). Could ozone impair volatile priming or eavesdropping in plants? Blande et al. (2010) addressed this question with lima bean plants *(Phaseolus lunatus)*, in which VOC priming by wounded leaves increases the production of extrafloral nectar (EFN)—a common mechanism for recruiting predators to the defense of plants is to provision them with sugar—in adjacent, nonwounded leaves (Heil and Silva Bueno 2007). Treatment with ozone concentrations of eighty parts per billion (ppb) reduced the functional distance of volatile priming of EFN from seventy to twenty centimeters, degrading the same herbivore-induced VOCs studied in cabbage ((E)-β-ocimene, DMNT, and so on), but not methyl salicylate or 2-butanone. However, to complicate things, the highest O_3 treatment (160 ppb) resulted in direct increase of EFN production, suggesting that high ozone stress directly induces the jasmonate pathway—the primary physiological response to plant wounding—independent of VOC priming. This result underscores the complex interplay between different environmental factors whose direct effects, including increased photosynthetic rates and carbon assimilation, decreased VOC emissions with elevated temperature and CO_2 (Vuorinen et al. 2004; Himanen et al. 2009) and ozone quenching by increased isoprene and monoterpene emission under high O_3, might cancel out physiologically under some conditions (Blande et al. 2010). The most urgent need at present is for behavioral and physiological experiments that integrate a full spectrum of realistically modeled ambient conditions and pollutants in the most naturalistic settings possible, for tritrophic and plant–pollinator systems.

Biotic Factors Affecting Information Content of VOC Cues (and the Plant's Adaptation to Them)

Just as the physical environment can affect the functionality of VOCs, the biotic environment also can affect the information content provided by an organism. The majority of compounds whose functions were discussed above as mediating interactions between plants and other organisms are common in nature. Their study necessitates an untargeted scientific approach in which the function of VOC emission is investigated as context-dependent and the roles of organisms as intended primary receivers of information (mutualists or dupes) or unintended eavesdroppers are considered to be interchangeable, depending on the community in which the interactions are played out. This approach widens the list of possible functions of VOC-mediated information transfer in each individual system but allows for specific predictions for the evolution of VOC-mediated information transfer to maximize positive ("intended") effects and minimize the potential of eavesdropping.

Volatile information transfer can be affected by additional organisms in a plant's community that alter the composition of scent blends. This can happen (1) through changes in VOC emission induced by third parties that affect information transfer between organisms engaged in an interaction within a plant, and (2) through VOC emissions from neighboring sources that directly or indirectly affect VOC information quality and so affect ongoing interactions.

The presence of a herbivore or pathogen that induces changes in plant metabolism and VOC emission can confound ongoing constitutive information transfer between plants and other organisms, as well as signaling induced by an already present herbivore. For example, herbivory in the wild tomato *Solanum peruvianum* induces changes in leaf and floral VOC emission (Kessler and Halitschke 2009). The pollinating bees in this system use floral VOC emission to find suitable flowers for pollen collection. Interestingly, the bees avoid VOC blends from flowers on herbivore-damaged plants, which results in significant reductions of plant fitness (Kessler et al. 2011). In a similar study on white turnip, *Brassica rapa*, herbivory reduced floral VOC emission and, in consequence, pollinator attraction. However, reduced pollinator attraction was compensated for by herbivore-induced production of more and earlier flowering as well as the increased attraction of parasitoids, so that plants did not suffer significant fitness costs of herbivory (Schiestl et al. 2014). In both cases, constitutive floral VOC-mediated plant–pollinator interactions were interrupted by herbivore-induced changes in plant metabolism, but the fitness effects for the plants were different, presumably reflecting ecological context.

Desurmont et al. (2014) reviewed cases in which insect species introduced into new habitats have significant negative effects on established VOC interaction networks and thus the fitness of participating organisms. Accordingly, information transfer within a native community is subject to diffuse selection that minimizes such negative third-party interference on VOC emission. Depending on the primary agent of selection on VOC traits, this could lead to relatively unspecific induced VOC emissions. For example, the big-eyed bug *(Geocoris pallens)*, a generalist predator of wild tobacco *(Nicotiana attenuata)*, is equally attracted to relatively unspecific VOC emissions induced by a diverse set of herbivores (Kessler and Baldwin 2001). Alternatively, in systems with multiple specialized receivers, induced VOC emission could potentially establish multiple private channels that are not easily compromised by simultaneous attacks. Evidence for such effects come from experiments where VOC emission is specifically induced by different herbivore species (Clavijo McCormick et al. 2012) or by parasitized versus nonparasitized individuals (Poelman et al. 2011), different densities of herbivores (Girling et al. 2011; Horiuchi et al. 2003), or developmental stages (Yoneya et al. 2009) of the same species that can inform very specific host search behaviors in predators and parasitoids (Poelman et al. 2011). These findings suggest that the complexity of an interaction network should be a major determining factor in the evolution of VOC-mediated specificity. The major prediction derived from this hypothesis is that the size of the interactions network mediated by herbivore-induced VOC emission should be strongly correlated with the specificity of induced responses. Comparative studies that combine interaction network analyses with assessments of VOC induction specificity can address this hypothesis.

Neighboring plants provide a different source of interference with VOC-mediated information transfer. A potential analogy for such effects is communication in a noisy environment, where background noise provides either false information or makes it difficult to extract (perceive) the information specific to an interaction. As discussed above, constitutive VOC emissions of a plant can be affected by the identity of the competing neighbor (Ormeño et al. 2007; Peñuelas and Llusià 1998) or by adsorption and reemission of VOCs of a neighbor (Himanen et al. 2010). Many cases of associational resistance or associational susceptibility are probably a result of VOC-mediated information transfer being compromised by emissions from biotic sources (Barbosa et al. 2009). In a recent study, altered headspace VOC emissions from endophytic fungus-infected *Lolium multiflorum* grass were interpreted to decrease recruitment of aphids to neighboring clover plants (*Trifolium repens*; García Parisi et al. 2014). Although in this case the emission conveyed associational resistance from a resistant grass to its neighbor, we expect the same VOC emission to compromise other VOC-mediated interactions of clover with its community. These examples illustrate that the outcome of VOC-mediated interactions depends not only on senders and receivers, but also on nonparticipating organisms providing the information landscape in which the interactions are played out. While there is need for more research in this direction, the available evidence suggests that VOC-mediated interaction networks evolve in the context of the surrounding community. We predict that different habitats (information landscapes) should potentially select for different VOC compositions and specificities for a given set of interactions. Furthermore, very "noisy" environments with respect to VOCs in the plant community should select for alternative interaction channels, leading to character displacement in signal quality and quantity.

From an applied perspective, understanding such associational effects can help to establish new pest control technologies that are based on manipulating the VOC-mediated information landscape (Pickett et al. 2014). One shining example is the "push–pull" technology developed to control stem-boring lepidopteran pests of maize (*Zea mays*) in East Africa (Hassanali et al. 2008). In this system, so-called "push plants" are intercropped between rows of maize. These push plants (e.g., *Desmodium*

uncinatum [Fabaceae], *Melinis minutiflora* [Poaceae]) constitutively produce VOCs such as (*E*)-β-ocimene and DMNT that maize would only
produce when damaged by caterpillars of stem-boring moths (e.g., *Bus-
seola fusca* [(Noctuidae], *Chilo partellus* [Crambidae]). This constitutive
repellent VOC emission manipulates the information landscape surrounding the maize plants, making the entire field repellent to gravid
adult moths (see above). Plants that are more attractive than maize, such
as the two native African grasses *Sorghum vulgare sudanense* and *Pennis-
etum purpureum*, are grown as borders, and so emphasize the contrast
between the repellent center and the attractive circumference of the field
(Khan et al. 2008). Our current understanding of such VOC-mediated
information networks suggests that solutions like the push–pull technology have to be specifically developed for each target crop species and,
most likely, for each region individually to maximize the positive effects
on crop production.

CONCLUDING THOUGHTS

Until recently, the study of plant volatiles has occurred in separate fields,
focused on economic aspects of essential oils, on large-scale emissions
and atmospheric processes, on floral physiology and pollinator behavior, on tritrophic interactions mediated by herbivore-induced plant
responses, or more recently on plant–plant interactions, both above and
below ground. Our own research programs have called attention to the
need to experimentally and conceptually integrate these phenomena into
a whole-plant context, in which the full spectrum of selective forces,
community interactions, and life history parameters are brought to bear
upon understanding how volatile communication impacts plant survival,
competitive ability, and reproductive success (Kessler and Halitschke
2009; Galen et al. 2011). Progress has been slowed in part by the nature
of identifying and quantifying volatile chemical traits, which requires a
scientific language (analytical chemistry) that is distinct from the study
of other plant characteristics and is inaccessible to many plant biologists.
Although such approaches provide reproducible patterns that hint at the
processes we have described above, they also generate jargon-rich data
sets that are harder to digest than measurements of morphology or
sensory traits related to the electromagnetic spectrum (Karban, this

book). In essence, chemical ecologists have had to name each pixel of a pointillist mural whose subject is only now coming into focus.

For these reasons, we chose in this chapter to survey these chemical pixels as words and phrases in a universal plant volatile language, not as formal linguists but as ecological tourists with a small and incomplete Berlitz guide. We have discovered some versatile chemical words in the form of conventional volatiles (e.g., methyl salicylate, caryophyllene, linalool), whose biological roles vary with dosage and context, as well as physical scale (Raguso 2008). These multifunctional compounds are akin to words like hot and cool, whose meanings vary richly across the geographic and historical spread of the former British Empire. We also have encountered chemical words whose meanings are more constrained, either because of their novelty (e.g., chiloglottones in sexual deception; Franke et al. 2009) or because their production is a reliable index cue of a specific state (e.g., volatile sulfides in brood-site deception; Jürgens et al. 2013). Yet, even specific signals can take on new meanings, as sulfides also are released as belowground herbivore-induced volatiles (Ferry et al. 2007; van Dam et al. 2010) or as attractants in nectar-rich bat-pollinated flowers (Raguso 2004). In a similar manner, neologisms such as the proper noun "Boojum" in Lewis Carroll's *Hunting of the Snark* have been co-opted by botanists to describe a distinctive desert tree *(Idria columnaris)* from Baja California (Humphrey 1974) and by physicists to describe the properties of liquid surfaces (Mermin 1981).

When should chemical neologisms evolve, and when should they maintain specific meanings? A novel VOC, like a key adaptation, might allow a plant to explore novel adaptive niches, which often leads to rapid diversification, as demonstrated for the evolution of nectar spurs in columbines *(Aquilegia)* and other plant lineages (Hodges and Arnold 1995). We suspect that adaptive shifts in plant volatiles may be driven by complex selective pressures through community interactions, and may result—as in defense chemistry (Futuyma and Agrawal 2009)—in ecological escape from eavesdropping herbivores and competitors, leading to subsequent diversification. The carrion- or fecal-mimicking flowers described above are well represented in African grassland and semidesert habitats (Jürgens et al. 2006) with large populations of antelopes and other herbivorous mammals providing potentially strong selective pressures. Lev-Yadun et al. (2009) question whether the evolution of such plants is driven by pollinator shifts or, rather, by a novel antiherbivore strategy—the smell of carnivore dung reliably indicates the presence of a predator—followed by subsequent adaptation to carrion flies as pollinators. The fact that one of these plants *(Hoodia gordonii)* not only produces fetid carrion flowers (Jürgens et al. 2006) but also is a commercial source of appetite suppressants marketed by the pharmaceutical industry (van Heerden et al. 2007) suggests that selective pressures by mammalian herbivores may have shaped the chemistry of these plants. It is also consistent with the "escape and radiate" key innovation idea that many plant lineages with carrion- or dung-mimicking flowers show high species diversity relative to their sister lineages (e.g., *Aristolochia, Rafflesia, Amorphophallus*; Davis et al. 2008).

Similarly, strong community-related selective pressure for chemical novelty could take the form of reproductive isolation or fruiting success. As discussed above, the majority of fig species studied thus far utilize specific ratios, or "bar codes," of terpenoids to attract pollinating wasps. In whole-plant and phylogenetic contexts, the terpenoid language of figs is not surprising, because they produce latex rich in nonvolatile terpenoids as a physical and chemical defense (Konno 2011). We predict reproductive character displacement to novel, nonterpenoid floral signals (e.g., methylanisol in *Ficus semicordata*) in communities packed with other, co-blooming fig species. This could be driven by the negative fitness consequences of hybridization due to female pollinator wasps laying eggs into the wrong host species (Herre et al. 2008) or with low fruiting success due to eavesdropping on shared fig volatiles by nonpollinating fig wasps (Proffit et al. 2009). Indeed, two races of *F. semicordata* pollinated by different wasp species in southern China show divergent chemistry of receptive figs, and their respective pollinators behave very differently before entering the fig through the ostiole (Wang et al. 2013). Conversely, when microhabitat differences allow sympatric *Ficus* species to share different populations of the same pollinator species, such selective pressures appear to be relaxed (Cornille et al. 2011).

These examples highlight the exciting direct and indirect roles that volatiles play in mediating important fitness-related aspects of plant life and provide the caveats that chemical communication in plants complements other sensory channels and their associated traits (Karban, this book) and that plant survival and reproduction are linked within a

whole-plant context. As the causes and consequences of volatile communication and its evolution become better understood, we look forward to a deeper, more predictive understanding of the lexicon of plant
volatile chemistry.

REFERENCES

Arab, A., J. R. Trigo, A. L. Lourenção, A. M. Peixoto, F. Ramos, and J. M. S.
Bento. 2007. "Differential Attractiveness of Potato Tuber Volatiles to
Phthorimaea operculella (Gelechiidae) and the Predator *Orius insidiosus*
(Anthocoridae)." *Journal of Chemical Ecology* 33:1845–55.

Arimura, G., R. Ozawa, T. Shimoda, T. Nishioka, W. Boland, and J. Takabayashi.
2000. "Herbivory-Induced Volatiles Elicit Defence Genes in Lima Bean
Leaves." *Nature* 406:512–15.

Arimura, G., K. Shiojiri, and R. Karban. 2010. "Acquired Immunity to Herbivory and Allelopathy Caused by Airborne Plant Emissions." *Phytochemistry*
71:1642–49.

Asai, N., T. Nishioka, J. Takabayashi, and T. Furuichi. 2009. "Plant Volatiles Regulate the Activities of Ca^{2+}-Permeable Channels and Promote Cytoplas-
mic Calcium Transients in Arabidopsis Leaf Cells." *Plant Signaling &*
Behavior 4:294.

Ayasse, M., J. Stökl, and W. Francke. 2011. "Chemical Ecology and Pollinator-
Driven Speciation in Sexually Deceptive Orchids." *Phytochemistry* 72:
1667–77.

Baldwin, I. T., R. Halitschke, A. Paschold, C. C. von Dahl, and C. A. Preston.
2006. "Volatile Signaling in Plant–Plant Interactions: 'Talking Trees' in the
Genomics Era." *Science* 311:812–15.

Barbosa, P., J. Hines, I. Kaplan, H. Martinson, A. Szczepaniec, and Z. Szendrei.
2009. "Associational Resistance and Associational Susceptibility: Having
Right or Wrong Neighbors." *Annual Review of Ecology, Evolution, and Systematics* 40:1–20.

Blande, J. D., J. K. Holopainen, and T. Li. 2010. "Air Pollution Impedes Plant-
to-Plant Communication by Volatiles." *Ecology Letters* 13:1172–81.

Bohman, B., R. D. Phillips, M. H. M. Menz, B. W. Berntsson, G. R. Flematti,
R. A. Barrow, K. W. Dixon, and R. Peakall. 2014. "Discovery of Pyrazines
as Pollinator Sex Pheromones and Orchid Semiochemicals: Implications
for the Evolution of Sexual Deception." *New Phytologist* 203:939–52.

Brodmann, J., R. Twele, W. Francke, G. Hölzler, Q.-H. Zhang, and M. Ayasse.
2008. "Orchids Mimic Green-Leaf Volatiles to Attract Prey-Hunting Wasps
for Pollination." *Current Biology* 18:740–44.

Bruce, T. J. A., L. J. Wadhams, and C. M. Woodcock. 2005. "Insect Host Location: A Volatile Situation." *Trends in Plant Science* 10:269–74.

Campbell, S. A., J. S. Thaler, and A. Kessler. 2013. "Plant Chemistry Underlies Herbivore-Mediated Inbreeding Depression in Nature." *Ecology Letters* 16:252–60.

Cardé, R. T., and M. A. Willis. 2008. "Navigational Strategies Used by Insects to Find Distant, Wind-Borne Sources of Odor." *Journal of Chemical Ecology* 34:854–66.

Carroll, M. J., E. A. Schmelz, R. L. Meagher, and P. E. A. Teal. 2006. "Attraction of *Spodoptera frugiperda* Larvae to Volatiles from Herbivore-Damaged Maize Seedlings." *Journal of Chemical Ecology* 32:1911–24.

Chen, C., and Q. Song. 2008. "Responses of the Pollinating Wasp *Ceratosolen solmsi marchali* to Odor Variation between Two Floral Stages of *Ficus hispida*." *Journal of Chemical Ecology* 34:1536–44.

Chen, C., Q. Song, M. Proffit, J.-M. Bessière, Z. Li, and M. Hossaert-McKey. 2009. "Private Channel: A Single Unusual Compound Assures Specific Pollinator Attraction in *Ficus semicordata*." *Functional Ecology* 23:941–50.

Chinta, S., J. C. Dickens, and J. R. Aldrich. 1994. "Olfactory Reception of Potential Pheromones and Plant Odors by Tarnished Plant Bug, Lygus lineolaris (Hemiptera: Miridae)." *Journal of Chemical Ecology* 12:3251–67.

Clavijo McCormick, A., S. B. Unsicker, and J. Gershenzon. 2012. "The Specificity of Herbivore-Induced Plant Volatiles in Attracting Herbivore Enemies." *Trends in Plant Science* 17:303–10.

Cornille, A., J. G. Underhill, A. Cruaud, M. Hossaert-McKey, S. D. Johnson, K. A. Tolley, F. Kjellberg, et al. 2011. "Floral Volatiles, Pollinator Sharing and Diversification in the Fig–Wasp Mutualism: Insights from *Ficus natalensis* and Its Two Wasp Pollinators (South Africa)." *Proceedings of the Royal Society B* 279:1731–39.

Cozzolino, S., and A. Widmer. 2005. "Orchid Diversity: An Evolutionary Consequence of Deception?" *Trends in Ecology and Evolution* 20:487–94.

Dafni, A. 1984. "Mimicry and Deception in Pollination." *Annual Review of Ecology and Systematics* 15:259–78.

Dani, F. R., G. R. Jones, S. Destri, S. H. Spencer, and S. Turillazzi. 2001. "Deciphering the Recognition Signature within the Cuticular Chemical Profile of Paper Wasps." *Animal Behaviour* 62:165–71.

Davis, C. C., P. K. Endress, and D. A. Baum. 2008. "The Evolution of Floral Gigantism." *Current Opinion in Plant Biology* 1:49–57.

De Moraes, C. M., M. C. Mescher, and J. H. Tumlinson. 2001. "Caterpillar-Induced Nocturnal Plant Volatiles Repel Conspecific Females." *Nature* 410: 577–80.

Denison, R. F., C. Bledsoe, M. Kahn, F. O'Gara, E. L. Simms, and L. S. Thomashow. 2003. "Cooperation in the Rhizosphere and the 'Free Rider' Problem." *Ecology* 84:838–45.

Desurmont, G. A., J. Harvey, N. M. van Dam, S. M. Cristescu, F. P. Schiestl, S. Cozzolino, P. Anderson, et al. 2014. "Alien Interference: Disruption of Infochemical Networks by Invasive Insect Herbivores." *Plant, Cell & Environment* 37:1854–65.

Dicke, M., and J. J. A. van Loon. 2000. "Multitrophic Effects of Herbivore-Induced Plant Volatiles in an Evolutionary Context." *Entomologia Experimentalis et Applicata* 97:237–49.

Dudareva, N., E. Pichersky, and J. Gershenzon. 2004. "Biochemistry of Plant Volatiles." *Plant Physiology* 135:1893–1902.

Engelberth, J., H. T. Alborn, E. A. Schmelz, and J. H. Tumlinson. 2004. "Airborne Signals Prime Plants against Insect Herbivore Attack." *Proceedings of the National Academy of Sciences* 101:1781–85.

Farré-Armengol, G., I. Filella, J. Llusia, and J. Penuelas. 2013. "Floral Volatile Organic Compounds: Between Attraction and Deterrence of Visitors Under Global Change." *Perspectives in Plant Ecology, Evolution and Systematics* 15: 56-67.

Ferry, A., S. Dugravot, T. Delattre, J.-P. Christides, and J. Auger. 2007. "Identification of a Widespread Monomolecular Odor Differentially Attractive to Several *Delia radicum* Ground-Dwelling Predators in the Field." *Journal of Chemical Ecology* 33:2064–77.

Franke, S., F. Ibarra, C. M. Schulz, C. Twele, J. Poldy, R. A. Barrow, R. Peakall, et al. 2009. "The Discovery of 2,5-Dialkylcyclohexan-1,3-Diones as a New Class of Natural Products." *Proceedings of the National Academy of Sciences* 106:8877–82.

Frost, C. J., H. M. Appel, J. E. Carlson, C. M. De Moraes, M. C. Mescher, and J. C. Schultz. 2007. "Within-Plant Signalling via Volatiles Overcomes Vascular Constraints on Systemic Signalling and Primes Responses against Herbivores." *Ecology Letters* 10:490–98.

Frost, C. J., M. C. Mescher, C. Dervinis, J. M. Davis, J. E. Carlson, and C. M. De Moraes. 2008. "Priming Defense Genes and Metabolites in Hybrid Poplar by the Green Leaf Volatile Cis-3-Hexenyl Acetate." *New Phytologist* 180:722–34.

Futuyma, D. J., and A. A. Agrawal. 2009. "Macroevolution and the Biological Diversity of Plants and Herbivores." *Proceedings of the National Academy of Sciences* 106:18054–61.

Galen, C., R. Kaczorowski, S. L. Todd, J. Geib, and R. A. Raguso. 2011. "Dosage-Dependent Impacts of a Floral Volatile Compound on Pollinators,

Larcenists, and the Potential for Floral Evolution in the Alpine Skypilot *Polemonium viscosum.*" *American Naturalist* 177:258–72.

García Parisi, P. A., A. A. Grimoldi, and M. Omacini. 2014. "Endophytic Fungi of Grasses Protect Other Plants from Aphid Herbivory." *Fungal Ecology* 9:61–64.

Gaskett, A. C. 2011. "Orchid Pollination by Sexual Deception: Pollinator Perspectives." *Biological Reviews* 86:33–75.

Gigord, L. D. B., M. R. Macnair and A. Smithson. 2001. "Negative Frequency-Dependent Selection Maintains a Dramatic Flower Color Polymorphism in the Rewardless Orchid *Dactylorhiza sambucina* (L.) Soò." *Proceedings of the National Academy of Sciences* 98:6253–55.

Girling, R. D., A. Stewart-Jones, J. Dherbecourt, J. T. Staley, D. J. Wright, and G. M. Poppy. 2011. "Parasitoids Select Plants More Heavily Infested with Their Caterpillar Hosts: A New Approach to Aid Interpretation of Plant Headspace Volatiles." *Proceedings of the Royal Society B* 278:2646–53.

Gouinguené, S. P., and T. C. J. Turlings. 2002. "The Effects of Abiotic Factors on Induced Volatile Emissions in Corn Plants." *Plant Physiology* 129: 1296–1307.

Hassanali, A., H. Herren, Z. R. Khan, J. A. Pickett, and C. M. Woodcock. 2008. "Integrated Pest Management: The Push–Pull Approach for Controlling Insect Pests and Weeds of Cereals, and Its Potential for Other Agricultural Systems Including Animal Husbandry." *Philosophical Transactions of the Royal Society B* 363:611–21.

Hebets, E. A., and D. R. Papaj. 2005. "Complex Signal Function: Developing a Framework of Testable Hypotheses." *Behavioral Ecology and Sociobiology* 57:197–214.

Heil, M., and R. Karban. 2010. "Explaining the Evolution of Plant Communication by Airborne Signals." *Trends in Ecology and Evolution* 25:137–44.

Heil, M., and J. C. Silva Bueno. 2007. "Within-Plant Signaling by Volatiles Leads to Induction and Priming of an Indirect Plant Defense in Nature." *Proceedings of the National Academy of Sciences* 104:5467–72.

Herre, E. A., K. C. Jandér, and C. A. Machado. 2008 "Evolutionary Ecology of Figs and Their Associates: Recent Progress and Outstanding Puzzles." *Annual Review of Ecology, Evolution, and Systematics* 39:439–58.

Himanen, S. J., J. D. Blande, T. Klemola, J. Pulkkinen, J. Heijari, and J. K. Holopainen. 2010. "Birch (*Betula* Spp.) Leaves Adsorb and Re-Release Volatiles Specific to Neighbouring Plants—a Mechanism for Associational Herbivore Resistance?" *New Phytologist* 186:722–32.

Himanen, S. J., A.-M. Nerg, A. Nissinen, D. M. Pinto, C. N. Stewart Jr., G. M. Poppy, and J. K. Holopainen. 2009. "Effects of Elevated Carbon Dioxide and

Ozone on Volatile Terpenoid Emissions and Multitrophic Communication of Transgenic Insecticidal Oilseed Rape (*Brassica napus*)." *New Phytologist* 181:174–86.

Hodges, S. A., and M. L. Arnold. 1995. "Spurring Plant Diversification: Are Floral Nectar Spurs a Key Innovation?" *Proceedings of the Royal Society B* 262:343–48.

Holopainen, J. K. 2004. "Multiple Functions of Inducible Plant Volatiles." *Trends in Plant Science* 9:529–33.

Horiuchi, J.-I., G.-I. Arimura, R. Ozawa, T. Shimoda, J. Takabayashi, and T. Nishioka. 2003. "A Comparison of the Responses of *Tetranychus urticae* (Acari: Tetranychidae) and *Phytoseiulus persimilis* (Acari: Phytoseiidae) to Volatiles Emitted from Lima Bean Leaves with Different Levels of Damage Made by *T. urticae* or *Spodoptera exigua* (Lepidoptera: Noctuidae)." *Applied Entomology and Zoology* 38:109–16.

Hossaert-McKey, M., C. Soler, B. Schatz, and M. Proffit. 2010. "Floral Scents: Their Roles in Nursery Pollination Mutualisms." *Chemoecology* 20:75–88.

Humphrey, R. R. 1974. *The Boojum and Its Home*. Tucson: University of Arizona Press.

Jersáková, J., and S. D. Johnson. 2006. "Lack of Floral Nectar Reduces Self-Pollination in a Fly-Pollinated Orchid." *Oecologia* 147:60–68.

Jersáková, J., S. D. Johnson, and P. Kindlmann. 2006. "Mechanisms and Evolution of Deceptive Pollination in Orchids." *Biological Reviews* 81:219–35.

Johnson, S. D., and A. Jürgens. 2010. "Convergent Evolution of Carrion and Faecal Scent Mimicry in Fly-Pollinated Angiosperm Flowers and a Stinkhorn Fungus." *South African Journal of Botany* 76:796–807.

Johnson, S. D., and L. A. Nilsson. 1999. "Pollen Carryover, Geitonogamy and the Evolution of Deceptive Pollination Systems in Orchids." *Ecology* 80: 2607–19.

Junker, R. R., and N. Blüthgen. 2010. "Floral Scents Repel Facultative Flower Visitors, but Attract Obligate Ones." *Annals of Botany* 105:777–82.

Junker, R. R. and D. Tholl. 2013. "Volatile Organic Compound Mediated Interactions at the Plant–Microbe Interface." *Journal of Chemical Ecology* 39:810–25.

Jürgens, A., S. Dötterl, and U. Meve. 2006. "The Chemical Nature of Fetid Floral Odours in Stapeliads (Apocynaceae–Ascelepiadoideae–Ceropegieae)." *New Phytologist* 172:452–68.

Jürgens, A., S.-L. Wee, A. Shuttleworth, and S. D. Johnson. 2013. "Chemical Mimicry of Insect Oviposition Sites: A Global Analysis of Convergence in Angiosperms." *Ecology Letters* 16:1157–67.

Karban, R., J. Maron, G. W. Felton, G. Ervin, and H. Eichenseer. 2003. "Herbivore Damage to Sagebrush Induces Resistance in Wild Tobacco: Evidence for Eavesdropping between Plants." *Oikos* 100:325–32.

Karban, R., and K. Shiojiri. 2009. "Self-Recognition Affects Plant Communication and Defense." *Ecology Letters* 12:502–6.

Karban, R., W. C. Wetzel, K. Shiojiri, S. Ishizaki, S. R. Ramirez, and J. D. Blande. 2014. "Deciphering the Language of Plant Communication: Volatile Chemotypes of Sagebrush." *New Phytologist* 204:380–85.

Karban, R., L. H. Yang, and K. F. Edwards. 2014. "Volatile Communication between Plants That Affect Herbivory: A Meta-Analysis." *Ecology Letters* 17:44–52.

Kariyat, R. R., K. E. Mauck, C. M. De Moraes, A. G. Stephenson, and M. C. Mescher. 2012. "Inbreeding Alters Volatile Signalling Phenotypes and Influences Tri-trophic Interactions in Horsenettle (*Solanum carolinense* L.)." *Ecology Letters* 15:301–9.

Kariyat, R. R., S. R. Scanlon, R. P. Moraski, A. G. Stephenson, M. C. Mescher, and C. M. De Moraes. 2014. "Plant Inbreeding and Prior Herbivory Influence the Attraction of Caterpillars *(Manduca sexta)* to Odors of the Host Plant *Solanum carolinense* (Solanaceae)." *American Journal of Botany* 101: 376–80.

Kessler, A., and I. T. Baldwin. 2001. "Defensive Function of Herbivore-Induced Plant Volatile Emissions in Nature. *Science* 291:2141–44.

———. 2004. "Herbivore-Induced Plant Vaccination: Part I. The Orchestration of Plant Defenses in Nature and Their Fitness Consequences in the Wild Tobacco *Nicotiana attenuata.*" *Plant Journal* 38:639–49.

Kessler, A., and R. Halitschke. 2009. "Testing the Potential for Conflicting Selection on Floral Chemical Traits by Pollinators and Herbivores: Predictions and Case Study." *Functional Ecology* 23:901–12.

Kessler, A., R. Halitschke, C. Diezel, and I. T. Baldwin. 2006. "Priming of Plant Defense Responses in Nature by Airborne Signaling between Artemisia tridentata and Nicotiana attenuata." *Oecologia* 148:280–92.

Kessler, A., R. Halitschke, and K. Poveda. 2011. "Herbivory-Mediated Pollinator Limitation: Negative Impacts of Induced Volatiles on Plant–Pollinator Interactions." *Ecology* 92:1769–80.

Kessler, A., and M. Heil. 2011. "The Multiple Faces of Indirect Defences and Their Agents of Natural Selection." *Functional Ecology* 25:348–57.

Kessler, D., C. Diezel, D. G. Clark, T. A. Colquhoun, and I. T. Baldwin. 2013. "Petunia Flowers Solve the Defence/Apparency Dilemma of Pollinator Attraction by Deploying Complex Floral Blends." *Ecology Letters* 16: 299–306.

Khan, Z. R., D. G. James, C. A. O. Midega, and J. A. Pickett. 2008. "Chemical Ecology and Conservation Biological Control." *Biological Control* 45:210–24.

Knudsen, J. T., R. Eriksson, J. Gershenzon, and B. Ståhl. 2006. "Diversity and Distribution of Floral Scent." *Botanical Review* 72:1–120.

Kolosova, N., N. Gorenstein, C. M. Kish, and N. Dudareva. 2001. "Regulation of Circadian Methyl Benzoate Emission in Diurnally and Nocturnally Emitting Plants." *Plant Cell* 13:2333–47.

Konno, K. 2011. "Plant Latex and Other Exudates as Plant Defense Systems: Roles of Various Defense Chemicals and Proteins Contained Therein." *Phytochemistry* 72:1510–30.

Lehtonen, J., and M. R. Whitehead. 2014. "Sexual Deception: Coevolution or Inescapable Exploitation?" *Current Zoology* 60:52–61.

Lev-Yadun, S., G. Ne'eman, and U. Shanas. 2009. "A Sheep in Wolf's Clothing: Do Carrion and Dung Odours of Flowers Not Only Attract Pollinators but Also Deter Herbivores?" *BioEssays* 31:84–88.

Long, S. R. 1996. "*Rhizobium* Symbiosis: Nod Factors in Perspective." *Plant Cell* 8:1885–98.

McFrederick, Q. S., J. D. Fuentes, T. H. Roulston, J. C. Kathilankal, and M. Lerdau. 2009. "Effects of Air Pollution on Biogenic Volatiles and Ecological Interactions." *Oecologia* 160:411–20.

McFrederick, Q. S., J. C. Kathilankal, and J. D. Fuentes. 2008. "Air Pollution Modifies Floral Scent Trails." *Atmospheric Environment* 42:2336–48.

Mermin, N. D. 1981. "E Pluribus Boojum: The Physicist as Neologist." *Physics Today*, April, 46–53.

Morrell, K., and A. Kessler. 2014. "Volatile-Mediated Information Transfer and Defence Priming in Plants: The Scent of Danger." *Biochemist* 36:26–31.

Murlis, J., M. A. Willis, and R. T. Cardé. 2000. "Spatial and Temporal Structures of Pheromone Plumes in Fields and Forests." *Physiological Entomology* 25:211–22.

Ollerton, J., and R. A. Raguso. 2006. "The Sweet Stench of Decay." *New Phytologist* 172:382–85.

Orians, C. 2005. "Herbivores, Vascular Pathways, and Systemic Induction: Facts and Artifacts." *Journal of Chemical Ecology* 31:2231–42.

Ormeño, E., C. Fernandez, and J.-P. Mévy. 2007. "Plant Coexistence Alters Terpene Emission and Content of Mediterranean Species." *Phytochemistry* 68:840–52.

Pellmyr, O, and C. J. Huth. 1994. "Evolutionary Stability of Mutualism between Yuccas and Yucca Moths." *Nature* 372:257–60.

Pellmyr, O., and L. B. Thien. 1986. "Insect Reproduction and Floral Fragrances: Keys to the Evolution of the Angiosperms?" *Taxon* 35:76–85.

Peñuelas, J., and J. Llusià. 1998. "Influence of Intra- and Inter-specific Interference on Terpene Emission by *Pinus halepensis* and *Quercus ilex* Seedlings." *Biologia Plantarum* 41:139–43.

Pickett, J. A., C. M. Woodcock, C. A. O. Midega, and Z. R. Khan. 2014. "Push–Pull Farming Systems." *Current Opinion in Biotechnology* 26:125–32.

Pinto, D. M., J. D. Blande, R. Nykänen, W. X. Dong, A.-M. Nerg, and J. K. Holopainen. 2007. "Ozone Degrades Common Herbivore-Induced Plant Volatiles: Does This Affect Herbivore Prey Location by Predators and Parasitoids?" *Journal of Chemical Ecology* 33:683–94.

Pinto, D. M., J. D. Blande, S. R. Souza, A.-M. Nerg, and J. K. Holopainen. 2010. "Plant Volatile Organic Compounds (VOCs) in Ozone (O_3) Polluted Atmospheres: The Ecological Effects." *Journal of Chemical Ecology* 36:22–34.

Pinto, D. M., A.-M. Nerg, and J. K. Holopainen. 2007. "The Role of Ozone-Reactive Compounds, Terpenes and Green Leaf Volatiles (GLVs) in the Orientation of *Cotesia plutellae*." *Journal of Chemical Ecology* 33:2218–28.

Pinto, D. M., P. Tiiva, P. Miettinen, J. Joutsenaari, H. Kokkola, A.-M. Nerg, A. Laaksonen, and J. K. Holopainen. 2007. "The Effects of Increasing Atmospheric Ozone on Biogenic Monoterpene Profiles and the Formation of Secondary Aerosols." *Atmospheric Environment* 41:4877–87.

Poelman, E. H., M. Bruinsma, F. Zhu, B. T. Weldegergis, A. E. Boursault, Y. Jongema, J. J. A. van Loon, et al. 2012. "Hyperparasitoids Use Herbivore-Induced Plant Volatiles to Locate Their Parasitoid Host." *PLOS Biology* 10:e1001435.

Poelman, E. H., S.-J. Zheng, Z. Zhang, N. M. Heemskerk, A-M. Cortesero, and M. Dicke. 2011. "Parasitoid-Specific Induction of Plant Responses to Parasitized Herbivores Affects Colonization by Subsequent Herbivores." *Proceedings of the National Academy of Sciences* 108:19647–52.

Proffit, M., C. Chen, S. Soler, J.-M. Bessière, B. Schatz, and M. Hossaert-McKey. 2009. "Can Chemical Signals, Responsible for Mutualistic Partner Encounter, Promote the Specific Exploitation of Nursery Pollination Mutualisms? The Case of Figs and Fig Wasps." *Entomologia Experimentalis et Applicata* 131:46–57.

Raffa, K. F. 2001. "Mixed Messages across Multiple Trophic Levels: The Ecology of Bark Beetle Chemical Communication Systems." *Chemoecology* 11:49–65.

Raffa, K. F., and E. B. Smalley. 1995. "Interactions of Pre-Attack and Induced Monoterpene Concentrations in Host Conifer Defense Against Bark Beetle-Microbial Complexes." *Oecologia* 102:285–295.

Raguso, R. A. 2004. "Flowers as Sensory Billboards: Progress Towards an Integrated Understanding of Floral Advertisement." *Current Opinion in Plant Biology* 7:434–40.

———. 2008. "Wake Up and Smell the Roses: The Ecology and Evolution of Floral Scent." *Annual Review of Ecology, Evolution and Systematics* 39:549–69.

Reddy, G. V., and A. Guerrero. 2004. "Interactions of Insect Pheromones and Plant Semiochemicals." *Trends in Plant Science* 9:253–61.

Renwick, J. A. A. 2002. "The Chemical World of Crucivores: Lures, Treats and Traps." *Entomologia Experimentalis et Applicata* 104:35–42.

Riffell, J. A., E. Shlizerman, E. Sanders, L. Abrell, B. Medina, A. J. Hinterwirth, and J. N. Kutz. 2014. "Flower Discrimination by Pollinators in a Dynamic Chemical Environment." *Science* 344:1515–18.

Rinnan, R., Å. Rinnan, T. Holopainen, J. K. Holopainen, and P. Pasanen. 2005. "Emission of Non-methane Volatile Organic Compounds (VOCs) from Boreal Peatland Microcosms—Effects of Ozone Exposure." *Atmospheric Environment* 39:921–30.

Rodriguez, E., and D. A. Levin. 1976. "Biochemical Parallelisms of Repellents and Attractants in Higher Plants and Arthropods." In *Biochemical Interaction between Plants and Insects*, edited by J. W. Wallace and R. L. Mansell, 214–70. New York: Plenum Press.

Runyon, J. B., Mescher, M. C., and C. M. De Moraes. 2006. "Volatile Chemical Cues Guide Host Location and Selection by Parasitic Plants." *Science* 313:1964–67.

Salzmann, C. C., S. Cozzolino, and F. P. Schiestl. 2007. "Floral Scent in Food-Deceptive Orchids: Species Specificity and Sources of Variability." *Plant Biology* 9:720–29.

Schaefer, H. M., and G. D. Ruxton. 2009. "Deception in Plants: Mimicry or Perceptual Exploitation?" *Trends in Ecology and Evolution* 24:676–85.

Schatz, B., R. Delle-Vedove, and L. Dormont. 2013. "Presence, Distribution and Effect of White, Pink and Purple Morphs on Pollination in the Orchid *Orchis mascula*." *European Journal of Environmental Sciences* 3:119–28.

Schiestl, F. P. 2005. "On the Success of a Swindle: Pollination by Deception in Orchids." *Naturwissenschaften* 92:255–64.

———. 2010. "The Evolution of Floral Scent and Insect Chemical Communication." *Ecology Letters* 13:643–56.

Schiestl, F. P., M. Ayasse, H. F. Paulus, C. Löfstedt, B. S. Hansson, F. Ibarra, and W. Francke. 1999. "Orchid Pollination by Sexual Swindle." *Nature* 399: 421–22.

Schiestl, F. P., and S. D. Johnson. 2013. "Pollinator-Mediated Evolution of Floral Signals." *Trends in Ecology and Evolution* 28:307–15.

Schiestl, F. P., H. Kirk, L. Bigler, S. Cozzolino, and G. A. Desurmont. 2014. "Herbivory and Floral Signaling: Phenotypic Plasticity and Tradeoffs between Reproduction and Indirect Defense." *New Phytologist* 203:257–66.

Schiestl, F. P., R. Peakall, J. G. Mant, F. Ibarra, C. Schulz, S. Franke, and W. Francke. 2003. "The Chemistry of Sexual Deception in an Orchid–Wasp Pollination System." *Science* 302:437–38.

Schiestl, F. P., and G. A. Wright. 2009. "Evolution of Floral Scent: The Influence of Olfactory Learning by Insect Pollinators on the Honest Signalling of Floral Rewards." *Functional Ecology* 23:841–51.

Seybold, S. J., D. P. W. Huber, J. C. Lee, A. D. Graves, and J. Bohlmann. 2006. "Pine Monoterpenes and Pine Bark Beetles: A Marriage of Convenience for Defense and Chemical Communication." *Phytochemistry Reviews* 5: 143–78.

Soler, C., M. Hossaert-McKey, B. Buatois, J.-M. Bessière, B. Schatz, and M. Proffit. 2011. "Geographic Variation of Floral Scent in a Highly Specialized Pollination Mutualism." *Phytochemistry* 72:74–81.

Stökl, J., J. Brodmann, A. Dafni, M. Ayasse, and B. S. Hansson. 2011. "Smells Like Aphids: Orchid Flowers Mimic Aphid Alarm Pheromones to Attract Hoverflies for Pollination." *Proceedings of the Royal Society B* 278:1216–22.

Sugimoto, K., K. Matsui, Y. Iijima, Y. Akakabe, S. Muramoto, R. Ozawa, M. Uefune, et al. 2014. "Intake and Transformation to a Glycoside of (Z)-3-Hexenol from Infested Neighbors Reveals a Mode of Plant Odor Reception and Defense." *Proceedings of the National Academy of Sciences* 111:7144–49.

Svensson, G. P., M. O. Hickman Jr., S. Bartram, W. Boland, O. Pellmyr, and R. A. Raguso. 2005. "Chemistry and Geographic Variation of Floral Scent in *Yucca filamentosa* (Agavaceae)." *American Journal of Botany* 92:1624–31.

Svensson, G. P., O. Pellmyr, and R. A. Raguso. 2011. "Pollinator Attraction to Volatiles from Virgin and Pollinated Host Flowers in a Yucca/Moth Obligate Mutualism." *Oikos* 120:1577–83.

Terry, L. I., R. B. Roemer, G. H. Walter, and D. Booth. 2014. "Thrips' Responses to Thermogenic Associated Signals in a Cycad Pollination System: The Interplay of Temperature, Light, Humidity and Cone Volatiles." *Functional Ecology* 28:857–67.

Theis, N., and L. S. Adler. 2011. "Advertising to the Enemy: Enhanced Floral Fragrance Increases Beetle Attraction and Reduces Plant Reproduction." *Ecology* 93:430–35.

Theis, N., M. Lerdau, and R. A. Raguso. 2007. "The Challenge of Attracting Pollinators While Evading Floral Herbivores: Patterns of Fragrance Emission in *Cirsium arvense* and *Cirsium repandum* (Asteraceae)." *International Journal of Plant Sciences* 168:587–601.

Urru, I., M. C. Stensmyr, and B. S. Hansson. 2011. "Pollination by Brood-Site Deception." *Phytochemistry* 72:1655–66.

Valone, T. J. 2007. "From Eavesdropping on Performance to Copying the Behavior of Others: A Review of Public Information Use." *Behavioral Ecology and Sociobiology* 62:1–14.

van Dam, N. M., B.-L. Qiu, C. A. Hordijk, L. E. M. Vet, and J. J. Jansen. 2010. "Identification of Biologically Relevant Compounds in Aboveground and Belowground Induced Volatile Blends." *Journal of Chemical Ecology* 36: 1006–16.

van Heerden, F. R., R. M. Horak, V. J. Maharaj, R. Vleggaar, J. V. Senabe, and P. J. Gunning. 2007. "An Appetite Suppressant from *Hoodia* Species." *Phytochemistry* 68:2545–53.

Vereecken, N. J., and J. N. McNeil. 2010. "Cheaters and Liars: Chemical Mimicry at Its Finest." *Canadian Journal of Zoology* 88:725–52.

Vereecken, N. J., and F. P. Schiestl. 2008. "The Evolution of Imperfect Floral Mimicry." *Proceedings of the National Academy of Sciences* 105:7484–88.

Vickers, N. J. 2000. "Mechanisms of Animal Navigation in Odor Plumes." *Biological Bulletin* 198:203–12.

Vuorinen, T., A.-M. Nerg, M. A. Ibrahim, G. V. P. Reddy, and J. K. Holopainen. 2004. "Emission of *Plutella xylostella*–Induced Compounds from Cabbages Grown at Elevated CO_2 and Orientation Behavior of the Natural Enemies." *Plant Physiology* 135:1984–92.

Wallin, K. F., and K. F. Raffa. 2000. "Influences of External Chemical Cues and Internal Physiological Parameters on the Multiple Steps of Postlanding Host Selection Behavior of *Ips pini* (Coleoptera: Scolytidae)." *Environmental Entomology* 29:442–453.

Wang, G., Compton, S. G., and J. Chen. 2013. "The Mechanism of Pollinator Specificity Between Two Sympatric Fig Varieties: a Combination of Olfactory Signals and Contact Cues." *Annals of Botany* 111:173–81.

Wong, B. B. M., and F. P. Schiestl. 2004. "How an Orchid Harms Its Pollinator." *Proceedings of the Royal Society B* 269:1529–32.

Yoneya, K., S. Kugimiya, and J. Takabayashi. 2009. "Can Herbivore-Induced Plant Volatiles Inform Predatory Insect about the Most Suitable Stage of Its Prey?" *Physiological Entomology* 34:379–86.

Zebelo, S. A., K. Matsui, R. Ozawa, and M. E. Maffei. 2012. "Plasma Membrane Potential Depolarization and Cytosolic Calcium Flux Are Early Events Involved in Tomato *(Solanum lycopersicon)* Plant-to-Plant Communication." *Plant Science* 196:93–100.

Zimmer, R. K., and C. A. Zimmer. 2008. "Dynamic Scaling in Chemical Ecology." *Journal of Chemical Ecology* 34:822–36.

THREE

Unraveling the "Radiometric Signals" from Green Leaves

Christian Nansen

WE ARE ALL VERY FAMILIAR WITH THE VISUAL DISPLAY, OR SIGNALING, of flowering plants. Through combinations of morphology, colors, color patterns, and scent (volatiles), flowers and their intricate relationship with pollinators represent some of the most extraordinary examples of coevolution between plants and animals. Extreme relationships between flowers and pollinators are found in plant species from the orchid family. It suffices to take a look at botanical (Hopper and Brown 2007) and pollination literature (PBS 2016) about hammer orchids and/or about sexual mimicry in a wide range of orchid species. Similarly extreme, species in other plant families, like the Nepenthes or the Venus flytrap, have converted their flowers into entomophagous (insect eating) and "reptiphagous" (reptile eating) machines (Adam 1998). In short, the combination of volatiles (scents), colors, and patterns of flowers represents a visual language about sex, resource allocation, rewards, and deception. And I have little doubt that the fundamental admiration for plants by most is linked to the appreciation and admiration for flowers and their derivatives (fruits). Or, to put it more bluntly, there are far less paintings, songs, or poems about plant roots or green leaves, and most people visit botanical gardens and go on wild flower trips during the plants' flowering periods. However, this chapter is exclusively about the green leaves of plants and the interpretation of radiometric signals they reflect, their reflectance profiles. That is, cameras and spectroscopy instruments can be used to analyze how green leaves are responding to growing conditions and possible stressors based on reflectance profiles. The main purpose of this chapter is to provide a review of the significance of basic and applied research discoveries into the "radiometric plant signaling" of green leaves. The word *signaling* is chosen instead of *language,* as a "language" would imply two-way communication between green leaves and us humans. Although this book chapter will unlikely spark a surge in the number of artistic contributions dedicated to green leaves, it is my hope that the information provided here about reflectance signals of green leaves will make you look differently at a green leaf the next time you encounter one.

Similar to flowers, green leaves also emit volatiles and visual signals that herbivores and other animals utilize to assess the quality of the plant as a food source and/or oviposition site. That is, the reflectance signals emitted from the surface of green leaves are, at least partially, associated with complex internal physiological processes inside the leaf. Insects and other herbivores with sensitive vision can "look" at a green leaf and use this information to determine the given plant's suitability as a food source or oviposition site. Even though the human eye may not be able to detect subtle differences in "greenness" among leaves, advanced imaging technologies and spectrometers can be used to analyze the reflectance spectra from green leaves and use this information to gain knowledge about a plant's stress level and about the possible stress factors involved. This type of plant reflectance research has gained considerable momentum from recent developments in computer processing, GPS technologies, imaging technologies, and unmanned aerial vehicles (UAVs; Figure 3.1). These technological improvements are opening the doors for new ways to interpret reflectance from green leaves

Figure 3.1. Example of an unmanned aerial vehicle (UAV) acquiring reflectance data from a canola field. (Photo credit: C. Nansen)

based on airborne remote sensing on a wide range of spatial scales and use this information to study plant responses to growing conditions.

What abiotic and biotic factors may affect the "greenness" or reflectance of a plant? How can we as humans utilize knowledge about the reflectance or greenness of plants to increase food production and to make it more sustainable?

Why Are Plant Leaves Green?

Without getting too far into optical physics and/or plant physiology and biochemistry, it is important to start out with a brief introduction to the "main characters" in this book chapter, "radiometric plant signals" and "green leaf." It was Isaac Newton who discovered that light could be separated into a spectrum of colors, and about one hundred years later James Clerk Maxwell discovered that light as we see it is part of a very wide radiometric spectrum (Raven et al. 1986). Radiometric energy refers to an oscillating electromagnetic field, in which the amount of energy is determined by the amplitude and the frequency of the oscillations. The frequency is inversely related to the spectral wavelength of the radiometric energy, so short wavelengths contain more energy than longer wavelengths. Radiometric energy is called "light," when it falls within the visible portion of the radiometric spectrum (between 380 and 700 nanometers [nm]), because those are the wavelengths that we humans can see (Davies 2004). Ultraviolet (UV) light (100–380 nm), near- and mid-infrared spectrum (700–3,500 nm), and thermal infrared spectrum (3,500–20,000 nm) are part of the radiometric spectrum in which human eyes have very low or no sensitivity, unless we use special cameras or spectroscopy equipment. Furthermore, the visible portion of the radiometric spectrum is divided into six basic light regions: "violet" (380–430 nm), "blue" (430–500 nm), "green" (500–560), "yellow" (560–600 nm), "orange" (600–650 nm), and "red" (650–700 nm).

Figure 3.2 shows the general reflectance profile acquired from a green leaf within the visible light spectrum, and it provides several important traits, which are essential for discussing plant reflectance as a set of radiometric signals to be analyzed and interpreted. First, the reflectance signal in individual spectral "bands" or narrow wavelength ranges (x-axis) is measured as "relative reflectance" (y-axis). It is relative,

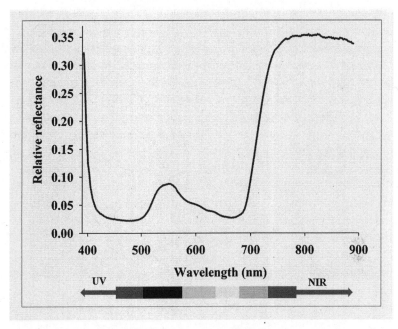

Figure 3.2. Typical reflectance profile from a green leaf. (Photo credit: C. Nansen)

because it is proportional to the reflection obtained from a white calibration surface. High relative reflectance means that comparatively less energy was absorbed by the plant leaf. Second, it is seen that, within the visible light, the reflection is lowest in the blue and red light spectra, and that it is highest in the green spectrum. The latter explains why plants look green to humans, and "dark greenness" basically means that less light is being reflected. Dark-green leaves absorb more radiometric energy, generally due to higher concentration of color pigments in leaves.

This chapter is not restricted to a discussion of plant reflectance signals in the visible light, as other parts of the radiometric spectrum may also be used to analyze and interpret reflectance profiles acquired from green leaves. Therefore throughout this book chapter, I will use the term *radiometric plant signal*, rather than *light*, to describe the wavelengths in which radiometric plant signals are investigated.

The second "main character" in this chapter is the green leaf. Plants perform "photosynthesis"—which means they produce or synthesize organic compounds (including sugars) based on energy derived from radiometric energy emitted by the sun. The absorption of radiometric energy involves highly complicated physiological processes inside chloroplasts, and color pigments, such as carotenoids and chlorophyll pigments, play a major role in the capture and conversion of radiometric energy into chemical energy. The visible portion of the radiometric spectrum is of particular relevance to reflectance studies of green leaves, because the photosynthesis process by plants is almost exclusively driven by radiometric energy from the blue and red light. At first glance, it may seem surprising that only a very small portion of the radiometric spectrum, the visible light, is being utilized by plants in photosynthesis. However, UV-light and even shorter wavelengths generally contain too much energy and are therefore potentially harmful to plant tissues, as they can cause breakage of chemical bonds or change in the structure of complex organic molecules (Raven et al. 1986). Plant leaves are therefore coated with cuticular waxes as part of their protective strategy against UV radiation (Shepherd and Griffiths 2006). For the same reasons, human skin also provides partial protection against UV light, as high-energy UV light causes dangerous changes to our skin cells. Due to climate changes and increases in UV-light-induced stress of crops in many agriculturally important regions, there is ongoing research into development and use of sunscreens for crop plants (Glenn and Puterka 2004). Regarding the near-infrared spectrum or longer spectral wavelengths, the amount of energy is simply too low to drive biochemical reactions, and the energy in those wavelengths is also absorbed by water (Raven et al. 1986), which is obviously one of the main constituents of growing plants. It is therefore clear that the biochemical processes in plant cells are uniquely suited to specifically utilize the radiometric energy in wavelengths from within the visible portion of the radiometric spectrum.

The positive correlation between concentration of chlorophyll and greenness is very important, because the concentration of color pigments is also tightly linked to many aspects of a plant's overall well-being. From experimental studies of green leaves, it has been demonstrated that important plant pigments have maximum absorption peaks at particular wavelengths (Dunagan et al. 2007; Jean-Philippe et al. 2012): chlorophyll a (430, 662, and 680 nm), chlorophyll b (448, 642 nm), and carotenoids

(448, 471 nm). In other words, there is clear evidence of reflectance at specific wavelengths to be directly linked to the chlorophyll content of leaves and their photosynthetic activity. Another important aspect of the correlation between concentration of chlorophyll and greenness is that it has been shown to follow both diurnal and circadian fluctuations in a number of plants, including tomato (Meyer et al. 1989), petunia (Stayton et al. 1989), tobacco (Paulsen and Bogorad 1990), and wheat (Nagy at el. 1987; Busheva et al. 1991). These fluctuations clearly demonstrate that a plant responds to temporal variations in growing conditions and attempts to adjust its "investment" in synthesis and maintenance of chloroplast and chlorophyll pigments. At a high spatial resolution, Busheva et al. (1991) studied the content of chlorophyll pigments a and b and the relative ratio between these two pigments in different portions of wheat leaves (basal, mid, and tip). They demonstrated that the most pronounced changes, both diurnally and seasonally, occurred in the basal leaf segments of the leaves. In other words, the reflectance profiles vary within a green leaf, and the different portions of a given leaf also show different levels of diurnal and seasonal fluctuations in their reflectance profiles. An even more extraordinary study showed that, in some plant species, the chloroplasts inside plant leaves show diurnal movement. Britz and Briggs (1976) studied an *Ulva* species and showed that during the day the chloroplasts were near the outer leaf surface, which resulted in high absorbance of radiometric energy; at night the chloroplasts were mainly located along the leaf sides, and the absorbance of radiometric energy was low. Britz and Briggs (1976) also demonstrated that the same mobility of chloroplasts was not found in closely related plant species, so that raises the questions of the main functions of this adaption and how it evolved in the first place. It is beyond the scope of this chapter to delve into these quite fundamental questions about chloroplasts, but the examples were included to emphasize how dynamic and variable reflectance profiles are dependent on (1) which portions of leaves the reflectance data were acquired from, and (2) when during the day and/or season they were acquired. This spatial and temporal variability underscores that plants have mechanisms to regulate (and possibly optimize) their investment in photosynthesis, but our current knowledge about such mechanisms is quite limited. The complex dynamics of leaf reflectance profiles underscore that they can provide valuable insight into a wide range of physiological processes in plants and their adaptability to different growing conditions.

The "Radiometric Signals" of Green Leaves

Remote sensing is defined as assessment of objects without involving physical contact, and use of airborne data (reflectance data acquired with imaging devices mounted on flying vehicles) has been an established research discipline for more than three decades (Billingsley 1984; Xie et al. 2008). To most people, *remote sensing* refers to such large-scale airborne studies, but it is actually also remote sensing to observe objects through a microscope. So all reflectance-based studies, irrespective of the spatial scale, may be considered remote sensing studies, because the identity and/or quality of objects (in the case of green plant leaves) is evaluated based on assessments of their "reflectance profiles." Such reflectance profiles represent series of reflectance values in narrow spectral bands or wavelengths. A normal digital camera collects reflectance data in three wide spectral bands: the red, green, and blue portions of the visible spectrum (and therefore also called an "RGB" camera). Thus, each pixel is associated with only three reflectance values, which is sufficient information to create an image similar to how an object looks to us humans.

A wide range of cameras and spectroscopy devices are used to acquire reflectance data from green leaves. These devices range in spectral properties, with some devices only collecting imaging data in five to one hundred spectral bands, while other devices acquire reflectance data in hundreds or even thousands of spectral bands. As would be expected, a classification of reflectance data acquired at a high spectral resolution (in hundreds of narrow spectral bands) is generally more likely to accurately detect a subtle stress response in a growing plant than an analysis of data acquired with a very coarse spectral resolution. However, the effect of spectral data resolution on classification accuracy of remote sensing data is not linear and is in itself a research topic with a range of implications (Nansen, Geremias, et al. 2013; Zhang et al. 2015). In addition to spectral resolution, reflectance data of green leaves are collected at a wide range of spatial resolutions—ranging from low spatial resolution satellite images with each pixel representing ten to one hundred square meters (Wikipedia 2016) to high-resolution studies with hundreds or thousands of pixels being acquired from single plant leaves (Nansen 2012; Nansen, Sidumo, et al. 2013). Remote sensing data acquired with a high spatial resolution (pixels smaller than the size of target leaves) are generally more likely to detect subtle stress responses in a growing plant, as it is possible to filter out (eliminate) reflectance signals from other objects (such as soil). Based on selections of spectral and spatial resolutions, an "image" can be produced. An image is a visual representation of information acquired from a particular object. In the current context, I will focus exclusively on reflectance-based imaging (not absorption). The goal is to provide a brief insight into how imaging of greenness can be analyzed or quantified to study and evaluate differences between plant species and the well-being of plant populations.

A Radiometric Signal—the Basics

Here, the term *signal* is defined as a way for us humans to interpret the identity (species identification), development (assess the plant's developmental stage), and stress level (different abiotic and biotic stressors) of a green leaf. Normally, signals are part of a "language" and therefore a means for two-way communication. I am deliberately not using the term *radiometric language* but using *radiometric plant signals,* as I am not proposing ways for plants and humans to establish two-way communication. Some readers may consider this clarification a bit pedantic, but it is becoming increasingly clear that plants both respond to sound and emit acoustic signals. Readers interested in plants and bioacoustics are encouraged to consult studies published elsewhere (Gagliano et al. 2012; Gagliano 2013).

The concept of a radiometric signal is here defined as a means to obtain insight into the well-being and ongoing physiological process within green leaves. As an analogue, consider how we use infrared thermometers to measure our internal body temperature. When humans become sick (stressed), we often get a fever as our body's immune system is mobilizing defensive strategies to protect us against the intruding pathogen. So, as part of a diagnostic assessment of our well-being, we associate an increase in internal body temperature with stress. We can use an infrared thermometer to measure the internal body temperature, but how does it actually work? Well, a radiometric energy source inside the thermometer emits a signal, which is reflected back from the skin and captured by a sensor. The difference in energy levels between the radiometric signal emitted and received is then converted into an estimate of the internal body temperature. The emitted radiometric energy only penetrates a few millimeters into our skin and elicits a radiometric signal in the near-infrared spectrum. So the use of this thermometer is based on the fundamental assumption that a signal from the outer body surface layers can be used to estimate the internal body temperature. And that is precisely what reflectance-based analysis of green leaves is also used for—using reflectance characteristics from the green leaf surface to make interpretations of the plant's well-being and the ongoing physiological process inside green leaves. In short, it is assumed that emerging signs of "stress" in growing plants can be identified based on reflectance data acquired from leaves.

SIGNALS OF GREENNESS

A simple way to demonstrate that plant cells within a leaf vary in greenness by responding differently to growing conditions is to quantify the "greenness" of pixels acquired with high spatial resolution from the surface of a leaf. Figure 3.3a shows photos of three bell pepper plants' leaves from plants subjected to different fertilizer treatments (Nansen, Geremias, et al. 2013).

The leaves were placed in similar horizontal positions and had "similar" greenness, although a slight decrease in brightness across phosphorus treatments may be apparent. Figure 3.3b shows the average reflectance profiles acquired from the three phosphorus treatments. It shows that: (1) high phosphorus treatment caused a considerable decrease in reflectance within the visible portion of the radiometric spectrum, and (2) low phosphorus treatment caused a considerable increase in reflectance in the near-infrared portion of the radiometric spectrum. The differences become more evident by using one average reflectance profile as "reference," and dividing the other average reflectance profiles ("test profile") by the reference profile. In this case, low phosphorus treatment was used as the reference profile. Using this approach, a value of one suggests that the average reflectance values at that wavelength were the

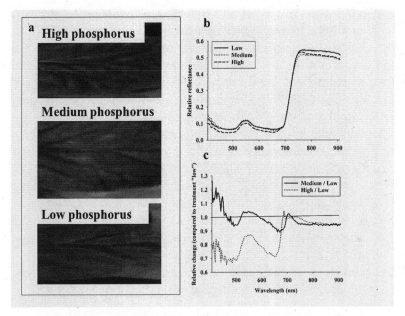

Figure 3.3. Reflectance data acquired from bell pepper leaves. *a*, Bell pepper plants in individual pots were grown under three different phosphorus treatments: low (0.008 g per pot), medium (0.016 g per pot), and high (0.032 g per pot). *b*, Average reflectance profiles from the three phosphorus treatments. *c*, Average reflectance profiles from medium and high phosphorus treatments divided by average reflectance profile from low phosphorus treatment as the reference. (Images: C. Nansen)

same; a value above one, that the average reflectance value in the test profile was higher than that in the reference profile; and a value below one indicates that the average reflectance value in the test profile was lower than that in the reference profile. The results from this simple visualization show that high phosphorus treatment caused about a 30 percent decrease in reflection of radiometric energy in the blue and red portions of the radiometric spectrum. This is likely explained by phosphorus treatment improving the plants' ability to synthesize color pigments, which results in a marked increase in photosynthetic activity (less reflection due to a higher level of absorption). The phosphorus treatment appears to have negligible influence on reflection profiles in the near-infrared portion of the radiometric spectrum. This observation is consistent with our understanding of how plants utilize radiometric energy.

The same comparison of average reflectance profiles could be applied to reflectance data acquired from the same plants at different time points, plants of the same species but grown under different water or temperature conditions, plants of the same species grown in different regions, different species of plants, and plants exposed to different biotic stressors, such as diseases and insects. Comparison of average reflectance profiles can be very helpful as an initial exploratory exercise to identify in which parts of the spectrum there appears to be a difference between the classes of reflectance data under investigation. Another simple approach is to ignore most of the reflectance information and only focus on that of a few spectral bands, which are then used to create specific indexes (Baghzouz et al. 2006), such as the normalized difference vegetation index (NDVI), the photochemical reflectance index (PRI), and the stress index (SI). Each of these indexes is used to calculate the ratio between reflectance values at particular spectral wavelengths for each pixel:

$$NDVI = (750 \text{ nm} - 705 \text{ nm}) / (750 \text{ nm} + 705 \text{ nm})$$
$$PRI = (531 \text{ nm} - 570 \text{ nm}) / (531 \text{ nm} + 570 \text{ nm})$$
$$SI = 693 \text{ nm} / 759 \text{ nm}$$

Often the same vegetation indexes are used to analyze aerial remote sensing data of vegetation, and greenness of grasses has been used to estimate their age (Knox et al. 2013). This use of well-established indexes to analyze reflectance data from a single leaf underscores that, even within leaves, plants may show specific responses to growing conditions in terms of their ability to capture radiometric energy—and therefore their reflectance profile. It should be pointed out that detection of subtle differences in reflectance data acquired from plants often requires analysis and classification approaches that are considerably more advanced than simple vegetation indexes. However, they illustrate the important point that reflectance data analysis can be used to identify and quantify differences among objects (in this case pixels of a canola leaf) that the human eye is unable to detect. Another important advantage of reflectance-based analyses is that they enable quantification of differences in visual appearance.

The Need for Detection of Unique Reflectance Responses by Plants

Above, I introduced the concept of plants tentatively "investing" in photosynthesis through synthesis and maintenance of color pigments. If a plant's well-being is compromised, for instance, due to lack of nutrients or presence of drought conditions, the plant may not be able to sustain its color pigment content. However, it also needs energy to compensate for the imposed stressor. In essence, a plant has to somehow "decide on" an "investment strategy" as a function of the environmental conditions under which it is growing. That is, it costs energy and resources to synthesize chloroplasts and chlorophyll pigments, but this is also what the plant needs in order to capture more radiometric energy and use it as an energy source for photosynthesis of sugars and other organic compounds. So, if the plant is too "conservative" in its life strategy and does not produce enough chloroplasts and chlorophyll pigments, then it will not have enough small engines to produce the energy and organic building blocks it needs to maintain its metabolic processes. I am fully aware that only a few people would suggest that a plant "makes actual decisions," but we are learning more and more about the adaptability and responsiveness of plants to growing conditions, and it is becoming increasingly obvious that plants have some level of cognitive capabilities (Gagliano 2015). The acknowledgment of plants' ability to communicate, learn, and respond to changing growing conditions has led to establishment of a new research discipline, plant neurobiology (Brenner et al. 2006). That is, something inside plants is detecting changes in the environment and is also communicating with other plants, and it leads to significant changes in resource allocations by plants as part of an optimized strategy. Typically, a plant growing under suboptimal conditions has reduced ability to utilize the radiometric energy as an energy source for the photosynthesis, so its "reflectance profile" is different from that obtained from a healthy (nonstressed) plant. The research literature is therefore full of examples of how the reflectance profiles from stressed plants are different from those acquired from nonstressed plants. Moreover, in their research on the exposure of a wide range of plant species to a diverse set of biotic or abiotic stressors, Carter and Knapp (2001) concluded that in most studies there appears to be an increase in reflectance in response to plant stress, and this increase is generally most noticeable near seven hundred nanometers. With the absorption of radiometric energy for photosynthesis being very efficient in that part of the visible light, it is not surprising that strong and consistent change in reflectance is detected in this part of the radiometric spectrum. In addition, it is intuitively logical that stress causes an increase in reflectance, because the plant's ability to utilize radiometric energy is partially compromised, so more energy is reflected back to the atmosphere from the leaf surface. From an academic standpoint, it is interesting that plants tend to have a quasi-universal stress reflectance response near seven hundred nanometers. However, it means that practical applications of reflectance-based stress detection systems may be limited, unless more unique reflectance features can be associated with different abiotic and biotic stressors (Nansen 2012). Under real-world or commercial conditions, plants will likely be adversely affected by several stressors simultaneously, and/or there may be important interactions between stressors. Thus, specific stress detection tools are needed, so that different stressors can be detected and quantified independently. Absorption of radiometric energy is not just determined by color pigment content in green leaves; a range of physiological processes interact, so different stressors are believed to cause slightly different changes to the reflectance profile. The latter statement is the fundamental assumption behind the concept of radiometric signals as indicators of plant stress.

How to Use the "Radiometric Signals" of Green Leaves—in Basic "Phenomics" Research

More than three decades of intensive molecular data mining have unraveled incredible amounts of knowledge about the expression of individual genes and gene families in plants, their importance in different biochemical pathways, and the complex regulation mechanisms involved. This type of research leads to molecular crop breeding, which is revolutionizing agriculture through development of better-adapted and higher-yielding crops, and to development of crops with specific qualitative traits, such as possessing certain nutritional traits or containing high content of compounds with certain commercial applications (Moose and Mumm 2008). Molecular plant data mining is also having a profound impact on our understanding of the evolutionary biology and systematics of plants. But most of these studies are conducted in cell cultures or highly experimental living plant materials, and much of this science has been based on the concept of "one gene, one function"—meaning that up or down regulation of one or a few proteins has been studied in response to manipulation of the expression of one or a few genes. Such studies are necessary for the discovery of the clear link between individual genes and specific phenotypic traits, but they also neglect the possible indirect effects of individual genes.

A new frontier in science is called "phenomics" (Houle et al. 2010), in which the roles of molecular and physiological modifications are assessed more broadly by characterizing their effects on the general performance on phenotypes and by also incorporating environmental effects into the analyses, without destroying the plant being investigated. In other words, phenomics is about noninvasive characterization and quantification of how individual organisms perform based on their genetic code and in response to environmental conditions. Due to the often-complex phenotypic responses by individuals, development of image-based systems to detect and characterize phenotypic responses has been vitally important to phenomics projects (Burleigh et al. 2013). This is best explained with an example. Rebolledo et al. (2012) studied forty-three rice genotypes grown under experimental drought stress conditions. Water availability in rain-fed rice fields can seriously limit harvested yields, so "drought stress" is a major challenge for effective rice production. The "phenomics" challenge in this study is to describe the plant traits (like plant growth characteristics and also metabolic variables) associated with drought response. Both genetic and environmental factors play a role, as some rice varieties are more drought resistant than others and because different metabolic variables have a wide range of effects on the growth of rice plants. Such a complex of genetic and metabolic variables is quite challenging to analyze because very few of the variables are independent (not affected by any of the other variables), so the "chicken–egg problem" is evident. However, the biggest challenge is that abiotic stressors, like drought, affect multiple metabolic (hexane, sucrose, and starch contents) and growth (number of leaves, size of leaves, shoot dry weight, and so on) variables, and the cumulative effect is therefore potentially not detected if each of the variables is examined individually.

Imaging technologies were not used in this particular study of drought stress of developing rice (Rebolledo et al. 2012), but that would have been a way to obtain quantifiable data on the overall effect of drought on the different rice genotypes. The reflectance response by the forty-three genotypes could have been categorized to identify two to five genotypes with either very high or low drought tolerance. Subsequently, the agronomic and metabolic variables within these extreme categories could have been analyzed to find the variables with the strongest association with drought tolerance. A very important benefit associated with the proposed reflectance-based approach is that metabolic analyses are often time-consuming and costly, so only conducting these analyses on genotypes with either very high or low drought tolerance would save both money and time. The tremendous potential of reflectance-based analyses as part of screening lines and genotypes in large crop breeding programs was highlighted and reviewed in detail by White et al. (2012).

How to Use "Radiometric Signals" of Green Leaves—in a Modern Crop-Farming Context

There are many potential applications of reflectance-based studies of green leaves in both natural and anthropogenic environments and across a wide range of spatial scales. Here, the main goal is not to review that literature but instead to provide a perspective on how powerful and widely applicable the technology is. And with a background in food production, I will attempt to provide this perspective through a discussion of some of the main challenges in modern crop farm management. Major trends in crop farm management over the last thirty years include (1) that the size of farm properties is increasing, and (2) that fewer people are seeking a career in agriculture and labor costs are increasing so it becomes harder to hire skilled workers (Nansen and Ridsdill-Smith 2013).

With more land to farm and with fewer workers, there is less time to monitor crop development during the growing season, and there is therefore an increasing risk of crops underperforming because of suboptimal management of inputs, such as applications of fertilizers and pesticides. A possible solution to this challenge is to develop and deploy systems to automate crop monitoring through use of imaging systems mounted on unmanned aerial vehicles (UAVs). That is, it may be possible to use small drones or multirotors to overfly crops and collect reflectance data at a sufficiently high spatial resolution to obtain data from individual crop plants and a sufficiently high spectral resolution to detect and accurately diagnose crop responses to a given stressor (i.e., a reflectance response to a disease or a nutrient deficiency). Once the reflectance data has been acquired and processed, it needs to be analyzed and converted into specific management recommendations that are conveyed to growers and their consultants. Such recommendations may be delivered as digital maps, which can be imported into the GPS-guided control systems of tractors, so that fertilizers and/or pesticides are applied based on where the reflectance data indicated that certain inputs were needed. In other words, the use of UAV-based acquisition of reflectance data from crop leaves makes it possible to optimize cropping inputs, so that they are only applied when and where they are needed.

For several decades, it has been a key driver in agricultural research to replace calendar-based and broadcast applications of agrochemicals with more targeted (precision agriculture) uses of cropping inputs. However, this development has so far been severely hampered by the lack of means to acquire monitoring data of growing crops that are both sufficiently sensitive to detect subtle levels of crop stress and acquired at a sufficiently high spatial resolution. But with continuous innovation in UAV technology, imaging systems, and ways to improve classification of reflectance data, it is almost certain that reflectance-based analysis of growing crops will become a major component of future crop management systems. Furthermore, the continuous improvement in acquisition and classification of reflectance data may lead to development of crop management practices with far less reliance on use of pesticides. That is, more detailed (both spatially and temporally) monitoring of crop status and development may enable a crop management approach similar to the concept of "preventive medicine" (Nansen, Sidumo, et al. 2013), in which the focus is on preventing crop stress rather than responding to its adverse impact on crop growth. Such an approach is becoming possible, because the ability to use reflectance-based technologies to detect early indications of crop stress has improved significantly in the last five to ten years.

To be more specific, it is well established that most crops become infested with pests when the crop is grown under some level of abiotic stress, say, drought stress or under nutrient-deficient conditions. The common approach is to suppress emerging or established pest outbreaks by applying pesticides. In cases where the pesticide is applied correctly, that will cause some level of control of the pest, but it does not remediate the cause of the infestation. So the pest is likely to reappear, especially if the pest is very mobile (and many economically important insect pests are). As an alternative, the preventive medicine approach is therefore to use reflectance-based technologies to monitor the crop and determine when and where crop plants appear to become stressed. Precision-targeted application of agrochemicals, such as fertilizers, is gaining growing interest both as part of optimizing input management and as part of developing "smart" pest-management practices (West and Nansen 2014). The key challenge regarding successful development of smart pest-management practices is to identify the stress-induced plant response accurately and early enough for the grower to have time to react and deploy necessary management practices, so that the crop does not become stressed and therefore more susceptible to pest infestations. Such improvements of future approaches to pest management in agriculture require a shift in emphasis from responsive applications of pesticides to early detection of crop stress, and use of plant reflectance profiling plays a role in that regard. The following examples provide some insight into why and how reflectance-based technologies can be used to develop innovative pest-management practices in cropping systems.

There is growing evidence that manipulating the potassium nutrition of field crops can be used to reduce the potential damage caused by insect pests feeding on crops. In a review of more than two thousand studies regarding effects of potassium (K) on pest and disease incidence in plants (Perrenoud 1990), the authors found that in 63 percent of these studies application of potassium led to a decrease in pest pressure. Potassium fertilization leads to increased leaf thickness, stronger epidermal cells, and decreases leaf nutrient concentrations of sugars and amino acids. Due to the physiological role and importance of potassium levels, it therefore seems reasonable to argue that outbreaks of insect and mite pests may be reduced through monitoring and precision applications of potassium fertilization. In other words, as an alternative to applying pesticides when pest outbreaks have occurred, it may be possible to prevent pest outbreaks through careful and continuous regulation of crop fertilization. Brennan and Grimm (1992) showed that losses of dry matter and seed yield of forage clover due to redlegged earth mites and blue-green aphids were markedly reduced by potassium fertilization. In a recent study, Nansen, Sidumo, et al. (2013) showed that potassium fertilization decreases attractiveness of maize plants to spider mites. Aphid survival was found to be negatively associated with potassium levels in soybeans (Walter and Difonzo 2007), wheat (Hasken and Poehling 1995), and canola (Sarwar et al. 2011). It has also been found that aphid population growth rates were negatively correlated with soil potassium (Myers and Gratton 2006). Thus, there appears to be strong evidence supporting the hypothesis that effective management of crop nutritional content may be a strategy to minimize crop susceptibility to pest infestations. With most field pests immigrating into crop fields during the growing season, the goal should be—based on carefully planned fertilizer applications—to make field edges proportionally more attractive and suitable as hosts than the central portion of the field. Under this scenario, scouting for emerging pest infestations is greatly simplified, because only field perimeters would need to be monitored for pest outbreaks, as they are the most likely portions to be infested. Thus, scouting and field crop management based on pest population thresholds can be conducted with a much higher level of precision and accuracy than today. This approach—the use of manipulative fertilizer applications to create "trap crops"—has already been investigated for a number of arthropod pest-crop systems, including diamondback moth infestations in canola (Staley et al. 2009; Badenes-Perez et al. 2010).

CONCLUDING REMARKS

The human eye may not detect all the nuances and details sequestered in the reflectance profiles emitted by green leaves, but they can be detected by cameras and spectroscopy sensors to elucidate complex plant responses to growing conditions. In other words, specific traits in these reflectance profiles can be interpreted as response signals, and careful interpretation of these signals can therefore be used to optimize crop management practices. Green leaf cells are factories of organic compounds needed to sustain all food webs, and their productivity is fueled by energy from a very narrow range of the radiometric spectrum, the visible light. Through careful analyses of reflectance profiles within the visible light spectrum, we can use reflectance data acquired from green leaves as a set of "radiometric signals" to study and manage plant performance. Such radiometric signals are based on basic assumptions about "healthy" (well-performing) plants reflecting radiometric energy in a particular way and that specific stressors or deficiencies cause unique changes to the reflectance profiles acquired from green leaves. Advanced imaging technologies and advanced classifications of reflectance data will undoubtedly play growing roles in both basic studies and management of plant populations (such as a growing crop) and communities, as we continue to learn more about what "plants can tell us" through their reflectance profiles.

I wish to thank Professor Emeritus Sten Bay Jørgensen for his thorough revision of an early version of this chapter.

References

Adam, J. H. 1998. "Reproductive Biology of Bornean *Nepenthes* (Nepenthaceae) Species." *Journal of Tropical Forest Science* 10:456–71.

Badenes-Perez, F. R., M. Reichelt, and D. G. Heckel. 2010. "Can Sulfur Fertilisation Improve the Effectiveness of Trap Crops for Diamondback Moth, *Plutella xylostella* (L.) (Lepidoptera: Plutellidae)?" *Pest Management Science* 66:832–38.

Baghzouz, M., D. A. Devitt, and R. L. Morris. 2006. "Evaluating Temporal Variability in the Spectral Reflectance Response of Annual Ryegrass to Changes in Nitrogen Applications and Leaching Fractions." *International Journal of Remote Sensing* 27:4137–57.

Billingsley, F. C. 1984. "Remote Sensing for Monitoring Vegetation: An Emphasis on Satellites." In *The Role of Terrestrial Vegetation in the Global Carbon Cycle: Measurement by Remote Sensing,* edited by G. M. Woodwell, 161–80. Chichester, UK: John Wiley & Sons.

Brennan, R. F., and M. Grimm. 1992. "Effect of Aphids and Mites on Herbage and Seed Production of Subterranean Clover (cv. Daliak) in Response to Superphosphate and Potash." *Australian Journal of Experimental Agriculture* 32:39–47.

Brenner, E. D., R. Stahlberg, S. Mancuso, J. Vivanco, F. Baluška, and E. Van Volkenburgh. 2006. "Plant Neurobiology: An Integrated View of Plant Signaling." *Trends in Plant Science* 11:413–19.

Britz, S. J., and W. R. Briggs. 1976. "Circadian Rhythms of Chloroplast Orientation and Photosynthetic Capacity in *Ulva*." *Plant Physiology* 58:22–27.

Burleigh, J. G., K. Alphonse, A. J. Alverson, H. M. Bik, C. Blank, A. L. Cirranello, H. Cui, et al. 2013. "Next-Generation Phenomics for the Tree of Life." *PLOS Currents* 5:ecurrents.tol.085c713acafc8711b2ff7010a4b03733.

Busheva, M., G. Garab, E. Liker, Z. Tóth, M. Szèll, and F. Nagy. 1991. "Diurnal Fluctuations in the Content and Functional Properties of the Light Harvesting Chlorophyll *a/b* Complex in Thylakoid Membranes: Correlation with the Diurnal Rhythm of the mRNA Level." *Plant Physiology* 95: 997–1003.

Carter, G. A., and A. K. Knapp. 2001. "Leaf Optical Properties in Higher Plants: Linking Spectral Characteristics to Stress and Chlorophyll Concentration." *American Journal of Botany* 88:677–84.

Davies, K. M. 2004. "An Introduction to Plant Pigments in Biology and Commerce." In *Plant Pigments and Their Manipulation*, edited by K. M. Davies, 1–22. Oxford: Blackwell.

Dunagan, S. C., M. S. Gilmore, and J. C. Varekamp. 2007. "Effects of Mercury on Visible/Near-Infrared Reflectance Spectra of Mustard Spinach Plants (*Brassica rapa* P.)." *Environmental Pollution* 148:301–11.

Gagliano, M. 2013. "Green Symphonies: A Call for Studies on Acoustic Communication in Plants." *Behavioral Ecology* 24:789–96.

———. 2015. "In a Green Frame of Mind: Perspectives on the Behavioural Ecology and Cognitive Nature of Plants." *AoB Plants* 7:plu075.

Gagliano, M., S. Mancuso, and D. Robert. 2012. "Towards Understanding Plant Bioacoustics." *Trends in Plant Science* 17:323–25.

Glenn, D. M., and G. J. Puterka. 2004. "Particle Films: A New Technology for Agriculture." In Horticultural Reviews, vol 31, edited by J. Janick, 1–44. Oxford, UK: John Wiley & Sons.

Hasken, K. H., and H. M. Poehling. 1995. "Effects of Different Intensities of Fertilisers and Pesticides on Aphids and Aphid Predators in Winter Wheat." *Agriculture, Ecosystems and Environment* 50:45–50.

Hopper, S. D., and A. P. Brown. 2007. "A Revision of Australia's Hammer Orchids (Drakaea: Orchidaceae), with Some Field Data on Species-Specific Sexually Deceived Wasp Pollinators." *Australian Systematic Botany* 20: 252–85.

Houle, D., D. R. Govindaraju, and S. Omholt. 2010. "Phenomics: The Next Challenge." *Nature Review Genetics* 11:855–66.

Jean-Philippe, S., N. Labbé, J. Damay, J. Franklin, and K. Hughes. 2012. "Effect of Mercuric Compounds on Pine and Sycamore Germination and Early Survival." *American Journal of Plant Sciences* 3:150–58.

Knox, N. M., A. K. Skidmore, H. M. A. van der Werffa, T. A. Groena, W. F. de Boerd, H. H. T. Prins, E. Kohid, and M. Peel. 2013. "Differentiation of Plant Age in Grasses Using Remote Sensing." *International Journal of Applied Earth Observation and Geoinformation* 24:54–62.

Meyer, H., U. Thienel, and B. Piechulla. 1989. "Molecular Characterization of the Diurnal/Circadian Expression of the Chlorophyll a/b-Binding Proteins in Leaves of Tomato and Other Dicotyledonous and Monocotyledonous Plant Species." *Planta* 180:5–15.

Moose, S. P., and R. H. Mumm. 2008. "Molecular Plant Breeding as the Foundation for 21st Century Crop Improvement." *Plant Physiology* 147:969–77.

Myers, S. W., and C. Gratton. 2006. "Influence of Potassium Fertility on Soybean Aphid, *Aphis glycines* Matsumura (Hemiptera: Aphididae), Population Dynamics at a Field and Regional Scale." *Environmental Entomology* 35: 219–29.

Nagy, F., S. A. Kay, and N.-H. Chua. 1987. "The Analysis of Gene Expression in Transgenic Plants." In *Plant Gene Research Manual,* edited by S. B. Gelvin and R. A. Schilperoort, 1–29. Dordrecht, Netherlands: Kluwer Academic Press.

Nansen, C. 2012. "Use of Variogram Parameters in Analysis of Hyperspectral Imaging Data Acquired from Dual-Stressed Crop Leaves." *Remote Sensing* 4:180–93.

Nansen, C., L. D. Geremias, Y. Xue, F. Huang, and J. R. Parra. 2013. "Agricultural Case Studies of Classification Accuracy, Spectral Resolution, and Model Over-fitting." *Applied Spectroscopy* 67:1332–38.

Nansen, C., and T. J. Ridsdill-Smith. 2013. "The Performance of Insecticides—a Critical Review." In *Insecticides: Development of Safer and More Effective Technologies,* edited by S. Trdan, 195–232. Rijeka, Croatia: InTech.

Nansen, C., A. J. Sidumo, X. Martini, K. Stefanova, and J. D. Roberts. 2013. "Reflectance-Based Assessment of Spider Mite "Bio-response" to Maize Leaves and Plant Potassium Content in Different Irrigation Regimes." *Computers and Electronics in Agriculture* 97:21–26.

Paulsen, H., and L. Bogorad. 1990. "Diurnal and Circadian Rhythms in the Accumulation and Synthesis of mRNA for the Light-Harvesting Chlorophyll a/b-Binding Protein." *Plant Physiology* 88:1104–9.

PBS. 2016. "Mimicry: The Orchid and the Bee." Evolution Library. Accessed September 11. http://www.pbs.org/wgbh/evolution/library/01/1/101101.html.

Perrenoud, S. 1990. *Potassium and Plant Health.* IPI Research Topics no. 3. 2nd rev. ed. Bern, Switzerland: International Potash Institute.

Raven, P. H., R. F. Evert, and S. E. Eichorn. 1986. *Biology of Plants.* 4th ed. New York: Worth Publishers.

Rebolledo, M.-C., M. Dingkuhn, A. Clément-Vidal, L. Rouan, and D. Luquet. 2012. "Phenomics of Rice Early Vigour and Drought Response: Are Sugar Related and Morphogenetic Traits Relevant?" *Rice* 5:1–15.

Sarwar, M., N. Ahmad, and M. Tofique. 2011. "Impact of Soil Potassium on Population Buildup of Aphid (Homoptera: Aphididae) and Crop Yield in Canola (*Brassica napus* L.) Field." *Pakistan Journal of Zoology* 43:15–19.

Shepherd, T., and D. W. Griffiths. 2006. "The Effects of Stress on Plant Cuticular Waxes." *New Phytologist* 171:469–99.

Staley, J. T., A. Stewart-Jones, G. M. Poppy, S. R. Leather, and D. J. Wright. 2009. "Fertilizer Affects the Behaviour and Performance of *Plutella xylostella* on Brassicas." *Agricultural and Forest Entomology* 11:275–82.

Stayton, M., P. Brosio, and P. Dunsnuir. 1989. "Photosynthetic Genes of *Petunia* (Mitchell) Are Differentially Expressed during the Diurnal Cycle." *Plant Physiology* 89:776–82.

Walter, A. J., and C. D. Difonzo. 2007. "Soil Potassium Deficiency Affects Soybean Phloem Nitrogen and Soybean Aphid Populations." *Environmental Entomology* 36:26–33.

West, K., and C. Nansen. 2014. "Smart-Use of Fertilizers to Manage Spider Mites (Acari: Tetrachynidae) and Other Arthropod Pests." *Plant Science Today* 1:161–64.

White, J. W., P. Andrade-Sanchez, M. A. Gore, K. F. Bronson, T. A. Coffelt, M. M. Conley, K. A. Feldmann, et al. 2012. "Field-Based Phenomics for Plant Genetics Research." *Field Crops Research* 133:101–12.

Wikipedia. 2016. "Remote Sensing Satellite and Data Overview." Accessed September 11. http://en.wikipedia.org/wiki/Remote_Sensing_Satellite_and_Data_Overview.

Xie, Y., Z. Sha, and M. Yu. 2008. "Remote Sensing Imagery in Vegetation Mapping: A Review." *Journal of Plant Ecology* 1:9–23.

Zhang, X., C. Nansen, N. Aryamanesh, G. Yan, and F. Boussaid. 2015. "Importance of Spatial and Spectral Data Reduction in Detection of Internal Defects in Food Products." *Applied Spectroscopy* 69:473–80.

FOUR

Breaking the Silence

Green Mudras and the Faculty of Language in Plants

Monica Gagliano

> There is a language older by far and deeper than words. It is the
> language of bodies, of body on body, wind on snow, rain on trees,
> wave on stone. It is the language of dream, gesture, symbol,
> memory. We have forgotten this language. We do not even
> remember that it exists.
> —Derrick Jensen, *A Language Older Than Words*

LANGUAGE IS OFTEN SAID TO BE ONE OF THE HALLMARKS OF BEING
human and is thought to have emerged from the interactions of three
adaptive systems, namely, individual learning, cultural transmission,
and biological evolution (Christiansen and Kirby 2003). The faculty of
language has provided humans with an effective tool for classifying
experiences, discriminating events, and communicating what has been
learned to others, thus steering adaptive adjustments to our way of being
in (and relating to) the environment. There are many theories about the
origins of language (Bolhuis et al. 2014), and while the truth is that we
simply do not know where, when, and how it did come about (Hauser
et al. 2002), the evolution of language in humans is considered to be one
of the most important events in the history of life on this planet (Maynard Smith and Szathmáry 1995). There is, of course, no denying that
language is essential to every aspect of our everyday lives. Through
words, gestures, and much more, we use language to inform others of
a multitude of feelings, desires, worries, questions, insights we experience in relation to the world that surrounds us. The question is whether
such use of language is truly an activity, a form of behavior, that makes
us unique and unlike other species.

The answer to this question clearly depends on how we define language and the criteria for its use. According to the *Oxford English Dictionary,* the formal definition of *language* includes: "a. The system of

spoken or written communication used by a particular country, people, community, etc., typically consisting of words used within a regular grammatical and syntactic structure"; "b. The vocal sounds by which mammals and birds communicate; (in extended use) any other signals used by animals to communicate"; "c. A means of communicating other than by the use of words, as gesture, facial expression, etc.; non-verbal communication." This definition readily acknowledges that nonhuman others share with our species a variety of cognitive and perceptual mechanisms that constitute "language" in its broad sense (see in-depth discussion on faculty of language in the broad sense [FLB] and in the narrow sense [FLN] and their components by Hauser et al. 2002). This definition denotes what language enables an organism to *do*, including both verbal and nonverbal communication, and makes no specific reference to language as a uniquely human trait. Strangely, however, an Internet search quickly reveals that it is a narrowly understood definition of language (FLN) and its tendency to assume human uniqueness that continue being indoctrinated in schools and universities worldwide as established scientific facts. The exceptionality of the human language is conceived by virtue of the unique capacity of recursion—a concept proposed with no explicit definition by Hauser et al. (2002), and subsequently defined by Pinker and Jackendoff (2005, 203) as "a constituent that contains a constituent of the same kind"; in other words, a phrase of a certain type can be found inside a phrase of the same type like a linguistic equivalent of a set of *matryoshka* dolls—and discrete infinity, the property by which a few finite elements produce an infinite number of expressions (see Chomsky 2000). The problem is that the empirical evidence does not support the theoretical assumption. For example, recursion may extend indefinitely *in principle,* but its actual usage is quite limited (e.g., Karlsson 2010; Laury and Ono 2010) and might even be absent in some human languages (Everett 2005). Additionally, recursive computations are not a unique feature of human language, because they are found in other capacities, such as numerical and spatial cognition, which are more broadly distributed among nonhuman species (e.g., Gentner et al. 2006; see also Arsenijević, 2008; Arsenijević and Hinzen 2010). This discrepancy needs addressing because our narcissistic way of thinking about the world perpetuates an attitude of disregard for the nonhuman world and, more generally, the natural environment.

To address the discrepancy, I make no attempt to be all-inclusive in my approach to this huge field of research on language. My inevitably selective treatment of the topic is primarily informed by emerging empirical evidence from the field of animal and plant behavioral ecology rather than the wide variety of perspectives offered by linguistics and psychology. It is my goal, however, to engage the transdisciplinary dialogue between the psychological and ecological understanding of language. I approach this task by bridging the gap between the human and nonhuman world in two ways. On the one hand, I bring the human world closer to nature by showing that much of human language is instinctual and by virtue of its very "materiality," closer than we think to the language of nonhuman others. On the other hand, by showing the greater complexity of nonhuman communication and thus elevating it closer to human language, I attempt to bring nature closer to the human world via, ultimately, the medium of a more universal understanding of language.

On the Human Superiority Complex

Based on the common view portrayed by a narrow definition of the faculty of language (FLN), human language is creative and unpredictable, while the nonhuman (generally restricted to some animal species) communication is believed to be stereotyped, instinctive, and predictable. But are we even sure that our use of language is truly unpredictable and creative? Based on linguistic evidence, Lakoff and Johnson (1980) showed that most of what we do every day and how we experience and describe the world around us is, in fact, instinctive; we think, act, and communicate with each other using a stereotyped language that is defined by our culture through a conceptual system of metaphors that we are normally not even aware of. Perhaps it is because this system of metaphors is unavoidable and typically unconscious (Lakoff 2010) that we come to conceive of our language as spontaneous, free, unpredictable, and creative, and, by logical extension, attribute each of these characteristics to our species as a whole, leading us into a trap of idolatry. Further, though there are many apparent differences between human and nonhuman language, it is sensible to question our long-standing practice of defining the essence of our humanness through the things that nonhuman others lack (Baker 2012). When we do so, we inevitably use the qualities and conditions of being nonhuman as oppositional foils. At best we fail to arrive at certain conclusions about our distinctiveness, and at worst we engender distorted beliefs about our own nature as well as that of non-human others—beliefs that then prove disastrous for our relationship both with our own natures and with those of others.

Because of its traditional foundation in human psychology and linguistics, the study of language assumes, to a greater or lesser extent, that human linguistic behavior constitutes the standard template for theorizing the issue. Clearly, taking human language as the diagnostic reference point to investigate what linguistic abilities are present in nonhuman others is inescapably anthropocentric and confines the interpretation of the communicative capacities of nonhuman others to the domain of human values and perception. To solve these problems, we need to envision an empirically tractable and phylogenetically neutral account of language (see examples from the evolution of signals and communication literature, such as Maynard Smith and Harper 2003; Jablonka and Lamb 2005; Skyrms 2010) that resists the temptation of looking for evidence of signaling systems in the nonhuman world that exhibit the various forms of signaling and communication that jointly make up human language. And even then, such a strategy is still anthropocentrically tinted, being inevitably bounded (as previously mentioned) to our conceptual system of metaphors, which by virtue of being a bundle of incomplete representations, favors one way of seeing while obscuring others (Lakoff and Johnson 1980).

In this chapter, I will propose a slightly different approach, one that deliberately embraces the human "filter" by employing the very same concepts, such as language, meaning, symbols, and culture (whose problematic character we have noted above), to then view language from a wider biological perspective in terms of what it lets us or any organism *do* (i.e., an embodied view of language; see review by Kelly et al. 2002; Clark 2006). By treating language as a real and perceivable feature of the whole organism–environment system, where linguistic information is used and comes to have meaning in the same basic way as in perceptual-action situations (see Wilson and Golonka 2013), we are able to consider language as a meaning-making *activity* at the core of every form of life, whether human or not.

Here, I offer an overview of the most recent empirically grounded advances in our understanding of the actual "gestures" and "utterances" of others, and particularly plants, which may assist in opening up a fresh dialogue about and with these nonhuman others. Ultimately, the overall aim is to encourage the emergence of new ways of understanding our experience of the world, and, in so doing, breathe life into a new narrative, where language is unbridled from human incarceration and its power is refocused toward a more integrated perception of the world.

The Making of Meaning

Humans use language to make sense of everything that surrounds them and to share their perceptions with others in order to reduce uncertainty about the world and possibly improve their capacity for survival. While our drive for language acquisition may be innate (Chomsky 1965; but see Tomasello 2003), its competent use in relation to a given cultural background is a meaning-making activity that is acquired through learning and inevitably bound to the context of its use (i.e., language is "meaning in use" à la Wittgenstein; see Baker 2012). In this sense, our language is about information, certainty, and survival, but, most profoundly, language serves us in our attempt to grasp the essence of what it is like to experience being human. As a vibrating link encoded with all the memories of our species and all the life forms we descended from, the making of meaning is our gateway to experience our deep history of connection to all others, humans and nonhumans. It is within this common milieu that humans and others use a language to represent events and things in their own environment, giving meaning to their individual experiences of being in the world. It is, then, accurate to say that (our) language is one of the hallmarks of being human; yet in its deeper connotation, such declaration is even more profound because it is equally true for all others.

As humans, bound as we are to our "dwelling-world" (see the notion of *Umwelt*; Uexküll 2010), is it even possible to conceive the making of meanings by others, to perceive the language of a humpback whale, a songbird, an ant, a sunflower, or a bacterium? Doing so could be a matter of attunement to the enormous vocabulary of bodily postures, elusive gestures, loud colorful displays, fleeting acoustic and chemical utterances, and barely palpable electromagnetic embraces used by all living forms, including us. Of course, this is no easy task. To start with, because of our innate anthropocentric bias (see discussion by Gagliano 2015), we are likely to recognize more readily (if at all) the existence of a language when this is primarily made of sounds (i.e., speech as the defining ability of using language to engage in dialogue and its absence used to denote the "silence" of nature; Vogel 2006). Yet this could be just the place where we start considering the language of nonhumans from a wider biological perspective based on available empirical evidences. Birds, for example, have a considerable repertoire of sounds, which develops into articulate and complex "songs" constituting part of their language and exists alongside their innate call system (equivalent to our human innate call system of screams, grunts, sighs, laughter). The avian language arises through learning earlier in life and it is fully developed by experience within the context of its use, so that birds know when to use certain vocalizations (e.g., a flight whistle is used only while the birds are flying; Rothstein and Fleischer 1987). The exquisite range of vocalizations, including a variety of distinct regional dialects that make up the avian language, is motivated by a need to communicate and imbued with meanings, such as declaring ownership of a territory, attracting the attention of a potential mate, or repelling an intruder, all of which are examples of interactive exchanges between two or more individuals (i.e., conversations). Moreover, a recent study by Flower et al. (2014) found that a species of African bird is in fact rather talented at using language, especially that of other species, to "cry wolf" in a bid to scare other animals away and steal their food. As true tacticians, these birds change their calls in response to the feedback they receive from their dupes, so that when one false alarm no longer works, the birds switch to another species' warning cry (Flower et al. 2014). Whether or not these birds are actually intuiting what others are thinking or adapting their behavior by reflecting upon their meaning-making activity, the fact remains that their mastery of language is, like our own, a demonstration of a very accomplished capacity for creating sounds that acquire meaning because of how the information is used (i.e., to direct and regulate the behaviors of themselves or others), which ultimately enhances their survival.

Exactly how, though, do we approach others whose languages we fail to notice because they appear silent to us? Plants seem to fall into this particular category because theirs is primarily (as currently understood) a silent language of shapes, colors, and scents (but see recent findings on the ability of plants to produce, detect, and respond to sound: Gagliano et al. 2012; Gagliano 2013; Appel and Cocroft 2014). Indeed, one of the primary ways through which plants interact and communicate within their environment is by a rich and complex bouquet of several thousand volatile chemicals (see Raguso 2008). Through their crafty use of this chemical language, plants are able to breathe out their message by encoding it with a single scented word that nonetheless conveys multiple meanings depending on the intended recipients. By adding just a bitter whisper (literally, only tiny amounts) of nicotine to their bouquet of nectar volatiles, for example, plants are able to discourage unwanted visitors such as florivores and nectar thieves; yet, this same nectar constituent is simultaneously used by the plant to manipulate the behavior of desirable floral visitors, such as hummingbirds, causing them to visit more flowers and ultimately increasing the plant's reproductive success (Kessler and Baldwin 2006; Kessler et al. 2008). Together with olfactory signals, visual displays including species-specific shapes, colors, and color patterns (see Raguso 2004) are the other prominent and well-known medium through which plants interact and communicate. More than 450 species of plants, for example, are able to change their color, position, and shape to advertise their trading hours and even promote further business deals if inadequately pollinated. Research by Willmer et al. (2009) describes the legume *Desmodium setigerum* as one such master shape-shifter, who will show off attractive lilac flowers at the start of the day to then turn them into a less eye-catching white and turquoise color as the flowers get pollinated over the course of the day. The color change is accompanied by a slower change in the shape of the flower as the upper petal folds downward over the reproductive parts, where the pollen is produced (anthers) and germinates (stigma). By rapidly changing the shape and color of its pollinated flowers and thus reducing their attractiveness, this plant directs pollinators to the unvisited flowers to be pollinated. If they received insufficient pollen, however, the flowers of this plant are unique in their ability to readvertise themselves as "open for business" by changing their shape to expose the stigma once again as well as shifting from the white and turquoise back to the attention-grabbing lilac color.

Like all language, these are gestures enriched with meaning, which plants use to invite, for example, the pollinators' behavior that is most beneficial at a specific time. Plants and pollinators, in particular, seem to have reached a mutual understanding of this interactive language, a kind of covenant where the meaning of colors, shapes, and scents together with the rules that govern these meanings are constantly negotiated, agreed upon, and embodied by the plant and its pollinators as part of their coevolutionary dialogue. In this sense, they do not differ considerably from any human gestures, including our spoken words; these, too, are perceptible material symbols "representing" the real things (or referents) of the world and to which a meaning is assigned and collectively agreed upon for a message to be shared among individuals. At this point, some may concede that plants display the faculty of language but nonetheless argue for human exceptionalism—our language is unique and distinctive because of the arbitrariness of the relationship between its signifiers and their referents (i.e., Saussure's theory of arbitrariness; Harris 1987), an arbitrariness that implies freedom and creativity. In other words, human language is special because of its power of abstraction, thanks to which the symbols we use are, in fact, not directly tied down to the physical things they symbolize. Yet, once again, how sure are we that our language truly is as subjective and not intrinsically connected to what it represents as we perceive? Even excluding the case of onomatopoeia and ideogramic languages (see discussion on this topic by Cuddon 2013), linguistic evidence shows that our words were originally much more rooted in a material connection to the things they referred to than we think, and abstraction of the symbol from the symbolized only developed over time (see Abram 1996, 99–102, 284).

Interestingly, even if we were to consider abstraction as an intrinsic and "required" property of human language, which allows a speaker to refer to something that is not present (as proposed by Vogel 2006), our exceptionalism remains illusionary. After all, if the case of human hunters mimicking partridge calls to lure the birds out of the bush is an example of abstraction (Vogel 2006), then the case mentioned above of the African bird that cries wolf using other species' warning calls is too. And so it is for orchids that dupe wasps into becoming unsuspecting pollinators by releasing a spectrum of chemicals very similar to what other plants emit when "crying for help" to summon predatory insects to feast on caterpillars (Brodmann et al. 2008) or even mimicking animal alarm (pheromone) calls such as those of honeybees, which are frequently captured by wasps to feed on their larvae (Brodmann et al. 2009). In other words, human language is not more exceptional than the very evolutionary process that shapes it—a natural dance by which the symbolic, yet material, gestures of all living species are perpetually moved into new and evolving habits of rendering meaning in the world.

The Faculty of Language: Learning, Culture, and Evolution in Plants

The versatility in the linguistic behavior of plants, as in other organisms, arises as a result of experience and, in turn, underpins the adaptive changes necessary to cope with novel and unpredictable challenges to ensure the survival and flourishing of individuals within their specific ecological, as well as social, settings. We now know that plant chemical language, for example, is endowed with true semantic flexibility, so that new meanings may be assigned to old chemical words and used in novel interactions and new contexts (see review by Holopainen 2004). Evolutionarily speaking, it is precisely through use and experience under different circumstances that this inventory of gestures and utterances is enriched with meaning and, most important, shared with others and across generations. From this perspective, the language of plants seem to emerge from the interactions of individual learning, cultural transmission, and biological evolution—the same three adaptive systems thought to be at the root of human language. Let's briefly consider how these three systems are expressed in the life of plants.

Even though, in the broadest sense, the evolution of language occurs at a species level over hundreds of thousands of years, the fundamental process of adaptation by which a linguistic behavior is acquired in the first place, performed by a number of individuals over time, and then passed from one generation to the next (i.e., on a cultural timescale of thousands of years), is driven by the individual's ability to learn during its lifetime (see discussion by Christiansen and Kirby 2003). In this regard, we have recently discovered that plants, like animals, can learn rapidly by acquiring novel behavioral features and change their behavior as a result of their individual experience (Gagliano et al. 2014); these learned behaviors can later become innate and potentially enhance evolutionary learning at the species level through the fundamental mechanism of genetic assimilation (i.e., the "Baldwin effect"). Based on what we currently know, it is in the context of plant defensive behavior against enemies that the linguistic ability of a plant to swiftly learn appears to be particularly significant. The volatile words plants whisper, for example, appear to be rather specific in what they convey (e.g., expressing useful and reliable information to the predators of their attackers). Moreover, by remembering past attacks, individual plants learn to adjust their volatile responses to be better prepared for future battles. And, as in humans, plants eavesdrop on the affairs of their neighbors and are able to use this vicariously acquired knowledge to mount a tactical defensive response even before they are attacked or damaged themselves (e.g., Paré and Tumlinson 1999; Karban et al. 2000; Heil and Ton 2008). Additionally, we know that the scented utterances of plants are more meaningful and effectively received among kin (i.e., genetically identical or related) than strangers (e.g., Karban and Shiojiri 2009; Karban et al. 2013). This finding and the recent discovery that recognition of relatives also involves specific leaf gestures and the perception of particular light signals (e.g., Crespi and Casal 2015) have demonstrated the occurrence of cooperative rather than competitive interactions among kin in plants (e.g., Dudley and File 2007; Murphy and Dudley 2009; File et al. 2012) as observed in a wide range of taxa. And, in general, this seems analogous to the process by which a group of individuals learns and passes on information to others, hence giving rise to what we call cultural transmission in humans and some animals.

Ultimately (and evolutionarily), being able to communicate with each other, form bonds of various kinds, and engage in teamwork are what makes language so important to us all, human and nonhuman alike. Consequently, I would like to invite the reader one last time in this chapter to interrogate the cultural validation of human linguistic uniqueness, in which humans strike us as "special" or "exceptional" when compared with nonhuman others. Whether human or not, the linguistic code of gestures and utterances unique to each species is designed as it is because those (and not any others) are the specific design features that have evolved to make for a useful communication system in the cultural context. Here, the term *culture* relates to the suite of behaviors and ways of living for that species or group of species. For example, the cultural background of plants refers to the interactions and relationships among plants as well as between plants, other organisms, and the environment. Therefore, the interpretation of these linguistic features is contingent on the knowledge of the appropriate cultural background in which the code is used. As such knowledge is understood and shared by members of a specific group (whether human or nonhuman), the code concurrently defines the communication system within that group and constructs the identity of those that use it. In humans, we can readily appreciate the extent to which speaking a language with a particular accent, for example, can easily give away one's geographical area of origin or even social status. And we appreciate that this is the case in other species too. The song patterns of humpback whales, for a well-known example, depend on where they live; populations inhabiting different ocean basins are known to sing distinctly different songs, and all males in a given population sing the same mating song until a catchier tune emerges as they encounter a migrating population, and then all males start singing the latest song (Noad et al. 2000; Garland et al. 2011). The degree of creativity associated with the composition of these songs and their sharing, as part of a cultural exchange among populations, is undeniable even if we still do not understand exactly how they are composed or what they communicate. To understand this "whale-specific" knowledge, there is no need to break into the "whale code" by translating it into human standards; it is essential, however, to appreciate the cultural background within which such "whale code" is used by whales. Somehow, we intuitively know that what whales are sharing using their language is solidly reposed within the nature of being a whale. And I believe this is exactly where the crux of the matter is rooted. The answer to my earlier question of whether it is possible for humans to conceive and appreciate the making of meanings by others may also be found here. What if language is a fundamentally natural and inevitable consequence of being that emerges as an organism makes meaning of its surroundings and, in turn, engraves the very identity of that organism and its physical embodiment in its world?

If we are to understand the language of plants, for example, the detecting, dissecting, and deciphering of their "code" (e.g., the identity of the chemicals a plant produces) need not be the main priority of our scientific endeavors. For example, how many more chemical compounds or light wavelengths do we need to detect and analyze before we can truly access, understand, and appreciate plant language? Much of the research in plant science is primarily focused on dissecting plants and deciphering their internal codes. However, the greatest advances in our understanding of plant language have been ingeniously delivered by a relatively small number of scientists who have adopted ecologically driven approaches, where the "cultural background" is taken into account. This is because meaning emerges during interactions among organisms; hence language is not a fixed property of that organism (e.g., a specific chemical compound) but rather a truly ecological, dynamic process of relationships by which meaning emerges to shape the production of behaviors that, in turn, shape new interactions for new meaning to emerge (e.g., a biology of language à la Maturana; see discussion on the topic by Kravchenko 2011). In practice, reducing such an active and ecologically vibrant process to a fixed and petri dish–like property has resulted in substituting the subjective material plant with an abstraction, the scientific idea of what an objectified plant is, does, and knows. Far from being trivial, the cultural construction of plants as *objects* of scientific exploration not only contradicts the emerging and expanded understanding of plant behavior, including matters of plant intelligence, agency, and intersubjectivity, but is also of ethical significance in the context of human–plant relations (Marder 2013). I suggest that an embodied conception of language could offer a valuable step toward the de-objectification of plants and the recognition of their subjectivity and inherent worth and dignity, renewing a sense of ecological intimacy and kinship with these nonhuman living *others* and, thus, promoting human care for nature (see Clayton and Meyer 2015).

Additionally, the task of redefining language through an embodied analysis could also be valuable to address the abstraction of our own species from nature. In fact, the same power of abstraction that has reduced living plants into lifeless objects over time has uprooted our own species from the physical connection we have to what we refer to.

Because the symbols we use have become themselves separate from (and only arbitrarily related to) what those symbols symbolize, the idea that *Homo sapiens* is separate from and dominating over all other species (hence special) is incarnated within the medium of communication itself, but nevertheless groundless. From this perspective, our abstractive power has resulted in the silencing (rather than the revealing) of the expression and faculties of ourselves as well as others, such as plants. Breaking this silence calls for a truly "cross-cultural" dialogue—a full immersion into the ecological context and beingness of humans, plants, as well as other nonhuman organisms with whom we share a common world. Such approach demands a new and enlarged sense of "cultural awareness" that appreciates both humans and plants (and any other organism) as embodied agents and recognizes that language always is contingent on the encounter of individual and subjective perceptual worlds and emerges at those points of contact and interaction in a truly ecological sense.

In today's globalized world, we are increasingly dwelling in highly cross-cultural environments, and an attitude of tolerance, respect, and appreciation for diversity is clearly fundamental for human societies to flourish in coexistence with all *others* (and, sadly, we know too well what happens when such attitude is not adopted). From this perspective, language is that open and tangible invitation to stop and smell the roses—as those plant volatile chemicals enter your nose, the cross-cultural dialogue is opened.

References

Abram, D. 1996. *The Spell of the Sensuous.* New York: Random House.

Appel, H. M., and R. B. Cocroft. 2014. "Plants Respond to Leaf Vibrations Caused by Insect Herbivore Chewing." *Oecologia* 175:1257–66.

Arsenijević, B. 2008. "From Spatial Cognition to Language." *Biolinguistics* 2:3–23.

Arsenijević, B., and W. Hinzen. 2010. "Recursion as a Human Universal and as a Primitive." *Biolinguistics* 4:165–73.

Baker, N. E. 2012. "The Difficulty of Language: Wittgenstein on Animals and Humans." In *Language, Ethics, and Animal Life: Wittgenstein and Beyond,* edited by N. Forsberg, M. Burley, and N. Hämäläinen, 45–64. New York: Bloomsbury.

Bolhuis, J. J., I. Tattersall, N. Chomsky, and R. C. Berwick. 2014. "How Could Language Have Evolved?" *PLOS Biology* 12:e1001934.

Brodmann, J., R. Twele, W. Francke, G. Hölzler, Q.-H. Zhang, and M. Ayasse. 2008. "Orchids Mimic Green-Leaf Volatiles to Attract Prey-Hunting Wasps for Pollination." *Current Biology* 18:740–44.

Brodmann, J., R. Twele, W. Francke, Y.-B. Luo, X.-Q. Song, and M. Ayasse. 2009. "Orchid Mimics Honey Bee Alarm Pheromone in Order to Attract Hornets for Pollination." *Current Biology* 19:1368–72.

Chomsky, N. 1965. *Aspects of the Theory of Syntax*. Cambridge, Mass.: MIT Press.

———. 2000. *New Horizons in the Study of Language and Mind*. Cambridge, UK: Cambridge University Press.

Christiansen, H. M., and S. Kirby. 2003. "Language Evolution: Consensus and Controversies." *Trends in Cognitive Science* 7:300–307.

Clark, A. 2006. "Language, Embodiment, and the Cognitive Niche." *Trends in Cognitive Science* 10:370–74.

Clayton S., and G. Meyer. 2015. *Conservation Psychology: Understanding and Promoting Human Care for Nature*. Chichester, UK: Wiley-Blackwell.

Crespi, M. A., and J. J. Casal. 2015. "Photoreceptor-Mediated Kin Recognition in Plants." *New Phytologist* 205:329–38.

Cuddon, J. A. 2013. *A Dictionary of Literary Terms and Literary Theory*. 5th ed. Chichester, UK: Wiley-Blackwell. Available at Blackwell Reference Online. http://www.blackwellreference.com/public/book.html?id=g9781444333275_9781444333275.

Dudley, S. A., and A. L. File. 2007. "Kin Recognition in an Annual Plant." *Biology Letters* 3:435–38.

Everett, D. L. 2005. "Cultural Constraints on Grammar and Cognition in Pirahã: Another Look at the Design Features of Human Language." *Current Anthropology* 46:621–46.

File, A. L., G. P. Murphy, and S. A. Dudley. 2012. "Fitness Consequences of Plants Growing with Siblings: Reconciling Kin Selection, Niche Partitioning, and Competitive Ability." *Proceedings of the Royal Society B* 279:209–18.

Flower, T. P., M. Gribble, and A. R. Ridley. 2014. "Deception by Flexible Alarm Mimicry in an African Bird." *Science* 344:513–16.

Gagliano, M. 2013. "Green Symphonies: A Call for Studies on Acoustic Communication in Plants." *Behavioral Ecology* 24:789–96.

———. 2015. "In a Green Frame of Mind: Perspectives on the Behavioural Ecology and Cognitive Nature of Plants." *AoB Plants* 7:plu075.

Gagliano, M., S. Mancuso, and D. Robert. 2012. "Towards Understanding Plant Bioacoustics." *Trends in Plant Science* 17:323–25.

Gagliano, M., M. Renton, M. Depczynski, and S. Mancuso. 2014. "Experience Teaches Plants to Learn Faster and Forget Slower in Environments Where It Matters." *Oecologia* 175:63–72.

Garland, E. C., A. W. Goldizen, M. L. Rekdahl, R. Constantine, C. Garrigue, N. D. Hauser, M. M. Poole, et al. 2011. "Dynamic Horizontal Cultural Transmission of Humpback Whale Song at the Ocean Basin Scale." *Current Biology* 21:687–91.

Gentner, T. Q., K. M. Fenn, D. Margoliash, and H. C. Nusbaum. 2006. "Recursive Syntactic Pattern Learning by Songbirds." *Nature* 440:1204–7.

Harris, R. 1987. *Reading Saussure: A Critical Commentary on the "Cours de linguistique générale."* La Salle, Ill.: Open Court.

Hauser, D. M., N. Chomsky, and W. T. Fitch. 2002. "The Faculty of Language: What Is It, Who Has It, and How Did It Evolve?" *Science* 298:1569–79.

Heil, M., and J. Ton. 2008. "Long-Distance Signalling in Plant Defence." *Trends in Plant Science* 13:264–72.

Holopainen, J. K. 2004. "Multiple Functions of Inducible Plant Volatiles." *Trends in Plant Science* 9:529–33.

Jablonka, E., and M. J. Lamb. 2005. *Evolution in Four Dimensions: Genetic, Epigenetic, Behavioral, and Symbolic Variation in the History of Life.* Cambridge, Mass.: MIT Press.

Karban, R., I. T. Baldwin, K. J. Baxter, G. Laue, and G. W. Felton. 2000. "Communication between Plants: Induced Resistance in Wild Tobacco Plants Following Clipping of Neighboring Sagebrush." *Oecologia* 125: 66–71.

Karban, R., and K. Shiojiri. 2009. "Self-Recognition Affects Plant Communication and Defense." *Ecology Letters* 12:502–06.

Karban, R., K. Shiojiri, S. Ishizaki, W. C. Wetzel, and R. Y. Evans. 2013. "Kin Recognition Affects Plant Communication and Defence." *Proceedings of the Royal Society B* 280:20123062. doi:10.1098/rspb.2012.3062.

Karlsson, F. 2010. "Recursion and Iteration." In *Recursion and Human Language,* edited by H. van der Hulst, 43–67. Berlin: De Gruyter Mouton.

Kelly, S. D., J. Iverson, J. Terranova, J. Niego, M. Hopkins, and L. Goldsmith. 2002. "Putting Language Back in the Body: Speech and Gesture on Three Time Frames." *Developmental Neuropsychology* 22:323–49.

Kessler, D., and I. T. Baldwin. 2006. "Making Sense of Nectar Scents: The Effects of Nectar Secondary Metabolites on Floral Visitors of *Nicotiana attenuata.*" *Plant Journal* 49:840–54.

Kessler, D., K. Gase, and I. T. Baldwin. 2008. "Field Experiments with Transformed Plants Reveal the Sense of Floral Scents." *Science* 321:1200–1202.

Kravchenko, A. 2011. "How Humberto Maturana's Biology of Cognition Can Revive the Language Sciences." *Constructivist Foundations* 6:352–62.

Lakoff, G. 2010. "Why It Matters How We Frame the Environment?" *Environmental Communication* 4:70–81.

Lakoff, G., and M. Johnson. 1980. *Metaphors We Live by*. Chicago: University of Chicago Press.

Laury, R., and T. Ono. 2010. "Recursion in Conversation: What Speakers of Finnish and Japanese Know How to Do." In *Recursion and Human Language*, edited by H. van der Hulst, 69–92. Berlin: De Gruyter Mouton.

Marder, M. 2013. *Plant-Thinking: A Philosophy of Vegetal Life*. New York: Columbia University Press.

Maynard Smith, J., and D. Harper. 2003. *Animal Signals*. Oxford: Oxford University Press.

Maynard Smith, J., and E. Szathmáry. 1995. *The Major Transitions in Evolution*. Oxford: Oxford University Press.

Murphy, G. P., and S. A. Dudley. 2009. "Kin Recognition: Competition and Cooperation in *Impatiens* (Balsaminaceae)." *American Journal of Botany* 96:1–7.

Noad, M. J., D. H. Cato, M. M. Bryden, M.-N. Jenner, and K. C. S. Jenner. 2000. "Cultural Revolution in Whale Songs." *Nature* 408:537.

Paré, P. W., and J. H. Tumlinson. 1999. "Plant Volatiles as a Defense against Insect Herbivores." *Plant Physiology* 121:325–31.

Pinker, S., and R. Jackendoff. 2005. "The Faculty of Language: What's Special about It?" *Cognition* 95:201–36.

Raguso, R. A. 2004. "Flowers as Sensory Billboards: Progress Towards an Integrated Understanding of Floral Advertisement." *Current Opinion in Plant Biology* 7:434–40.

———. 2008. "Wake Up and Smell the Roses: The Ecology and Evolution of Floral Scent." *Annual Review in Ecology, Evolution, and Systematics* 39: 549–69.

Rothstein, S. I., and R. C. Fleischer. 1987. "Vocal Dialects and Their Possible Relation to Honest Status Signalling in the Brown-Headed Cowbird." *Condor* 89:1–23.

Skyrms, B. 2010. *Signals: Evolution, Learning, and Information*. Oxford: Oxford University Press.

Tomasello, M. 2003. *Constructing a Language: A Usage-Based Theory of Language Acquisition*. Cambridge, Mass.: Harvard University Press.

Uexküll, J. von. 2010. *A Foray into the World of Animals and Humans: With a Theory of Meaning*. Minneapolis: University of Minnesota Press.

Vogel, S. 2006. "The Silence of Nature." *Environmental Values* 15:145–71.

Willmer, P., D. A. Stanley, K. Steijven, I. M. Matthews, and C. V. Nuttman. 2009. "Bidirectional Flower Color and Shape Changes Allow a Second Opportunity for Pollination." *Current Biology* 19:919–23.

Wilson, A. D., and S. Golonka. 2013. "Embodied Cognition Is Not What You Think It Is." *Frontiers in Psychology* 4:58.

II. Philosophy

FIVE

To Hear Plants Speak

Michael Marder

HUMANS ARE SOCIAL BEINGS ANXIOUS TO COMMUNICATE WELL beyond the confines of our own time and place—with future generations, with hypothetical extraterrestrial civilizations, or with other animals, such as dolphins, who interact through sonar media. In each case, the expansion of the linguistic model is faced with the problem of encoding a message intended for others (be they human or not) and deciphering possible messages emitted by them. But only very rarely does language itself become a problem.

Plants have not escaped the human desire for universal communicability. Consider the following, relatively recent, examples. In 2008, a café owner in the Japanese town of Kamakura decided to connect electric sensors to a plant, translating its sensitivity to light, heat, and so forth into Japanese with the help of an algorithm. The result was a widely publicized blog by *Midori-san*, or Mr. Green, the nickname of the *Hoya kerrii* specimen housed in the café and plugged into global telecommunication networks.[1] A more serious instance of the same impulse is evident in seismologists' increasing reliance on trees' sensitivity to changes in ultralow-frequency geomagnetic pulsations in a bid to anticipate earthquakes.[2] In the capacity of biotic measuring devices, trees now serve as the "networking elements of the Internet of Things."[3]

We want to "hear" plants "speak." What kind of hearing must we resort to and what sort of speech corresponds to it? How to translate the language, or the languages, of plants into terms that are intelligible within the scope of our human languages? What is lost in this transposition, above all when it strives to make everything about the plants' communication with themselves, with other plants, with insects or other animals transparent? What are the conditions of possibility for a cross-kingdoms translation and what is the place of the untranslatable in it? And, in the first place, is the expression "the language of plants" defensible?

From the outset, it should be clear that the word *language*, conjugated with plants, does not *merely* point in the direction of a metaphor, an analogy, or a judgment of nonhuman forms of life based on human phenomena. We are sure not to hear plants speak if we continue to hold onto the anthropocentric prejudice that sees in our intelligence, cogitation, and languages the gold standards of intelligence, cogitation, and language as such. Protagoras's sophistic insight, reported by Sextus Empiricus—"The human is the measure of all things *[pantōn khrematōn einai metron ton anthrōpon]*: of things that are, that they are; of things that are not, that they are not"[4]—still overshadows our thinking millennia after its formulation. While, traditionally, Protagoras has been read as an advocate of extreme relativism in human knowledge (a reading that is not too fanciful, given that the sophist himself insisted on the changeability of truth depending on what appears to each individual), today his statement has an opposite ring to it, if it is heard in a cross-species context. The human *(to anthrōpos)* as human has now become the measure of all things, so every form of life appears to be deficient by comparison with it. Thanks to humanity's alienation from the rest of the natural world, it has turned into an exception that, starting from itself, dictates the rule.

We must, by all means, resist this piece of contemporary sophistry that passes for common sense. A serious consideration of the nature of language cannot afford to ignore the question about its broader meaning, outside a purely human enunciation. This is what Walter Benjamin called "language as such" or "the language of things," to which we shall return. Similarly, in discussing the language of plants, we are invited to reexamine the prevalent and highly impoverished view of language as a tool for transmitting or sharing information. At the very least, it will be necessary to supplement this view with the symbolic dimension of language, the mechanisms of expression and articulation, and the extended material–substantive side of vegetal biosemiotics.

In this chapter, I discuss the question of the language of plants as one of translation into the more or less familiar frameworks of human discourse. I outline four possible modalities of such translation: (1) the symbolic "language of flowers" that announces itself in writers as diverse as St. Augustine and Sigmund Freud, Novalis and James Joyce; (2) the figure of "talking trees" that pervades various cultures, from the sacred grove of Dodona in ancient Greece, through vegetalized human beings in Dante's *Inferno,* to the Persian Trees of the Sun and the Moon; (3) biochemical communication between plants (as well as between plants and animals), quantified in plant sciences; and (4) the plants' participation in the language of things, with its spatial nexuses and articulations. I argue that to hear plants speak we must learn to listen to the lacunae and silences of language, leaving plenty of room for the untranslatable (and, hence, the unspeakable) in these practices of translation. To the extent that each of the four modalities refuses to impose specifically human voices, categories, and discursive molds onto the vegetal world, it accomplishes this task and gets in touch with the self-expression of plants.

THE SYMBOLISM OF THE "LANGUAGE OF FLOWERS"

When a plant is converted into a symbol, its own language all but disappears underneath a shroud of meanings humans throw over it. It is no longer possible to discern the self-signification of vegetal life, because the symbolic plant does not refer to itself but turns into a token for something else entirely, thrown or cast (*ballein,* "to throw" at the root of *sym-bol*) outside itself. More often than not, the symbolic deployment of vegetation proceeds in the name of spirit—of spirit's separation from the world of matter, which is treated as the substratum for the impression of spiritual marks or signatures. A prevalent strategy in this regard is to locate the ultimate and, indeed, the only source of meaning in nonmaterial reality, divesting existence here below of any inherent sense of its own. Plant life, too, is presumed to be senseless in and of itself, unless it is put in the service of spirit. At the same time, spirit lays heterogeneous claims on various parts of plants; if flowers have been the most symbolic among these, it is because they have been framed, in the history of Western thought, as the closest to the spiritual realm. This is particularly evident in Goethe's botanical theory, according to which the pure colors of petals are the objective embodiments of light. A refined leaf, the petal is the pinnacle of the plant's spiritual ascent toward sexual difference, or "propagation through two genders."[5] Hence, instead of the language of plants, philosophers and poets resort to what they univocally designate as "the language of flowers."

St. Augustine excelled in the symbolic interpretations of texts and phenomena that permitted him to distance his theology and Christianity as a whole from Judaism. He resisted as much the vestiges of paganism (the full title of his 426 C.E. magnum opus is *The City of God against the Pagans*) as Jewish literalism, with its obedience to the letter of the divine law, rather than its spirit. And plants—particularly, flowers and fruit—were at the forefront of this endeavor. Let me give you an example. When it came to God's commandment to Adam and Eve to "be fruitful and multiply," Augustine refuses to take it literally because physical reproduction is a capacity humans share with plants and animals. For him, it would be absurd to contend that this should be a distinctly human virtue, because other kinds of creatures actually excel in it more than humans. "I might further say," he writes in his *Confessions*, "that this blessing, had I found it bestowed upon trees and plants and land animals, belongs to those kinds which are propagated by reproduction; but for plants and trees and beasts and serpents, there is no mention of 'Increase and multiply.'"[6] The way out of this conundrum is to spiritualize fruitfulness, rendering it symbolic; the true meaning of "fruit" is linked to the "works of mercy"—a spiritual and exclusively human activity that erases its vegetal (and sexual) origin, that is, physical reproduction.

Note that the association of plants with sexuality is the staple feature of "the language of flowers." Where a direct discourse on matters sexual is forbidden, invocations of vegetal sexuality abound as substitutes for the open outpouring of human desire. Thus, in the seventeenth and eighteenth centuries, botany was "the most explicit discourse, in the public domain, on sexuality."[7] In the eyes of Augustine, such connotations of plant life were central to pagan fertility cults, represented by the goddess Flora, who governed everything that bloomed and who was honored in the annual *ludi florales,* or floral games, replete with theatrical performances by naked actresses and prostitutes, some of whom were dramatically transubstantiated into Flora. Augustine calls worship of such deities "wholly wanton, impure, immodest, wicked, and unclean."[8] In order to purify the "shameful propitiation" of paganism, he sublimates plant parts into the spiritual sphere of good works, distancing them from their natural and literal origins. It follows, then, that the Augustinian vegetal symbolism was twice removed from the plants themselves and their languages: while pagan cults already took flowers to be the symbols of sexuality, Augustine recodified this association, making it fit the nascent lexicon of Christian spirituality.

What the author of *The City of God* found especially abhorrent was the flourishing of nonreproductive sexualities, of flowers that did not give rise to fruit. Reacting to the myth of Attis, the Phrygian god of vegetation, who in a temporary frenzy castrated himself, died, and was resurrected, Augustine mocks ancient Greek philosopher Porphyry for thinking that "Attis symbolizes the flowers, and that his mutilation signifies the fact that the flowers must fall before the fruit comes." In a curious inversion of the symbolic operation, whereby a humanized figure (or its organs) betokens a plant (or its parts),

> it is not the man himself, therefore—or the symbolic man whom they call Attis—whom they have compared to flowers, but his male organs. These, indeed, fell while he was still alive; or, rather, they did not fall, nor were they plucked, but were clearly torn off. But when the flower was lost, there was no subsequent fruit; rather, barrenness followed. What, then, is the meaning of this remnant . . . ? Do not the Greeks' fruitless attempts to find an interpretation simply persuade us that what rumor reports and writing records is no more than the tale of a castrated man?[9]

Augustine cannot tolerate a situation whereby a human symbolizes a plant. Nor does he think that flowers have any inherent value unless they give way to fruit or, more precisely, to the products of spiritual activity. Although it appears that his solution to the story of Attis is to take it literally, as "no more than a tale of a castrated man," the ironic meta-symbolism of this episode is that Attis denotes the fruitless nature of Greek thought itself, unacquainted with the truth of Christianity.

Centuries later, in *The Interpretation of Dreams* (1900), Freud will avow the symbolic bond uniting the language of flowers to human sexuality and the fabric of the unconscious. Here, Freud's report on his "Dream of the Botanical Monograph" is of special significance: "I had written a monograph on a certain plant. The book lay before me and I was at the moment turning over a folded colored plate. Bound up in each copy there was a dried specimen of the plant, as though it had been taken from a herbarium."[10] Through a variety of unconscious associations, Freud establishes that the words *botanical* and *monograph* "constituted 'nodal points' upon which a great number of the dream-thoughts converged."[11] This, in turn, illustrates his theory of "condensation" and "overdetermination," which, postulating complex, nonlinear, psychic causality, accounts for the emergence of any given dream-image in sleep. Needless to say, the associations Freud records are suffused with sexual innuendo: he reproaches himself for forgetting to bring flowers to his wife; recalls a clinical case, in which Frau L. rightly interpreted her husband's forgetting to give her flowers for her birthday as a sign of his cooling passion; alludes to the "blooming looks" of Professor Gärtner (Gardener) and his wife, and so on.[12]

Even more explicit are the interpretations of dream-images involving live flowers, as opposed to dry ones. When a neurotic young woman reports that in her dream she "arrange[s] the center of a table with flowers for a birthday," Freud hears in that "an expression of her bridal wishes: the table with its floral center-piece symbolized herself and her genitals; she represented her wishes for the future fulfilled, for her thoughts were already occupied with the birth of a baby; so her marriage lay a long way behind her." Once specified, the flowers included in the arrangement were ascribed their precise symbolic functions: the lily of the valley concurred with "its popular sense as a symbol of chastity," while the violet, said and heard in English, conveyed "the dreamer's thoughts on the violence of defloration." [13] We could further multiply these examples, but the conclusion to which they lead us is already obvious: when it comes to the language of flowers, or, perhaps, to the question of symbolism generally construed, psychoanalysis is at its least critical and self-critical. The analytic drive stops at the level of "popular symbolism," accepting the authority of received ideas. Admittedly, this acceptance is partly unavoidable, insofar as it echoes the "truths" held self-evident from the standpoint of the analysand her- or himself. Yet, the analysand's own interpretation should only present materials for psychoanalysis to elaborate upon, not ready-made conclusions for analytic practice.

The formal framework of Freudian associations also appears to be identical to that of Augustine: plants are the symbols for nonvegetal reality, namely, the sphere of human sexuality. Nevertheless, we cannot miss the fact that Freud's "language of flowers" does not translate them into the vocabulary of abstract spirit but into that of the unconscious, the wholly embodied, material, extended part of the psyche. A psychoanalytic *symbol* is a *symptom* of psychic repression or, at least, of sublimation, which analytic dream-work obstinately traces back to its roots. No doubt, his translation of the language of flowers into that of the unconscious fails to respect the inherent meaningfulness of vegetal life, with the plants' own lived interpretations of the world, wherein they grow, orient themselves, and interact. But it is equally certain that Freud's materialism, his resistance to an artificial separation between the mind and the body, or spirit and matter, is more faithful to this life than Augustine's harnessing of flora as an allegory for spiritual reality. Freudian flower symbolism charts the way back to plant life *through* the human psyche, for which sexuality (the classical reproductive faculty that originates in plants and is shared by all living beings) is constitutive. Cast away from itself as a symbol, the plant comes back to itself, having engulfed much of what we consider to be the human. That is why, in *Plant-Thinking,* I pointed out that the "object of *psycho*-analysis, wherein we might detect a vegetal approach to the psyche, is no longer 'a soul of another genus' but an extended psychic thing entwined with the body itself—a somatic, and thus divisible, soul akin to that of a plant."[14]

Across the spectrum of symbolic transcriptions and translations it motivates, the literal plant, the plant itself, remains untranslatable. There are no plants growing on the fields of metaphor, in the gardens of allegory, or in the forests of symbols—except of the metaphorical, allegorical, and symbolic kinds. When I say or write "the plant itself," I do not mean a vegetal version of the Kantian thing-in-itself, foreclosed to human knowledge. I have in mind the language of the plants themselves, irreducible to our superimposition of meanings upon them. Anything but theoretical hairsplitting, this distinction intimates that what we are dealing with is not the classical relation between a signifier—that is, the word *plant,* or the symbolic structure it participates in—and the corresponding signified, prelinguistic concept, but an interchange between two languages: the biosemiotics of vegetal life and human signification. Whether we are mindful of this or not, whatever we know about plants is due to a successful translation from the former to the latter, a translation that never exhausts or depletes but, on the contrary, enriches what it translates (just as no translation, say, from Bulgarian to Swahili, exhausts either the source or the target languages).

Georges Bataille calls psychoanalytic attention, precisely, to the problem of the preanalytic symbolism of plants in his short essay "The Language of Flowers." Rejecting outright the search for a secret meaning of vegetal life, he attacks, without overtly naming Freud, the "fortuitous and superficial character" of this language.[15] Although "love can be posited from the outset as the natural function of the flower," Bataille prefers to shed light on the imperfections, obscene sights, and smells of decay that form the reverse of flower symbolism.[16] "The flower is betrayed by the fragility of its corolla," he writes, and, "thus, far from answering the demands of human ideas, it is the sign of their failure. In fact, after a very short period of glory the marvelous corolla rots indecently in the sun, thus becoming, for the plant, a garish withering. Risen from the stench of the manure pile—even though it seemed for a moment to have escaped it in a flight of angelic and lyrical purity—the flower seems to relapse abruptly into its original squalor: the most ideal is rapidly reduced to a wisp of aerial manure."[17] Bataille couples flower symbolism with the other type of Aristotelian movement, namely, decay, which, alongside growth and metamorphosis, is proper to plants. His flowers bespeak death, the ever-present shadow and source of meaning for earthly existence, be it vegetal or human. With all its gory, unsightly attributes that cannot be easily idealized, the symbolism of decay does not respond to the demands of spirit in its flight from the messiness of matter. It is, therefore, symbolism at the limit of the symbolic, nearly letting a finite life express itself in the language of finitude.

In philosophy and in psychoanalysis, contestations of "the language of flowers" slide on a scale that stretches between idealism and raw materialism; in literature, a corresponding continuum extends between Romanticism and irony. Consider Novalis and Joyce. The hero of Novalis's unfinished novel (1802), Henry von Ofterdingen, embarks on a quest for the idealized and nonexistent "blue flower," *blaue Blume*, which he saw in a dream. In anticipation of its symbolic association with his future beloved, Mathilda, he notices that "the flower leaned towards him and its petals displayed an expanded blue corolla wherein a delicate face hovered."[18] When he finally meets Mathilda, he asks himself: "Do I not feel as I did in that dream when I saw the blue flower? What strange connection is there between Mathilda and that flower?"[19] The connection is, after all, not that strange in the context of Romantic imagination, with its double predilection for the dream, the impossible, the nonexistent, on the one hand, and Goethe's natural language of symbols based on "the symbolic language of nature" itself,[20] on the other. No wonder that Novalis's blue flower grew to be the symbol not only of Mathilda or of metaphysical longing but of the entire German Romantic movement.

The traditional imputations of purity and innocence to the language of flowers that persist in Novalis are a far cry from its utilization by Leopold Bloom, the protagonist of Joyce's *Ulysses* (1922), who adopts the pseudonym "Henry Flower" in the course of an erotic correspondence with his admirer, Martha Clifford. Accompanied by a dry and scentless yellow flower, Martha's letter of longing to meet Bloom implies that he is *her* "blue flower": unreachable and largely indifferent. "Henry Flower" transforms these epistolary yearnings in his stream of consciousness under the heading "the language of flowers," though not before a mocking meditation on such a language, ironically resonating with his pseudonym, which is a synonym of his "real" family name: "Language of flowers. They like it because no-one can hear." His rendition of the letter is even more biting: "Angry tulips with you darling manflower punish your cactus if you don't please poor forgetmenot how I long violets to dear roses when we soon anemone meet all naughty nightstalk wife Martha's perfume."[21] Bloom hints that the muteness of the language of flowers is propitious to launching veiled sexual insinuations under the cloak of civility. He proceeds to unveil the obscenities that lurk in this symbolic language by intermingling real and imaginary flower names with portions of Martha's text. We obtain, consequently, a series of charming phytophallic images, from the sadomasochistically tinged "punish your cactus" to the suggestive "naughty nightstalk."

In the works of Bataille, Joyce, and, to a lesser extent, Freud, the materialist and ironic inversions of the language of flowers internally undermine the spiritual and idealist thrust of symbolic expression. The instability of flower symbolism at the limit of its aspiration to ideality reveals a certain noncoincidence between the symbolizing and the symbolized: reproductive parts of plants can be the signs of purity and innocence *or* obscenity; magnificent growth *or* decay; an unfulfillable yearning *or* sadomasochistic fantasy . . . In other words, this instability indicates that there is much more to the language of plants than meets the eye in the superficiality of its symbolic dimension.

The Ethics of "Talking Trees"

The figure of "talking trees" traverses different cultures and seems to belong to the Jungian archetypes of the collective unconscious. In India, the rustling of leaves on a pipal tree is thought "to be done by supernatural beings, who speak by that means."[22] Herodotus reports in *The Histories* that the cult of the oracle of Zeus at Dodona in northern Greece, also mentioned in Homer's *Iliad* and by Socrates in Plato's *Phaedrus*, was centered on an oak grove.[23] The rustling of sacred oaks foretold the future, as interpreted by the oracle-priests of the Selli tribe: "The Selli listened to the rustling of the sacred oak and thought they could hear voices. A question spoken with a loud voice in the wind found an answer in the sighing, whispering, and rustling of the oak trees."[24] According to Buddhist traditions, trees that supplied the materials for the statues of Buddha could be observed "speaking, crying, or producing other miraculous signs."[25] Alexander the Great came across prophesying talking trees in Persia; known as Iskandar there, he received predictions about his demise from the male cypress Mithra, the Tree of the Sun, and the female cypress Mao, the Tree of the Moon: "In a letter to his mother, Iskandar describes how the Tree's male heads told him by day that he would soon die, and how the female heads informed him by night that he would not return home."[26] In the fables of Zhuangzi, in Dante's *Inferno,* in Tolkien's *The Lord of the Rings* with its Ents, and the like plants are endowed with anthropomorphic features, which invariably include the ability to speak.

The very partial list I have compiled above is meant to give but a foretaste of the prevalence of talking plants in world religions and literatures. As a rule, I see in this figure an improvement over the symbolic impositions of human meanings and languages onto plants. Unlike the flowers we have encountered in the previous section of this chapter, the trees themselves speak—for themselves and, usually, with knowledge that is superior to that of human beings. But they do not all speak uniformly, in the same manner, through the same media, or in the same kinds of languages. The crucial difference between their ways of speaking has nothing to do with the (at times forced) divergence between literary and religious representations; rather, it concerns the narrow identification of speech with vocalization, as opposed to its comprehension in terms of giving signs.

The assumption that to have a language is to be able to speak is both erroneous and unethical. Its mistaken premise is thoroughly anthropocentric, insofar as it ties linguistic phenomena to the voice, which only humans possess, and grants other creatures the right to speak only on the condition that they ventriloquize quasi-, proto-, or post-human voices. When trees converse, as they do in Dante's *Divine Comedy*, the silence of vegetal life, which does not preclude the plants' self-expression, is broken and disrespected. The sentient trees inhabiting his *Inferno* weep, moan, and emit "these loud, unhappy voices."[27] From a broken twig, "words and blood" *(parole e sangue)* gush forth together.[28] And the bush Dante addresses says in response: "Indeed your sweet words break me / Away from silence *[non posso tacere]*: let it not weigh on your ears / If I am enticed to prattle a bit, for your sake."[29] We learn that these infernal talking plants are the souls of those who, having committed suicide, are punished with vegetal incarnations for taking their lives. If these plants are allowed to speak, it is because their bodies are but the vehicles for human spirits, pieces of sentient matter that trap the souls of sinners and cause them pain and suffering: "in the poisoned shade of the bushes we are."[30] Simply put, hell, for Dante, is human thought, speech, and sensation trapped in the body of a plant.

In lieu of translating the language of trees into its anthropic equivalent, we witness a substitution of vegetal communication with the patently logocentric aspects of human behavior. Moreover, this act of substitution does not recognize the origin it has erased: it does not register the existence of plant languages it has automatically denied and replaced with speech. Not every author imagining speaking trees succumbs to such blindness, however. The cult of Dodona inspired James Howell's 1645 *Dendrologia: Dodona's Grove, or The Vocall Forrest*, which narrates the history of England and Continental Europe with the help of talking trees. Howell's poetic opening sentence is worth citing: "It fortun'd not long since that Trees did speake, and locally move, and meet one another: Their ayrie whistlings and soft hollow whispers became articulate sounds, mutually intelligible, as if to the soule of *vegetation* the sensitive faculties and powers of the intellect also had beene co-infused into them."[31] The similarity between trees and people was, in Howell's world, a two-way street. While the former spoke like the latter, early human discourse bore a semblance to the rustling of leaves: "in the nonage of the world, mens voyces were indistinct and confus'd; and sojourning chiefly in the Woods, by a kind of assimulation and frequent impressions in the eare, they resembled those soft susurrations of the Trees wherewith they conversed."[32] The hypothetical convergence of vegetal and human discourses evinces, at the same time, respect for the potentialities of plant language and modesty with regard to the capacities of the early representatives of *Homo sapiens*. Howell does not endeavor to erase the vegetal origin in an act of thoughtless substitution but, on the contrary, attempts to construct a richer and more fluid version of plant existence that reverberates with other forms of life. Still, the markers of intelligence in his account are drawn from a set of largely unquestioned human values, ranging from locomotion to abstract thought and, of course, vocalization. Given the anthropocentric parameters of Howell's world, then, it makes no sense to delve into the difficulties of translating plant languages into their human counterparts, and vice versa, not to mention to care about what remains untranslatable in this carryover.

Although Master Zhuangzi's tale about Carpenter Shi and an oak, dating back to the late Warring States period in China (third century B.C.E.), also puts words in the "mouth" of a tree, it is acutely aware that plants, as well as all other living beings, live in excess of the value systems we thrust upon them. The carpenter's complaint about an old oak he encounters near a village shrine is that it is useless—neither yielding edible fruit nor timber that could stay afloat if used to build a boat. When the oak appears in his dream, it queries him about the value of value, understood in the sense of usefulness. The utility of fructiferous trees, it states, "makes life miserable for them, and so they don't get to finish out the years Heaven gave them but are cut off in mid-journey. . . . As for me, I've been trying for a long time to be of no use. . . . If I had been of some use, would I ever have grown this large? Moreover, you and I are both of us things. What's the point of this—things condemning things? You, a worthless man about to die—how do you know I am a worthless tree?"[33] The lesson of the tree is that all existence is worthless from the standpoint of instrumental rationality and that its real worth lies in this absence of utility. More than that, human beings cannot presume to know that their scales of value extend to all creatures, including plants. The thingly nature of the carpenter and the oak implies their finitude, which simultaneously unites and separates them, whether ontologically, epistemologically, or axiologically. Now, all this is said in a speech that, delivered by a tree, subtly undermines the prominence of spoken discourse, which, too, is a disproportionately inflated human value. After all, the determinations of the worth or worthlessness of plant languages must contend with the analogous dilemma of the divergent axiologies, or valuations, corresponding to different modes of life.

Perhaps the most ethically attuned among the narratives of talking trees are those that do not conflate speech with vocalization. When ancient Hindus listened to the rustling of pipal tree leaves, trying to discern the conversations of celestial nymphs *(apsarās)* said to inhabit them, and when ancient Greeks engaged in fortune-telling based on the same sound produced by the wind as it passed through an oak grove, they imbued the interaction of plants and the elements with a secret meaning. To distill meaningful signs from the sounds that leaves, branches, and currents of air emit in concert is not to force a human semantic form onto the language of plants, but to attend to them in the places of their growth. It is also to acknowledge the untranslatable and the indiscernible, what cannot be picked up by the human ear and what exceeds our every interpretative venture, oriented toward the communication (if not the communion) of plants and the elements. Not by chance, the absence of speech from this language parallels the desire to glimpse a temporal absence, namely, the future. Since the future is not a fixed, substantive object of knowledge, a suitable approach to it cannot rely on the rigid mechanisms of comprehension and identification, built into human cognition and speech. We must strain to listen to the inaudible, if we are to be in touch with the future and with the language of plants themselves.

Talking trees are the conduits to another possible relation to plants, perceived no longer as mute objective surfaces for the inscription of meaning that has originated outside them but as nonhuman subjects in their own right. Does contemporary plant science grasp and elaborate upon this possibility? Does it prepare the ground for an ethics of vegetal life? Or does it provide yet another inflexible method of translating the language of plants into a system of human conceptuality, oblivious to the untranslatable?

THE SCIENCE OF PLANT COMMUNICATION

One virtue of a scientific approach to the language of plants is that it is highly resistant to the groundless attribution of symbolic meanings to vegetal life, on the one hand, and the anthropomorphic temptation pervading stories about talking plants, on the other. Plant scientists work with biochemical substances and electrical signaling that are part and parcel of the plants' life activity. It is from these processes and products of vegetal existence that they deduce the language of plants. A tomato plant, whose leaves have been damaged by a pathogen, can send warning signals concerning the impending attack to nearby undamaged plants through shared mycorrhizal underground networks. Under controlled experimental conditions, the undamaged plants demonstrated a sharp increase in defense enzymes, such as peroxidase, in response to the communication they had received.[34] While the damaged tomato specimen recruits the fungi that symbiotically grow around its roots to transmit this information to its neighbors, the biochemical message it sends is inseparable from its own life, in all its materiality. The only remaining problem seems to be the accuracy of the scientific transcription and interpretation of such communications.

To backtrack a little, it bears mentioning that plant science (like all science, for that matter) is not interested in the materiality of "plant-being" but preoccupied with the information transmitted between plant cells, tissues, specimens, and across biological kingdoms (i.e., between plants and insects). As the word itself shows, information is a very formal notion, quite removed from the life activity of the organisms that share it. In order to arrive at this highly mediated result, scientists must first operationalize their terms and quantify their research, such that whatever they observe would be countable and measurable. Even if plant science does not transform the language of plants into a set of symbols, it does render it in numbers and numeric codes. As I wrote in *Through Vegetal Being*, "More often than not, the scientific codification of reality resorted to numbers, totally indifferent to what they quantified. Sensible qualities dissolved in these empty universals that aimed to supplant, among other things, the elements, within which and thanks to which life unfolded."[35] Quantitative measurements equally apply to a tomato plant and a birch tree, a marigold and a cactus, a chainsaw and the trunk of the pine it is about to cut, a plant and a planet . . . In the universal language of science, the oft-incomparable languages, dialects, and idioms of living beings and extended objects are irretrievably lost. Moreover, convinced that everything in material reality can be, in principle, quantified and operationalized, scientists come to believe in the possibility of a total translation of the world into their code. They lose sight of the untranslatable, the nonquantifiable, and nonoperationalizable—things that, for all intents and purposes, do not exist within the bounds of the scientific worldview. The language of plants is robbed of its specificity, its unique purchase on and expression of life, when it is absorbed without its untranslatable remainder into the research conducted by plant scientists.

Another problem with the scientific construction of the language of plants has to do with the way of operationalizing this concept. For scientists, "language" means "communication," which, in turn, signifies "transfer of information" in the shape of biochemical, electrical, and other kinds of signals. As a result, plants are conceived of as emitters and receivers of information,[36] following the computational standard of intelligence. To be sure, we must be cognizant of the theoretical background for this reductive approach to the language of plants: in the twenty-first century, it is virtually impossible to find scientific formulations of human language and cognition, let alone other-than-human linguistic phenomena, that do not depend upon information exchanges and computational techniques. So entrenched is this paradigm that, nowadays, the untranslatable, in the case of plants, animals, and humans alike, refers to the incommunicable expression and the deep, noninformational stratum of linguistic reality. Martin Heidegger's famous definition of language as "the house of Being"[37] would be a good start for rethinking the alternatives to its current hegemonic operationalization, provided that its scope would extend beyond humans (so that plant languages would be the house of plant being, animal languages the house of animal being, and so forth).

The controversy raging in the scientific community around the idea of plant behavior and intelligence is a moot point, compared to a notion of language that would be qualitatively different from signal emission, processing, and sharing. In 2007, thirty-three plant scientists signed a letter, published in *Trends in Plant Science,* in which they attacked the emerging area of study called "plant neurobiology."[38] Their chief argument was that, absent the brain and nervous system, plants cannot by any stretch of the imagination be considered intelligent. The signatories to the letter omit the fact that, despite lacking such organs (indeed, any organs whatsoever), computers not only fall under the current category of intelligence but also serve as its prototypes. At its best, science demonstrates the primacy of functions over the organs that execute them, which is why a brain is not absolutely indispensable for intelligence any more than an actual eye is necessary for vision. The presence of photoreceptor cells, detecting light, the direction from which it emanates, and where it fits in the color spectrum, is sufficient to qualify for vision—a capacity that plants also possess. *Arabidopsis* alone has "at least eleven different photoreceptors,"[39] allowing it to see without eyes. The same is true for language: it is not necessary to have the organs for vocalization to exercise this particular function, above all, within the paradigm of communication as a sharing of information.

The tragedy of twenty-first-century plant sciences is that, on the one hand, they prepare the ground for a dispensation to plants of their language and intelligence, while, on the other hand, they make plants, as coherent living beings with their own modes of expression, disappear. Once again, this slight of hand is typical of all contemporary science, to the extent that it breaks its objects of study down to such minute molecular, atomic, or subatomic components that they are no longer recognizable as entities we experience in everyday life. The primacy of functions over structures has a role to play in this process, which, applied to plants, dissolves them into their chemical components, hormones, or cellular action potentials. Today, there is no more botany, which used to study plants in their integrity, underscoring their morphology and place in systems of classification, itself quite far from the language of the plants themselves. Instead, we have "plant sciences," including an interdisciplinary mix of biochemistry, biophysics, molecular and cell biology, evolutionary theory, and ecology, that happen to converge on the same type of living being.

Curiously enough, the decoupling of language from spoken discourse in the communicative model of Western sciences has much in common with the deconstruction of phonocentrism (that is, of the privilege of speech over writing) in the work of Jacques Derrida. In general, deconstruction may be thought of as a hypercritical deformalization and destabilization of existing meaning clusters. Yet, Derrida also welcomes scientific formalization by means of mathematization, "whose progress is in absolute solidarity with the practice of a nonphonetic inscription."[40] The language of plants is an excellent illustration of "the practice of nonphonetic inscription," a trace or a mark written in biochemical substances and transmitting meaning to other plants and animals. Having said that, Derrida is not ready to give up the legacy of Edmund Husserl, who bemoaned the crisis of the European sciences. According to the founder of phenomenology, this crisis was caused by the separation of formalized and quantified thinking from the "lifeworld." In keeping with Husserl's critical philosophy, meant to accompany scientific research and bring it back to the level of existence obscured underneath layer upon layer of abstraction, he writes: "It seems to me that critical work on 'natural' languages by means of 'natural' languages [travail critique sur les langues 'naturelles' au moyen des langues 'naturelles'], an entire internal transformation of classical notation, a systematic practice of exchanges between 'natural' languages and writing should prepare and accompany such formalization."[41] Akin to phenomenological reduction, deconstructive critique must shadow scientific translation and codification, while nurturing whatever resists this carryover and prevents nonphonetic inscription from lapsing into sheer abstraction. Plant science, even in its most courageous and pioneering instantiations, requires such a deconstructive supplement.

THE PHENOMENOLOGY OF VEGETAL SELF-EXPRESSION

If I were asked to provide a succinct definition of the language of plants, I would characterize it as "an articulation without saying." We would do well to recall that the word *articulation* has two, apparently unrelated, senses: expressing in words and joining two or more things in space. It is one of those rare locutions that combines—indeed, articulates—the ideal and the material strata of language, uniting the Cartesian *res cogitans* and *res extensa*. But what exactly do plants articulate in their language devoid of words?

First of all, themselves: as they proliferate by means of modular growth, reiterating their already existing morphological units, branching out in all directions, they reaffirm vegetal being, which, through them, becomes more spatially pervasive. In other words, plants articulate themselves with themselves, as they join the semiautonomous growths of which they consist. Second, plants articulate the burgeoning emergence, or self-generated appearance, that distinguishes the Greek conception of nature, or *phusis*. Their growth provides a palpable image of nature as a growing whole, encompassing everything that exists. Third, if plants articulate water, air, fire, and earth, it is because they are rooted not only in the earth but in all four elements, including the heat and light of the sun, the atmosphere that they enrich with oxygen, and the moisture they require for their flourishing. Plants are the first living bridges between the elements that, thanks to them, become livable for animals and humans. The connections they forge are nothing short of the language of life itself.

With regard to the first sense of vegetal articulation, we might say that it has much in common with the Benjaminian "language of things," or "the language as such." Still, holding on to the idea of the perfection of human language, compared to that of things, Benjamin argues that "language itself is not perfectly expressed in the things themselves. This proposition has a double meaning, in its metaphorical and literal senses: the languages of things are imperfect, and they are dumb. Things are denied the pure formal principle of language—namely, sound. They can communicate to one another only through a more or less material community. This community is immediate and infinite, like every linguistic communication."[42] The community of things, comparable to that of plants, is created thanks to the environmental articulation of things with other things, plants with other plants and with themselves. Both communities are essentially silent (though "plant bioacoustics" is a cutting-edge area of research, which may still prove this assertion wrong),[43] and it is this muteness that, in Benjamin's view, denies them the perfection of ideal expression, which, since Aristotle, has been tied to the vocal medium, hearing, and sounds. The members of the two communities communicate by contiguity, through the spatial relations they establish with other things, plants, and the elements. Now, things are not merely deposited in empty space; in their mutual articulations, they create a sense of place, a habitable world that accretes around them. This principle is further refined in plants that actively articulate organic and inorganic facets of their milieu, form ecosystems and microclimates, influence the atmosphere, and transform the places of their growth into receptacles for other kinds of life. Their self-affirmation is, simultaneously, the affirmation of their others, be they the classical elements, animals, or humans. And, just as sound is extraneous to the functioning of their language, so their intentionality need not connote a *purposeful* extension of hospitality to the other. Rather, vegetal hospitality is one and the same with the very heteronomous being of plants.

Usually circumspect on the subject of the capacities of plants, Augustine nevertheless admits the possibility of vegetal intentionality, understood as the desire to be known by humans. "In the case of trees," he suggests, "their nourishment and generation have some resemblance to sensation. Yet, though they and all corporeal things have causes which lie hidden in their nature, they do display their forms, which give shape to the visible structure of our world, for perception by our senses; and so it seems that, even though they themselves cannot know, they nonetheless wish to be known."[44] Between the lines of the Augustinian text, we may detect the second level of vegetal articulation and self-expression, related to the ancient notion of *phusis* as a burgeoning emergence. The causes of plants may be hidden, but parts of plants give themselves to our sight and other senses, lending themselves to perception in the course of germination, whereby the seed or the seedling leaves the darkness of the soil, to which it remains attached. The same is true for nature as a whole (particularly according to Plotinus, who exerted enormous intellectual influence on Augustine): in perceivable, natural things, *phusis* silently comes into the light, as the aboveground portion of the plant does, and presents itself to our senses, while remaining rooted in itself. The order of phenomenality—of the surfaces available for perception— and that of nature—of self-generated growth toward the light—blend with each other in vegetal growth. Whereas the human variety of language *(logos)* repeats and formalizes the self-presentation of phenomena, in the languages of things and of plants, a silent *logos* and sunlit phenomenal surfaces (whether they are static in inanimate objects or proliferating in plants) belong together in an uninterrupted unity. Thus, in addition to exemplifying the growth of nature as a whole, plants stand for the principle of a material living expression as such, demonstrating how a being can come into the light, appear, and signify itself.

I have already touched upon the third vegetal articulation of the classical elements and of organic and inorganic worlds. It is crucial to understand that, at this level, the language of plants means that they *form* a world, by which they, themselves, are shaped. The world is neither an empty container for reality nor the seemingly unlimited stuff it contains; rather, its character is deeply relational, dependent on the interconnections between the beings that populate it. The world is what happens in between. To insist, as I am doing here, that plants form a world is simply to emphasize that they institute relations of lived and living significance between things. Take gravitropism: as a result of sensing gravity in the area of root tips, plants are able to send their roots down into the earth and instruct stems to grow upward.[45] "Above" and "below" are spatial determinations that were meaningless and unrelated to each other before their articulation in vegetal growth. Language fosters and develops such meaningful relationality. Although we, humans, habitually enunciate the relations between things (and so form *our* world) in semantic units that we string together in speech, the groundwork of sense lies in the practical articulation and integration of the disparate parts of our environment. The closer we come to actual existence, the more our languages and articulations resemble those of plants.

As for the untranslatable, in the phenomenology of self-expression it entails the tension and the opacity between the two meanings of articulation: the verbal and the spatial. A charitable reading of Benjamin's unwavering commitment to the perfection of human language would reveal that the latter alone manages to synthesize the material and the ideal types of relationality. Unlike plants, humans can signify both by speaking and by altering our bodily positions (i.e., by resorting to "body language"). Granted, meaningful articulations are more varied in our case, but, by and large, we disregard the kind of extended self-expression, which we share with plants, and, by the same token, fail to consider their language qua language. It would be a mistake, however, to conjecture that vegetal self-expression is a less complex, if not a primitive, form of language and, therefore, of being. Everywhere, we are surrounded by plants and even bear a trace of their orientation toward the world in our subjectivities. Nonetheless, the world *of* plants and the language through which it is constructed are opaque to us, because we tend to translate it, without further ado, into the ideal medium of semantic articulation. Besides universal communicability, humans espouse the ideal of an equally universal translatability of all other languages into that "of man." Paradoxically, the greatest opacity stems from the dream of total transparency, which does not permit the dreamers—every single one of us—to encounter other-than-human beings on a nonanthropocentric turf. Only once the languages of things, plants, and animals are given their due, in a mixture of ontological fidelity and justice, will this encounter be possible. It is, perhaps, then that we will finally hear plants speak.

NOTES

1. "Midori-san or the Slightly Potty Blog," *CORDIS,* October 24, 2008, http://cordis.europa.eu/express/20081024/finally_en.html.

2. A. C. Fraser-Smith, "ULF Tree Potential and Geomagnetic Pulsations," *Nature* 271 (1978): 641–42.

3. Francis daCosta, *Rethinking the Internet of Things: A Scalable Approach to Connecting Everything* (New York: Apress, 2013), 109.

4. Sextus Empiricus, *Outlines of Pyrrhonism,* 1.216–19.

5. Johann Wolfgang von Goethe, *The Metamorphosis of Plants* (Cambridge, Mass.: MIT Press, 2009), 6.

6. Augustine, *Confessions* 13.24.35.

7. Alan Bewell, "'On the Banks of the South Sea': Botany and Sexual Controversy in the Late Eighteenth Century," in *Visions of Empire: Voyages, Botany, and Representations of Nature,* ed. David Philip Miller and Peter Hanns Reill (Cambridge, UK: Cambridge University Press, 2011), 173–93 [174].

8. Augustine, *The City of God against the Pagans* 2.27.

9. Ibid., 7.25.

10. Sigmund Freud, *The Standard Edition of the Complete Psychological Works of Sigmund Freud,* 24 vols., trans. and ed. James Strachey (London: Vintage, 2001), 4:169.

11. Ibid., 283.

12. Ibid., 169–71.

13. Ibid., 5:374.

14. Michael Marder, *Plant-Thinking: A Philosophy of Vegetal Life* (New York: Columbia University Press, 2013), 44. In continuation of this passage, I write: "Post-metaphysical thought, such as that of psychoanalysis, no longer believes in the fiction of the indivisible and immortal soul 'of another genus.' Psychic divisibility becomes the destiny of humanity that, perhaps without knowing it, sets for itself an infinite task: that of recuperating its vegetal heritage."

15. Georges Bataille, *Visions of Excess: Selected Writings, 1927–1939,* ed. Allan Stoekl (Minneapolis: University of Minnesota Press, 1985), 10.

16. Ibid., 11.

17. Ibid., 12.

18. Novalis, *Henry von Ofterdingen: A Novel,* trans. Palmer Hilty (New York: Continuum, 1992), 17.

19. Ibid., 104.

20. Brad Prager, *Aesthetic Vision and German Romanticism: Writing Images,* Studies in German Literature, Linguistics, and Culture Series (Rochester, N.Y.: Camden House, 2007), 150.

21. James Joyce, *Ulysses* (London: Wordsworth, 2010), 69.

22. Frederick J. Simoons, *Plants of Life, Plants of Death* (Madison: University of Wisconsin Press, 1998), 47.

23. Herodotus, *The Histories* 2.55–56; Homer, *Iliad* 16.234; Plato, *Phaedrus* 275b.

24. Philipp Vandenberg, *Mysteries of the Oracles: The Last Secrets of Antiquity* (London: I. B. Tauris, 2007), 27.

25. Fabio Rambelli, *Vegetal Buddhas: Ideological Effects of Japanese Buddhist Doctrines on the Salvation of Inanimate Beings* (Kyoto: Scuola Italiana di Studi sull'Asia Orientale, 2001), 52.

26. John Renard, *Islam and the Heroic Image: Themes in Literature and the Visual Arts* (Macon, Ga.: Mercer University Press, 1999), 156.

27. Dante Alighieri, *The Divine Comedy,* Northwestern World Classics Edition, trans. Burton Raffel (Evanston, Ill.: Northwestern University Press, 2010), part 1, *Inferno,* canto 13, lines 25–30.

28. Ibid., 40–45.

29. Ibid., 55–60.

30. Ibid., 105–10.

31. James Howell, *Dendrologia: Dodona's Grove, or The Vocall Forrest,* 3rd ed. (Cambridge: Humphrey Moseley, 1645), i.

32. Ibid., ii.

33. Zhuangzi, *The Complete Works of Zhuangzi,* trans. Burton Watson (New York: Columbia University Press, 2013), 30–31.

34. Yuan Yuan Song et al., "Interplant Communication of Tomato Plants through Underground Common Mycorrhizal Networks," *PLoS ONE* 5, no. 10 (2010), doi: 10.1371/journal.pone.0013324.

35. Luce Irigaray and Michael Marder, *Through Vegetal Being: Two Philosophical Perspectives* (New York: Columbia University Press, 2016), 191.

36. Refer, for instance, to Hirokazu Ueda, Yukito Kikuta, and Kazuhiko Matsuda, "Plant Communication: Mediated by Individual or Blended VOCs?," *Plant Signaling & Behavior* 7, no. 2 (2012): 222–26.

37. Martin Heidegger, *On the Way to Language,* trans. Peter D. Hertz (New York: Harper & Row, 1971), 22.

38. Amedeo Alpi et al., "Plant Neurobiology: No Brain, No Gain?," *Trends in Plant Science* 12, no. 4 (2007): 135–36.

39. Daniel Chamovitz, *What a Plant Knows: A Field Guide to the Senses* (New York: Scientific American/Farrar, Straus and Giroux, 2012), 22.

40. Jacques Derrida, *Positions,* trans. Alan Bass (Chicago: University of Chicago Press, 1972), 34.

41. Ibid., 34–35; French edition, *Positions* (Paris: Les Éditions de Minuit, 1972), 47.

42. Walter Benjamin, "On Language as Such and on the Language of Man," in *Selected Writings,* vol. 1, *1913–1926,* ed. Marcus Bullock and Michael W. Jennings (Cambridge, Mass.: Belknap Press of Harvard University Press, 1996), 62–74 [67].

43. Cf., for instance, Monica Gagliano, "The Flowering of Plant Bioacoustics: How and Why," *Behavioral Ecology* 24, no. 4 (2013): 800–801.

44. Augustine, *City of God* 11.27.

45. Robyn M. Perrin et al., "Gravity Signal Transduction in Primary Roots," *Annals of Botany* 96, no. 5 (2005): 737–43.

SIX

What the Vegetal World Says to Us

Luce Irigaray

THE VEGETAL WORLD IS THE MOST ANCIENT ENVIRONMENT, AND even dwelling place, of humanity. Unfortunately, above all in the West, it is more and more considered to represent the memory of a primitive and wild life that our culture has permitted us to overcome. For a long time, it, at least, remained a background of our infancy, which accompanied our discovery of life and our growth. We came into a world of trees, bushes, grass, and flowers, which helped us find our own breathing and, little by little, reach autonomy with respect to our mother, whose blood initially brought us the oxygen that we needed. Furthermore, these vegetal surroundings, beyond the fact that they contributed to the awakening of all our senses, were populated with a crowd of various insects and other animals, which accustomed us to coexisting with living beings different from those of our family and of our human neighborhood. We grew up in the midst of a living world that introduced us to universal exchanges and sharing in life.

Then, sharing with the vegetal world was quite natural. We did not exist without communicating, in a way being in communion, with it. And this was so immediate, simple, and going without saying that we did not ponder over what was passed on to us and so participated in our existence.

But we wandered away from the vegetal world as we moved away from our living being. We neglected and finally forgot what being alive presupposed. And our planet and the entirety of living beings needed to be put in danger so that we would come back to thinking about what living means and how we have to care about this preliminary condition of our existence.

TEACHING LIFE

The first gesture that life requires is breathing. We were born thanks to breathing by ourselves, and death will result from stopping breathing.

As such, the vegetal world is the best partner of our existence. Indeed, by purifying the atmosphere, by transforming carbon dioxide into oxygen, it brings the most precious help to our life. Even our heart, not to speak of other bodily functions, cannot act as a vital and psychic or spiritual center of our organism without us inhaling pure air—even loving cannot happen without breathing.

Furthermore, breathing does not amount only to an individual process but introduces us into a universal sharing in which we have to assume a specific role. For lack of considering the importance of breathing by ourselves, our cultural and social norms have turned our human world into a universe of conflicts between people struggling for their survival, before any master–slave struggle at an economic level. Instead of uniting their breathing toward an increase and sharing of life, people are, without knowing it, fighting one against the other in order to appropriate air and vital energy that they need so that they would not die, and it is only in a vegetal environment that they, finally, look quiet and happy.

In such places, they also enjoy views, sounds, smells, sometimes tastes that nourish their senses, together with the warmth of the sun or the caresses of the wind touching their skin. They are put back in themselves, furthermore, in accordance with seasonal rhythms, all things allowing them to experience with joy what being alive means. It suffices to gaze at the exuberant behavior of little children, and also of domestic animals, when they pass from the town to the country, to note such a fact, that we do not always make time to check as far as we are concerned. Now we ought not only to take the time to enjoy this gift of the natural surroundings but also to meditate about it. We will, then, discover that the help they provide us regarding our life is not limited to that example.

The vegetal world is also an irreplaceable teacher about the cultivation of life. It tells us that life cannot develop without endurance, patience, and faithfulness to oneself. No doubt, this is more complicated for us than for a tree. We have neither simple nor fixed roots, and the return to our original being is not obvious. However, our manner of wandering farther and farther away from our living belonging, of becoming what a constructed environment imposes on us as norms and customs, and of being fragmented into the objects of our needs, our desire, and our duties, little by little removes us from a possible living growth. And this probably explains why the majority of humans stop growing so quickly, whereas a tree continues growing until its death. The communion with vegetal life is a phenomenon that we can also observe, especially in Eastern traditions, in a few wise persons, who remain close to nature and cultivate their breathing and bodily energy instead of thinking that we can only decline and fade once we have reached a presumed physical maturity. This indicates that we are not able to harmonize, each time, being and letting be, as a tree does. We remain passive with respect to our natural development; we barely take charge of its blossoming; whereas we actively work toward constituted forms to which we subject our natural forms, instead of cultivating them until their complete realization. A tree reminds us how much this can lead to a beautiful and fruitful result, before or outside any human intervention.

However, the vegetal world also teaches us that we cannot continue being alive without becoming. Now, we generally imagine, at least in the West, that we received life once and for all at birth, and we have only to maintain it without caring about its development. This way, we become torn between a mother who gave natural life to us and a father whose role would be to introduce us into a cultural or spiritual life. But such a split has been paralyzing our human growth for centuries. If we want to make a future for ourselves and for the world possible, we must urgently realize that our cultural and spiritual life has to correspond to the blossoming and flowering of our natural belonging, and define a culture that allows for a continuity between these presumably separate lives. All the more so since this artificial partition has damaged our relations to and with other living beings, whether human or nonhuman, who can neither communicate nor share with us because of the constructed forms that we have interposed between us.

More surprisingly, the vegetal world gives us some advice about an appropriate manner of behaving toward our surroundings. It says to us that it is not fitting to take advantage of the environment in which we live without making our contribution to it. But if, for its part, it cares about breathable air, we, as humans, behave like consumers who, furthermore, pollute, without being really concerned with our contribution to the atmosphere and our living surroundings, be they vegetal, animal, or human.

The vegetal world provides us with other teachings, and the remarkable thing is that it teaches without any articulated speech, as some masters, especially of the East, do. It thus reminds us of the value of silence, notably for cultivating life and having living relations with others different from us. What is more, in this way, it lets us choose to hear or not to hear its teaching.

A tree teaches us in silence. This corresponds to its own destiny as a vegetal being. Perhaps, our destiny as humans is to make appear what the language of a tree says. This requires that we pay attention to it and to our relation to and with it. For example, breathing consciously ought to lead us to think about the gift of pure air coming from the vegetal world and to express our gratitude for it. And this ought to happen for everything that we receive from the plant world: for our survival, our blossoming, and our spiritual becoming. We must consider and show with our own language how the vegetal word contributes to the discovery and the accomplishment of us, humans, as living beings coexisting with other, human or nonhuman, beings.

A World without Words

The vegetal world speaks a language without words, which, as terms, both exhaust meaning and paralyze it. Plants talk without articulating and naming—as life does. They do not use language as a tool, or a technique, in order to express themselves. They say through shaping their own matter. And the separation of signifier from signified is not operative here. Ultimately, the connection between the two would be inverted: the signifier being what the plant lets be of the signified, outside of any uptake or mastery by a word. The signifier is now the appearing of the action of the signified, but no term creates a division and a parallelism between what is saying and what is said. It is the growing of the plant that is the source of what appears of its existence. Motion is not at the service of a finality external to it; motion includes its finality that cannot be extrapolated from it. So, the flower or the fruit are not the end of a plant, but a moment of its growth, which cannot be separated from the complete motion without removing the plant from its life. Now we, as humans, do that in theory as well as in practice, which prevents us from listening to vegetal teachings.

In a way, for the vegetal world, saying is doing or acting, to quote the title of J. L. Austin's book *How to Do Things with Words* and that of John R. Searle's *Speech Acts* that treat the question of performative language. However, doing and even acting are now endowed with a meaning that is closer to being than to merely embodying an intention. And being also has a meaning different from that which is usually attributed to this term. It is the word, or the wording, of life itself that involves both being and letting-be as elements and moments that are inextricably linked to one another in vegetal existence. Growing is the way of existing of vegetal being. And it cannot be expressed in a partial mode—for example, as or through a term—because it always amounts to a comprehensive word. A word that says the whole or says nothing about vegetal life. A word that also sums up the past in the present existence without any forgetting or resentment. What exists today is the becoming of what the vegetal is originally without the possibility that it wastes its essence through artificial and wrong existence, as it can happen for us, especially due to our language.

A plant says what it is, and its way of growing is the word of its existence. It can neither wander nor imitate nor lie—save if a human intervenes in its becoming. But is it, then, still a vegetal being? As such, it has a structure of its own, one could say a syntax of its own, that does not tolerate being distorted by the intervention of an operation or a term extraneous to it. The vegetal world remains faithful to its roots without a possible appropriation of another essence, which would expropriate it of its own origin, notably through a borrowed term or syntactic structure. Its achievement corresponds to the actualization of its initial potential without the possible implantation of another existence. A vegetal being becomes the one that it is. Its becoming is more or less flourishing according to the space and the environment in which it grows, but it does not need hybridization for its natural blossoming and reproduction, as is the case of us humans. Resorting to it, or to grafting, can provoke an additional production, but not contribute to the accomplishment of its being. It is a human technique that constrains it to produce more, as does the manner of pruning it today, especially in cities, in order to ensure its survival in spite of an unfavorable atmosphere. But can we still speak of a vegetal world? Or of a human fabrication that has already removed the plant from its own essence, existence, and truth? And is not vegetal being, then, deprived of its own word? A word through which it says itself without being able to say anything but itself.

Plant language is not selfish or egocentric. Its wording corresponds to a celebration of what it received from the elements and from its comprehensive environment. It does not keep this gift for itself, but expresses its gratitude by growing, coming into flowers, bearing fruits—a way of celebrating, or giving thanks, and of sharing what it got, not without having increased it at least tenfold. The vegetal world is probably the best contributor in the economy of universal exchanges. And, once more, its discourse is action, its word is becoming, without fixing it in any terms, which could interrupt the motion of growing. Its specific incarnation, taking shape, or producing from what it appropriated of the elements, of the world, is acknowledgment, in every sense of the word.

As such, to stop for gazing at a tree can grant us a soul, in the original sense of the soul, and that experience can bring us back to a living soul, made of energy, breath, memory, gratitude, which thus corresponds both to a gathering with ourselves and being in communion with the living. A thing that renders us capable of finding our place in the world among other living beings. This way, we can dwell on the earth and try, for our part, to embody what suits our human destiny.

The soul, then, is a place that can provide us with an internal earth or humus, which is both material and spiritual, and in which we can safeguard and care about the development of that which we must keep and cultivate, thanks to a memory that allows for deferring an immediate actualization. Our contribution to the universal economy of the living is, thus, different from that of the vegetal world. However, this reminds us of being faithful to our duty to play our role in this economy, if we take the time to perceive what it communicates to us, without merely appropriating it. The invisibility of the sap that maintains it alive and ensures continuously its becoming can awaken or revive in us a living soul, which is at once attention, memory, reserve, and source of life and love that we must preserve and share. A gesture that becomes possible if we renounce the appropriation of the other, spendthrift and sterile passion, as well as infinite wanderings in abstract elaborations, in order to patiently and responsibly work on our human accomplishment as that of one specific living being in the midst of others, upon which it is dependent for its survival, but also for its growth and happiness.

A PREDICATION WITHOUT OBJECT

Our language, our logic are based on the connection between a subject and an object. In a way, the one does not exist without the other, as Hegel brilliantly demonstrated. However, this supposedly necessary link is per-haps the result of a certain conception of truth that is not universally valid either for us, as humans, or for other living beings.

For example, if I try to analyze the predicate in the wording of the vegetal world, I will note that it is without any object. The verbs that can express it are *to live, to grow, to blossom, to flower, to spring up again, to die,* and so on—a process that cannot aim at or end in a specific object or objective but living, which amounts to being. Unfortunately, we lack words for saying the becoming or the moving of being as living.[1] As it cannot be split between the signifier and the signified, vegetal language cannot be split between enunciation and utterance. And this is perhaps why we, as Westerners, do not hear its word, because it does not obey the same logic as ours. But we could question our logic and wonder whether it does not result from a contempt for and finally a forgetting of a word capable of saying life as such. This cannot happen through separate and discontinuous utterances that do not express the various manifestations of a single truth, a single being, irreducible to any other. Are we not, as humans, constantly losing our life in a way of existing cut off from our living resources, not to say from our being, given the traditional meaning that we attribute to the word? However, without rooting our existence in a personal and unique being, what results for our life and for our living sharing with others? Are they not going to nothing? And does not our favoring having over being fragment and dis-sipate our living energy until its exhaustion?

Obviously, being cannot for all that signify a motionless and unchanging state, and no more does it refer to an unattainable ideal. From our conception, our being is existing as a potential—the genome— that we have to actualize and to develop. It is something that we have not yet taken into account seriously enough as humans, ignoring our natural determinations—be they physical, relational, or contextual—and imagining that we could be universal and neuter individuals.

Vegetal teaching invites us to take another path, especially through showing us that it would be advisable to learn how to say ourselves instead of putting the stress of our discourses on objects and favoring the talking about something. No doubt, we always do that, but without being conscious of the fact. For example, what could be our claim to speak in a universal, objective, and neutral way, as we are all the time telling ourselves, even in our most theoretical discourses? We cannot avoid expressing our particularities as a syntax acting stealthily on our officially admitted grammar. If it was not the case, then truth would not evolve historically, and all of our discourses would not be sexuated, even without our knowing (cf. the numerous surveys that I have carried out on this topic, also with various collaborators, in *I Love to You*; *Je, Tu, Nous*; *Luce Irigaray: Key Writings*; *Le partage de la parole*, and others).

Before aiming to predicate something about something, or even someone, conforming this way to our traditional logic, we would be well advised to be concerned with predicating ourselves, which would clarify our relation to truth, but also to all sorts of human or nonhuman others, allowing space for their own existence and for possible meetings with them.

In a way, the vegetal world shows us how to predicate ourselves and indicates the necessity of a return to the middle voice in order to make such a process and word possible. Does not εἶναι (einai) always involve ἔμμεναι (emmenai)? Now, we have lost this morphological form in most of our current grammars, as we lost the meaning of dialectics as a predication, especially of ourselves, which goes all over the domain, the subject, it tries to say: δια-λέγεσθαι (dia-legesthai).

This saying has more to do with being than with having, and neglecting our living being little by little led to the forgetting of this wording. Then we entered into an epoch when our subjectivity became defined by possessions and quantity more than by transformation and qualities—with the concomitant evolution of the meaning of the verb *to appropriate*, which gradually related to having more than to being. With the latter acceptation, becoming means "to turn into" and not "to acquire something": for example, the caterpillar turns into a butterfly. We are thus mistaken when we affirm that a tree has leaves and flowers; it would be more correct to say that it is presently leaving or flowering. The tree becomes what it is, remains faithful to its being that takes one form or another according to the natural, and not the constructed, space and time in which it is.

The vegetal world unveils something about the destiny of our culture by questioning the reduction of our language to a certain logic, due to which our relation to being as living got lost. If our Western history has gradually neglected the importance of the vegetal world, it likewise forgot what "to be" means. It assimilated this word to an idea extrapolated from any existence—as "Being" could be heard—instead of interpreting it as the specific origin and the determined development of each living being. This way of understanding "to be" has rendered possible the dissociation between being and existing, with all the human wanderings and drifts in appearances, artificial behaviors, and constructed realities that have removed us from our life, its achievement, and its possible sharing.

In Conclusion: On the Path toward Being

It is to a return of our existence to the blossoming of our living being that the vegetal world invites us. Our removal from the vegetal world has been accompanied by the loss of a language that serves the accomplishment and sharing of life, especially through the cultivation of breath and energy, together with the contemplation of the way of growing and blossoming, more generally of appearing, for example, of a tree. It is then the question of a language that lets, and even gives, each one its own being, and, in a way, entrusts to each one the responsibility for its destiny. Now, instead of a living being reaching its appearing only thanks to a human thinking, supported by a *logos*, it is its appearing as disclosure of life that sets us thinking.

It is thus no longer a "that" or a "what" that encourages us to think, but a "who" awakens in us questions, inviting us to meditate on them. I allow myself to use "who" to refer to all living beings capable of appearing and providing their meaning by themselves without a human intervention or fabrication. "Essence," now, alludes to the natural roots that develop forms corresponding to their original potential, and not to a "Being," which, supposedly, grants us a possible, but already constructed, perception of the existence of living beings. Every essence is, this way, endowed with a specific existence, that is, a wording of its own, before any formulation about it coming from us. Contemplating a tree can thus become a means to leave our past metaphysics without falling into a worse nihilism, because it awakens in us a wording foreign to our traditional *logos* and predicative logic. No doubt, such a gesture is not sufficient for making our path as humans, but it can reopen the cultural horizon within which we have been trapped and pave the way toward an existence that achieves our living being, instead of subjecting it to a culture, which contributes to sterilizing life more than making it blossom. It can especially lead us to stop before another human in order to contemplate his or her "who" as living and wonder about how to clear a way for rendering possible a meeting and a dialogue between us with respect for our mutual differences.

Opening and lingering on the crossroads between two existences differently rooted in nature can send each one back to their original being and invite them to remember and cultivate it in order that a sharing would happen. Such an event could fertilize the development of each life, but also contribute toward the weaving of the horizon of a new cultural postmetaphysical era, in which the spring of transcendence does not depend on values but on consideration for the irreducible difference between living beings.

NOTES

For a better understanding of this text, it would be advisable to refer to my work on language (see, for example, Luce Irigaray, *To Speak Is Never Neutral* [London: Continuum, 2002]; and *Le partage de la parole* [Oxford, UK: Legenda, 2001]); to Martin Heidegger, *What Is Called Thinking?* (New York: Harper & Row, 1968); and to the book I coauthored with Michael Marder, *Through Vegetal Being* (New York: Columbia University Press, 2016).

1. Cf. on this topic Martin Heidegger, "A Dialogue on Language: Between a Japanese and an Inquirer," in *On the Way to Language* (San Francisco: Harper & Row, 1982), 1–54.

SEVEN

The Intelligence of Plants and the Problem of Language

A Wittgensteinian Approach

Nancy E. Baker

THERE IS THE LANGUAGE *OF* PLANTS AND OUR LANGUAGE *FOR* plants. It is the latter that will occupy me here, but, of course, our language *for* plants will include how we talk about the language *of* plants. Science, needless to say, is essentially concerned with empirical questions or problems, namely, with facts or, said differently, with the truth or falsity of statements about phenomena. Occasionally, however, science runs up against purely conceptual issues, such as those that arise in the development of a new paradigm. But even in a settled, working paradigm there can also be issues around the meaning of particular words or concepts and disagreements about their applicability to certain phenomena. This is particularly true in the case of research involving animals and plants.

Current scientific work in the field of plant biology, particularly in what has become known as "plant neurobiology," has led some scientists to apply mental concepts such as *intelligence, consciousness, knowledge, learning, memory, choice,* and *intention* to the behavior of plants. Unsurprisingly, this has caused quite a negative reaction on the part of other plant scientists, who accuse the users of this language of being "anthropomorphic"—and worse. Plant biologist Lincoln Taiz says that the writings of plant neurobiologists suffer from "over-interpretation of data, teleology, anthropomorphizing, and wild speculations."[1] According to biologist Clifford Slayman, "'Plant intelligence' is a foolish distraction, not a new paradigm."[2] Even the concept of *behavior* is thought by some not to apply to plants. The charge of anthropomorphism is the same criticism leveled against those animal behavior scientists, who see a continuity between animal and human behavior, and apply to animals concepts assumed by some to be strictly applicable to human beings.

For various reasons, including enormous progress in the area of research into animal behavior, more continuity with human behavior is being noted, studied, and shared with the general public—at least in the case of so-called higher animals.

In the field of philosophy, the metaphysics of the human–animal divide has been called into question by various contemporary continental philosophers, but it is only the twentieth-century philosopher Ludwig Wittgenstein who has done this in linguistic terms. Though he explicitly considers our language for the behavior of animals and not that of plants, what he shows us in general about our concepts and the criteria for their application can be useful in addressing some of the issues raised in the debates concerning "plant neurobiology" and the "intelligence" of plants. I am a city gardener, not a scientist, or even a philosopher of science. Rather, as a philosopher, I have concentrated on the later work of Wittgenstein, in particular his philosophy of mind, and have considered at some length his very frequent mention of animals.[3] What I propose to do here is not to solve the problem of the plant biologists' language for plants or to present arguments for or against a particular position, but rather, with the help of Wittgenstein, to remove some confusions about mental concepts and their application, and to show what is and is not relevant in considering whether the concept *intelligent* applies to plants.

Before moving on to Wittgenstein, I would like to quote a description of plant neurobiology that contains the language some other plant biologists object to so much:

> The new view . . . is that plants are dynamic and highly sensitive organisms, actively and competitively foraging for limited resources both above and below ground, and that they are also organisms which accurately compute their circumstances, use sophisticated cost–benefit analysis, and that take defined actions to mitigate and control diffuse environmental insults. Moreover, plants are also capable of a refined recognition of self and non-self and this leads to territorial behavior. This new view considers plants as information-processing organisms with complex communication throughout the individual plant. Plants are as sophisticated in behavior as animals but their potential has been masked because it operates on time scales many orders of magnitude slower than that operation in animals.[4]

Here we see used for plant behavior various concepts assumed to be strictly applicable to human beings and some higher animals by the critics of plant neurobiology, for example, *sensitive, competitive, refined recognition of self and nonself,* and so on.

Concepts, Criteria, and Context

Wittgenstein, considered by many to be the greatest twentieth-century philosopher, was primarily interested in language and our temptation to misuse it in ways that end up confusing us about the nature of the mind, knowledge, philosophy, and language itself. This included his showing us how philosophers, as well as what he called "the philosopher in us," have misunderstood not only the concept of *meaning* but also the concept of *concept* itself and have, as a result, ended up misusing language and leaving us with many unsolved problems and paradoxes. According to Wittgenstein, the empirical scientist is also "subjected to the temptations of language like everyone else, he is in the same danger and must be on guard against it."[5] In what is known as his "later work," Wittgenstein painstakingly showed that language is far more complex than we realize. The meaning of a word, for example, is not an object, event, or action to which the word refers, but is rather a function of how the word is used in various contexts, as well as of its complex connections to other words—if there were no cups in the world the concept *saucer* would have a different meaning. In addition, he cured us of the temptation to see the sole function of language as that of describing reality. Description is only one of what he called our "language-games." There are many others—cursing, translating, exclaiming, praying, telling jokes, explaining, expressing, and so forth. He thought the misuse of words—and, again, not just by the philosopher—was due to our "craving for generality," namely, to our tendency to "assimilate" to each other unlike cases for which there is one term. What makes this assimilation seem possible is the presupposition of the one essence they all supposedly share. In the history of Western thought we can see this tendency to "generalize" in the creation of the animal–human divide, where animals are used as an oppositional foil to define and elevate the human being. Several things result from this polarization: (1) all animals are lumped together under one unexplored term, *animal,* their multiplicity and particularity completely ignored; (2) they become simply objects without any subjectivity or milieu of their own; and (3) they are treated from this anthropocentric point of view as somehow incomplete or lacking what human beings have. The overcoming of this divide, in the view of some plant scientists, has resulted in a zoocentrism[6] where, at least, "higher" animals are brought over to the human side of the divide. But, then, "lower" animals and plants continue to be excluded, made into objects, and treated as somehow incomplete, while the divide and its framework remain. Wittgenstein himself was guilty of this: "It would almost be strange if there did not exist animals with the mental life of plants. I.e., lacking mental life."[7]

In their essay "Cognition in Plants," which makes use of the model of "embodied cognition," Calvo and Keijzer warn against exactly the kind of generalization to which Wittgenstein was referring: "Within embodied cognition, the notion of cognition—which is based on perception and action—is used to make sense of a wide range of behaviors exhibited by 'simple' animals, like nematodes or flies. The message is clearly that we should avoid *generalized dismissive intuitions* concerning such animals and attempt a more empirically informed approach" (italics mine).[8] Trewavas and Baluška, in their essay "The Ubiquity of Consciousness," give us many examples of the "empirically informed approach" that reveals the sophisticated behavior of slime mold, amoeba, and bacteria.[9] In other words, this approach avoids "generalized dismissive intuitions" and uncovers the multiplicity of intelligent behaviors in so-called "simple animals." In his book, *Plant-Thinking*, Michael Marder has done this deconstruction of generalization and objectification ontologically. Wittgenstein does it linguistically. In each case, it is multiplicity and particularity that are brought to light and emphasized.

Wittgenstein counters the linguistic tendency to "generalize" with what in many circles is known as antiessentialism. He shows us that in the case of our ordinary, nonmathematical concepts there is no essence, no single thing that all referents of the same word or concept have in common. The example he chooses to illustrate this is the concept of *game*. To put it in different terms, there are no necessary and sufficient conditions for something's being a game, as there are in the case of stipulated mathematical concepts. Those who object to the concept *intelligent* being applied to plant behavior, or, indeed, to the concept of *behavior* being applied to plants seem to have a mathematical model of "definition" in mind. In the case of games, there are, after all, board games, card games including solitaire, tennis games, football, and so on. Every time we try to find the essence of all games, namely, the one thing they all have in common, or to come up with a hard-and-fast definition of the word *game*, we find exceptions. If this is true, then we might well ask what holds all games together as games? Wittgenstein had two metaphors for this. One was the notion of a "family resemblance": if we lined up all the members of a family so that the first and last one had no features in common we could still tell they were members of the same family because of all the overlapping resemblances in between—two with the same nose, one of those with the same chin as another, and so on. The other metaphor was that of a thread that gets its strength not from one single strand running through the whole thing but from all its overlapping fibers. One can easily see this at work in the courtroom of the common law tradition, where cases are decided on the basis of precedent. Is the particular case before us murder or manslaughter? The case is unique and the decision is made on the basis of the degree of its resemblance to what has happened before. The concept *intelligent* is surely like this. There is not one essence common to all uses of the term.

All this is also true of the concept of *language* itself:

> You talk about all sorts of language-games, but have nowhere said what is essential to a language-game, and so to language: what is common to all these activities, and makes them into language or parts of language. . . .
>
> And this is true.—Instead of pointing out something common to all that we call language, I'm saying that these phenomena have no one thing in common in virtue of which we use the same word for all—but there are many different kinds of *affinity* between them. And on account of this affinity, or these affinities, we call them all "languages."[10]

In the case of the example of *game* Wittgenstein responds to the objection to this by saying, "Don't say: 'They *must* have something in common or they would not be called "games"'—but *look and see* whether there is anything common to all.—For if you look at them, you won't see something that is common to *all,* but similarities, affinities, and a whole series of them at that. To repeat: don't think, but look!"[11] In his "look and see" Wittgenstein anticipates Calvo and Keijzer in their admonition against "generalized dismissive intuitions" and Trewavas and Baluška in their reporting of actual empirical research.

In the case of the concept of *behavior*, we can see that it is very broadly used to apply to living as well as nonliving things. We do, after all, speak of the behavior of the stock market, of chemical reactions, of planets, of electrons, and so on. There is in all these cases some form of movement or change, but when it comes to animals and human beings Wittgenstein reminds us that behavior involves very particular kinds of movement and change: "If you want to act like a robot—how does your behavior deviate from our ordinary behavior? By the fact that our ordinary movements cannot even approximately be described by means of geometric concepts."[12]

Another way he puts it is the following: "Our attitude to what is alive and what is dead is not the same. All our reactions are different.—If someone says, 'That cannot simply come from the fact that living beings move in such-and-such ways and dead ones don't,' then I want to suggest to him that this is a case of the transition from 'quantity to quality.'"[13] Being not "robotic" is as well a characteristic of facial expressions, an important part of behavior in humans and even in many animals, as Darwin knew. Even in a dog's face we can see what Wittgenstein calls "a lack of stiffness."[14] Human facial expressions are forms of behavior and can even be intentionally used as forms of communication, for example, "making a face" in response to a question, raising the eyebrows quizzically or disapprovingly, nodding in agreement, and so on. Even speaking a language is an activity, a form of behavior: "The word 'language-game' is used here to emphasize the fact that the *speaking* of a language is part of an activity, or of a form of life."[15] So is looking at something or staring at something or trying to make out what that is in the distance. Plant biologist Stefano Mancuso tells Michael Pollan that it is "our tendency to equate behavior with mobility" that prevents us from seeing rooted plants as "behaving."[16] But it must be the essentializing plant scientist and, perhaps, "the philosopher in us" who have that tendency, resulting from having in mind a hard-and-fast, and hence exceedingly narrow and rigid, word-object definition of *behavior*. Our ordinary use of the word *behavior* involves all the above, and even though we are constantly observing, understanding, and interpreting behavior, we might not use the word very often in everyday speaking. Some other examples might be "His behavior was atrocious," said of someone "rooted" in his chair and speaking with inappropriate hostility, or "Her behavior was very strange—she stood stock-still and looked at that flower for over twenty minutes." What to call a particular kind of behavior is usually the answer to the question "What is she doing?" in ordinary life, as well as in the sciences of human psychology and animal behavior. Insofar as facial expressions, even speaking, are activities or forms of behavior, we can see from what Wittgenstein shows us that behavior goes far beyond physical mobility in the sense of change of location. It may go beyond any kind of obvious change and motion as well. Examples might be pretending to be asleep or unconscious, which would involve only the movement of breathing.

When we see time-lapse videos of plants moving in their different ways, the question "What are they doing?" is perfectly appropriate, perfectly meaningful. The answer will be the *naming* or *describing* of a particular behavior. The fact that they are rooted does not seem relevant at all. Most important is that how to *explain* the behavior chemically, biochemically, or evolutionarily is the answer to a different question, for example, "What is the mechanism that makes this behavior possible?," and has nothing to do with the answer to the question "What are they doing?" Even when we detect some kind of biochemical change in a plant in response to an environmental phenomenon, it makes perfect sense to ask, "What is it doing?" The answer could be "defending against" something or "communicating to the other trees in the forest" or "choosing " or "recognizing kin" or "distinguishing self from nonself." Another way the question can be asked is "What is the meaning of this behavior? What is it for?" The problem with this that some would see is that this kind of response seems like a passive reaction, namely, a stimulus–response form of behavior. The research on plants, however, has uncovered many examples of complex learning, memory, and even a kind of planning ahead in plant behavior. Distinguishing between a mere passive reaction to a stimulus and a true response requires taking the surroundings, both spatially and temporally, into consideration.

This brings us to the importance of context. Concepts have criteria for their application. We decide on the basis of the presence or absence of those criteria whether a concept is applicable or not to an entity or situation. When it comes to mental concepts, bodily behavior and context are all-important. If we ask what kinds of bodily behavior count as criteria for the application of the concepts *sad* or *happy,* say, we might come up with certain postures, tones of voice, or speed of movement. But context as well is all-important here. Does the individual to whom we are applying the concept *sad* on the basis of his slightly bent over posture and slowness of *movement* actually have a backache or sore feet instead? When we discover he has just won the lottery and is still recovering from a car accident two weeks earlier, we might not be tempted to see him as sad at all, even though his behavior satisfies *some* of the criteria for the application of the concept *sad.* He could also be pretending to be sad or rehearsing for a part in a play, but being thrilled at having won the lottery. In a simpler example Wittgenstein says: "Pain-behaviour and the behaviour of sorrow.—These can only be described along with their external occasions. (If a child's mother leaves it alone it may cry because it is sad; if it falls down, from pain.) Behaviour and kind of occasion belong together."[17]

Because plant life occurs in a very different time dimension from our own, time-lapse photography shows us much that we would not perceive otherwise. In the case of Mancuso's time-lapse film of the growing vine aiming for a pole, we see a context different from a vine moving in the same way and in the same direction, but being blown by the wind with no pole in sight. Time-lapse photography and the context of the presence of a pole open up the possibility of applying concepts like *searching for, choosing, intending* in answer to the question "What is it doing?"

Wittgenstein also tells us that there may be very different kinds of behavior: "Forms of behaviour may be incommensurable. And the word, 'behaviour,' as I am using it, is altogether misleading, for it includes in its meaning the external circumstances—of the behaviour in the narrower sense. / Can I speak of one behaviour of anger, for example, and another of hope?"[18] The incommensurability of forms of behavior and the important role of surroundings or context, temporally and spatially, open up the possibility of treating certain biochemical changes in a plant as forms of behavior. This gets us away from thinking of behavior solely in visualizable, bodily behavioral terms. After all, Wittgenstein shows us that even speaking a language, which includes what we actually say, is a form of behavior.

ORDINARY LANGUAGE

For Wittgenstein, the way the philosopher, the generalizing "philosopher in us," and even the scientist go wrong is in distorting or misusing our *ordinary language*. But, we may well ask, why should the plant scientist pay any attention to ordinary language? After all, words like *electron, quantum,* or *electromagnetic* are not ordinary. Here we must notice the difference between the physical sciences and the life sciences. As Michael Marder puts it, "In different ways, Henri Bergson and Hans Jonas argued that, unlike physics or chemistry, biology and evolutionary theory do not obey 'objective' laws, because life introduces a fair degree of indeterminacy into matter. . . . The irreducible indeterminacy of biology implies that every form of life is not a totally predictable object of study, but a subject in its own milieu."[19] Plants, animals, and human beings are not inanimate objects whose behavior is predictable according to the various laws of nature. They are, instead, adaptive subjects that respond to their indeterminate complex surroundings, that is, they behave in complex and unpredictable ways. As Wittgenstein puts it, "The concept of a living being has the same indeterminacy as that of a language."[20] What counts as "a living being"? What is the *definition* of a living being? If what are the necessary and sufficient conditions for something being a living being is meant, the answer is that there are none. Unlike mathematical concepts, *living being* is an indeterminate concept. As Wittgenstein puts it, in our desire for a determinate definition we fail to see something like the following: "It's as if we imagined that the essential thing about a living being was the outward form. Then we made a lump of wood in that form, and were abashed to see the stupid block, which hasn't even any similarity to life."[21]

The reason ordinary language is so important here is that all our mental concepts have their origin in the indeterminate behavior of living beings, in particular, the behavior of human beings. We naturally extend these terms to other forms of life: "only of a living human being and what resembles (behaves like) a living human being can one say: it has sensations; it sees; is blind; hears; is deaf; is conscious or unconscious."[22] There is also the reverse extension: we pig out, rat on our friends, wolf down our food, feel sheepish. These are not metaphors; they are embedded in our language. There is no need for "as if."

DEVELOPMENT

But how can we use the same word for the behavior of a plant and the behavior of a person? Michael Pollan describes the behavior of the vine headed for the pole as "effortful and striving," but later he says that "striving," "knowing," and "noticing" are all "metaphorical in the case of plants," even though, as he acknowledges, it looks exactly like striving in the videos.[23] To help us out of the anthropocentric view that words like these are metaphors, Wittgenstein reminds us that we quite correctly use the same word for behaviors that occur at different developmental stages. Unlike most philosophers, he had a strong sense of development, both of the human individual and of the evolution of species. He often used examples of children and animals to show various things about how our language works. In the case of the application of certain concepts to children, for example, that of *reading,* the criteria will be different from those we use for the literate adult. If we heard an educated adult sounding out words like a first grader we would think he was seriously dyslexic. On the other hand, if we came upon a first grader just beginning to sound out words, we would say "He's reading!" and be quite pleased. As Wittgenstein says, "The word 'to read' is applied *differently* when we are speaking of the beginner and of the practiced reader."[24] Context changes, therefore criteria change. The same is true of a word applied at different evolutionary levels. Take the concept *recognize*: A plant recognizes kin; an amoeba recognizes nutrients; a dog recognizes its master; a chimp recognizes a new sign it has learned; a botanist recognizes a new species of orchid. Part of the relevant context in each case is the range of possible behaviors. Maybe we are wrong that what we see in the botanist's behavior is the recognition of a new species of orchid. Maybe his behavior is a response to seeing a gorgeous flower. As far as we know, the number of possibilities in this case is greater than those in the case of the dog or the chimp or the amoeba. It does not matter. The important thing is that all these behaviors are instances of some kind of discrimination, and thus recognition. What other word is there for the amoeba consistently choosing nutrient over nonnutrient under a variety of conditions? In "The Ubiquity of Consciousness," Trewavas and Baluška describe some of those conditions and conclude: "This and other observations on the behaviour of Amoeba led Walker (2005) to conclude that 'Amoeba perceives, recognizes, chooses and ingests a variety of prey that is not much short of the choice of higher animals, it recognizes its own kind and engages in cooperative behaviour,' particularly in cooperative hunting."[25] Because of the increasing complexity of context and criteria, we might say that the word *recognize* means something slightly different at each stage of development—but all these uses are held together by a "family resemblance," which is why we have the same word for the behavior at all these different levels. The plant recognizes kin and does not compete; the plant recognizes self in its roots as opposed to not-self in the roots of another plant; the plant recognizes the threat of drought, and so on—and in ways that are not predetermined. There is no need for "as if." It would sound silly to say "The plant is behaving as if it is recognizing such and such."

MENTAL CONCEPTS REQUIRING LANGUAGE AND THEIR PRIMITIVE ROOTS

It is, however, true that to recognize a new species of orchid *as* a new species of orchid requires language. To be able to hope requires language:

> One can imagine an animal angry, fearful, sad, joyful, startled. But hopeful? And why not?
>
> A dog believes his master is at the door. But can he also believe his master will come the day after tomorrow?—And *what* can he not do here?—How do I do it?—What answer am I supposed to give to this?
>
> Can only those hope who can talk? Only those who have mastered the use of a language. That is to say, the manifestations of hope are modifications of this complicated form of life. (If a concept points to a character of human handwriting, it has no application to beings that do not write.)[26]

> One does not say that a suckling hopes that . . . , but one does say it of a grown-up—Well, bit by bit daily life becomes such that there is a place for hope in it.[27]

Given the fact that some concepts can be applied only to language users, how could the amoeba's behavior have anything in common with the botanist's? We have already seen that Wittgenstein points us to different levels of development for which we would use the same word, just as we use the same word for many different kinds of games. In addition, he introduces the notion of "primitive roots" to show how a language-dependent behavior grows out of prelinguistic behavior: "But what is the word 'primitive' meant to say here? Presumably that this sort of behaviour is *pre-linguistic*: that a language-game is based *on it,* that it is a prototype of a way of thinking and not the result of thought."[28]

Mental phenomena such as sensations, emotions, even cognitions are in many cases accompanied by typical kinds of behavior, what Wittgenstein calls "natural expressions" that we share with animals: "What is the natural expression of an intention? Look at a cat when it stalks a bird or a beast when it wants to escape."[29]

In addition to sharing the natural expressions of intention with animals, we human beings can also express intention linguistically, a behavior based on the more "primitive" prelinguistic natural expression. It turns out that we share certain natural expressions with plants as well. The effortful, striving behavior of the vine is a perfect example: we recognize the similarity to our behavior. On the other hand, Marder rightly says, "The intelligence of plants is not merely a shadow of human knowing, and their behavior is not a rudimentary form of human conduct."[30] On the surface, this seems to take issue with the notion that developmentally prior behaviors are "primitive," but Wittgenstein's word *primitive* does not mean "rudimentary" in this sense. This can be seen in what he shows us about completeness.

COMPLETENESS AND LACK

Just as mice do not lack the ability to fly and we humans do not lack the ability to "breathe" under water, so plants do not lack any abilities we humans have. Nor do we humans lack any of the extraordinary abilities of plants. The world and capacities of each form of life are complete in themselves. The amoeba's recognition of nutrient is not a "rudimentary," namely, an incomplete version of the botanist's recognition of a new species of orchid, but rather a "primitive," namely, simpler and hence complete version of a more complex version. The first grader's reading is complete in itself. We do not say it lacks the essence of adult reading and therefore we should use another word for it. There is only lack from an adult-anthropocentric point of view, namely, a generalizing, objectifying, polarizing point of view. We can see this in the following remarks of Wittgenstein: "A treatise on pomology may be called incomplete if there exist kinds of apples which it doesn't mention. Here we have a standard of completeness in nature. Suppose on the other hand there was a game resembling that of chess only simpler, no pawns being used in it. Should we call the game incomplete?"[31] No, of course not. In the case of human "disabilities" (an interesting word), because the majority of us are a certain way, we tend to use that way as a standard of completeness, but, as Wittgenstein points out, "a colour-blind man is in the same situation as we are, his colours form just as complete a system as ours do; he doesn't see any gaps where the remaining colours belong."[32] If most people were color-blind, that is, if they made up the whole society, we would not think of color blindness as incomplete. The same is true of feeblemindedness: "One imagines the feeble-minded under the aspect of the degenerate, the essentially incomplete, as it were in tatters. And so under that of disorder instead of a more primitive order (which would be a far more fruitful way of looking at them). We just don't see a *society* of such people."[33] What is being discovered, or uncovered, about plant behavior puts plants in their own "society," in other words, in their own complete world manifesting their own remarkable forms of behavior. Seeing this contributes to the ending of anthropocentric comparing, objectifying, and ranking. As Wittgenstein says, "The difficulty of renouncing all theory: One has to regard what appears to be obviously incomplete, as something complete."[34] And, "theory" here is the generalizing "without looking" or the "generalized, dismissive intuitions" of Calvo and Keijzer.

CONCEPTUAL CONFUSIONS BEHIND THE OBJECTIONS TO PLANT INTELLIGENCE?

Those who object to the use of the word *intelligent* for plant behavior seem to have some particular definition in mind and forget that we all apply the word to the behavior of animals and children and that its application is indeterminate. There is a metaphysics underlying the assumption that maintains that human beings are the standard of everything. It includes that we are individual solitary selves, with an inner life not accessible by others; that we are divided into mind and body; that body has nothing to do with the mental, or, on the other side, that mind is brain; that cognition necessarily involves representation of some kind, and so on. This leads to serious misunderstandings about the criteria for the application of the word *intelligent* to plant behavior. These criteria have nothing to do with brains or neurons. The ancient Greeks, the nonliterate Papua New Guinea hunters, and the ten-year-old chess expert watching a stunning move and calling it "really intelligent" had and have no trouble correctly applying the concept *intelligent* while knowing nothing about brains. The debate among plant biologists about whether it is appropriate to use the language of neurons for the mechanisms that make certain plant behaviors possible is not at all the same as the debate about whether the behavior of a plant without a nervous system can be called "intelligent." The latter is about behavior. We do not need to know anything about brains in order to call a particular behavior in a certain context "intelligent." The debate among plant biologists about whether the language of "neurons" is appropriate or useful for the mechanisms that make the behavior at issue possible is a theoretical, and ultimately empirical, debate, not a conceptual one about what to call the behavior in question.

What is called "the homunculus fallacy" adds another aspect to this mistake, namely, the application of behavioral terms to brains. To think that plant behavior cannot be intelligent because plants do not have brains is to imply that brains are intelligent. Brains are not "intelligent," they do not "think," "talk," or "recognize kin," whether in the human or the plant way. Persons, animals, and plants do. What the homunculus fallacy amounts to is the notion that there has to be another "little man" inside us human beings running the show, the pilot of the ship, as it were. The fallacy part is that there would then have to be another little man inside that one ad infinitum. Wittgenstein, interested in meaning rather than argument, simply shows us what the relevant criteria are for the application of mental concepts. And brains or neurons have nothing to do with it. Pollan quotes Taiz making this mistake: "Lincoln Taiz has little patience for the notion of plant pain, questioning what, in the absence of a brain, would be doing the feeling. He puts it succinctly: 'No brain, no pain.'"[35] Not only is this an example of the homunculus fallacy, because it is persons and animals who feel pain, not their brains, but also, interestingly, there is no such thing as feeling in brain tissue anyway.

There are human and some animal behaviors made possible by brains. There are plant behaviors made possible by very different mechanisms. The mechanisms are what explain the behavior in terms of what makes them possible. But the mechanisms are not the behavior, nor do they provide the context that contributes to the meaning of the behavior, namely, the criteria for what to call it. Once we see that brain states and neurons in no way provide criteria for the application of a concept such as *intelligent,* it becomes clear that the new plant biology is not discovering or inventing new criteria for our concepts but rather discovering contexts and behaviors we did not see before and that, once seen, we recognize as the criteria for the application of those concepts. In the animal realm there are many examples of this. To the dismay of anglers, it turns out that fish feel pain, but it took a lot of research actually to see their avoidance behavior, one of the main criteria of a living being feeling pain.

Another subtler mistake is to assume that behavior is just an external manifestation of the real thing and that we can only infer on the basis of behavior and the similarity of it to our own behavior while we are feeling something like pain, that the other is feeling what we feel. Our inference, however, can never be certain on this view. This is the famous problem of other minds that Descartes left us with.[36] It is not the place here to go through Wittgenstein's devastating deconstruction of this, but I will mention two things I see in the plant literature. One is the notion that intelligent behavior is a manifestation of intelligence or some kind of faculty, as if these were two separate things. Pollan writes that, although Slayman "doesn't think that plants possess intelligence, he does believe they are capable of intelligent behavior."[37] Another version of this is that, since intelligent behavior is the *result of* intelligence and since plants lack intelligence, their behavior cannot be called "intelligent." The other residue of Descartes's influence I see is an emphasis on the lack of interiority in plants, as if this is what defines intelligence. Well, not only do dogs, slime molds, toddlers, and probably most of us lack interiority while engaging in intelligent behavior, but, again, like the brain, interiority has nothing to do with the criteria for the application of the concept *intelligent.* Here is Wittgenstein on the seeming importance of the inner in the case of human beings: "The 'inner' is a delusion. That is: the whole complex of ideas alluded to by this word is like a painted curtain drawn in front of a scene of the actual word use."[38]

We might say it is like a painted curtain drawn in front of a scene of actual behavior and its context. Wittgenstein asks whether a cat waiting by a mouse hole and an experienced robber waiting for his victim have to be thinking about the victim in question, namely, have to be experiencing something inner for us to see that they are lying in wait for them?[39] No, we know the context and behaviors for "lying in wait for" in both cases.

Pollan writes: "'Memory' may be an even thornier word to apply across kingdoms, perhaps because we know so little about how it works. We tend to think of memories as immaterial, but in animal brains some forms of memory involve the laying down of new connections in a network of neurons. Yet there are ways to store information biologically that don't require neurons."[40] Here we see both mistakes—conflating memory with how it works, in this case with brain states, and memories with a kind of Cartesian immaterial mind.

How Great Does the Resemblance Have to Be?

As we have seen, Wittgenstein shows us that our human mental terms apply only to what "resembles (behaves like) a living human being." This sounds like recognizable behavior in the case of animals with faces and arms and legs, which includes for Wittgenstein a "wriggling fly." But given how much we may not have seen with the naked eye, for example, pain behavior in fish, and given how much context matters in deciding what to call a behavior, it seems like no stretch at all to call many forms of plant behavior *intelligent*. In considering whether the concept *thinking* can be applied to animals Wittgenstein says that when the "*rhythm* of work, play of expression etc. was like our own, . . . there is no deciding *how* close the correspondence must be to give us the right to use the concept 'thinking' in their case too."[41] He also very importantly says, "It is unnatural to draw a conceptual boundary line where there is not some special justification for it, where similarities would constantly draw us across the arbitrarily drawn line."[42] Again, what the new plant science is uncovering is what we have not been able to perceive before.

We already know the criteria for the application of concepts like *behavior* and *intelligent* and all the variability and indeterminacy of contexts—and, like the jury in the common law tradition, we know how to extend the concept through a family resemblance.

Notes

1. Cited in Michael Pollan, "The Intelligent Plant: Scientists Debate a New Way of Understanding Flora," *New Yorker*, December 23 and 30, 2013, 92–105 [94].

2. Cited in Pollan, "The Intelligent Plant," 94.

3. See Nancy E. Baker, "The Difficulty of Language: Wittgenstein on Animals and Humans," in *Language, Ethics, and Animal Life: Wittgenstein and Beyond*, ed. Niklas Forsberg, Mikel Burley and Nora Hämäläinen (New York: Bloomsbury, 2012), 45–64. In this essay I show that, in spite of being associated with the importance of language, Wittgenstein is in no way anthropocentric, but on the contrary sees the continuity between animals and humans more than any other philosopher I can think of. Part of the reason he can do this is that he knows that language itself is a form of behavior. The essay is primarily concerned with his text.

4. František Baluška et al., "Neurological View of Plants and Their Body Plan," in *Communication in Plants: Neuronal Aspects of Plant Life*, ed. František Baluška, Stefano Mancuso, and Dieter Volkmann (Berlin: Springer-Verlag, 2006), 19–35 [31].

5. Ludwig Wittgenstein, "Notes for Lectures on 'Private Experience' and 'Sense Data,'" in *Philosophical Occasions, 1912–1951*, ed. James C. Klagge and Alfred Nordmann (Indianapolis: Hackett, 1993), 274–75.

6. Pollan, "The Intelligent Plant," 94.

7. Ludwig Wittgenstein, *Culture and Value* (Chicago: University of Chicago Press, 1980), 57.

8. Paco Calvo and Fred Keijzer, "Cognition in Plants," in *Plant-Environment Interactions: From Sensory Plant Biology to Active Plant*, ed. František Baluška (Berlin: Springer-Verlag, 2009), 247–64 [248].

9. Anthony J. Trewavas and František Baluška, "The Ubiquity of Consciousness: The Ubiquity of Consciousness, Cognition, and Intelligence in Life," *EMBO Reports* 12, no. 12 (2011): 1221–25.

10. Ludwig Wittgenstein, *Philosophical Investigations*, rev. 4th ed. (Oxford: Wiley-Blackwell, 2009), §65.

11. Ibid., §66.

12. Ludwig Wittgenstein, *Remarks on the Philosophy of Psychology*, 2 vols. (Chicago: University of Chicago Press, 1980), vol. 1, §324.

13. Wittgenstein, *Philosophical Investigations*, §284.

14. Ludwig Wittgenstein, *Last Writings on the Philosophy of Psychology*, vol. 2 (Oxford, UK: Blackwell, 1992), 65.

15. Wittgenstein, *Philosophical Investigations*, §23.

16. Pollan, "The Intelligent Plant," 94.

17. Ludwig Wittgenstein, *Zettel* (Berkeley: University of California Press, 1967), §492.

18. Wittgenstein, *Remarks on the Philosophy of Psychology*, vol. 1, §314.

19. Michael Marder, "What Is Plant-Thinking? Botany's Copernican Revolution," *Los Angeles Review of Books*, March 17, 2013, http://lareviewofbooks.org/article/what-is-plant-thinking-botanys-copernican-revolution.

20. Wittgenstein, *Zettel*, §326.

21. Ludwig Wittgenstein, *Philosophical Grammar* (Berkeley: University of California Press, 1974), §85.

22. Wittgenstein, *Philosophical Investigations*, §281.

23. Michael Pollan, "Do Bean Plants Show Intelligence?," *New Yorker*, December 20, 2013, video, http://www.newyorker.com/tech/elements/video-do-bean-plants-show-intelligence.

24. Wittgenstein, *Philosophical Investigations*, §156.

25. Trewavas and Baluška, "The Ubiquity of Consciousness," 1223–24.

26. Wittgenstein, "Philosophy of Psychology," in *Philosophical Investigations*, §1.

27. Wittgenstein, *Remarks on the Philosophy of Psychology*, vol. 2, §15.

28. Wittgenstein, *Zettel*, §541.

29. Wittgenstein, *Philosophical Investigations*, §647.

30. Marder, "What Is Plant-Thinking?"

31. Ludwig Wittgenstein, *The Blue and Brown Books* (New York: Harper Torchbooks, 1965), 19.

32. Wittgenstein, *Zettel*, §257.

33. Ibid., §372.

34. Wittgenstein, *Remarks on the Philosophy of Psychology*, vol. 1, §723.

35. Pollan, "The Intelligent Plant," 102.

36. In his search for the foundations of knowledge, the enormously influential seventeenth-century philosopher René Descartes found those foundations in his own subjectivity, in that which could not be doubted. This, along with his splitting off mind and body from each other into two conceptually separate realms, left him (and us) with the famous problem of other minds. If all I know for certain is my own subjectivity or mind and all I perceive of you is

your body, how do I know what is mentally going on in you or whether you even have a mind?

37. Pollan, "The Intelligent Plant," 102.

38. Wittgenstein, *Last Writings on the Philosophy of Psychology,* 2:84.

39. Wittgenstein, *Remarks on the Philosophy of Psychology,* vol. 1, §829.

40. Pollan, "The Intelligent Plant," 100.

41. Wittgenstein, *Zettel,* §102.

42. Wittgenstein, *Remarks on the Philosophy of Psychology,* vol. 2, §628.

EIGHT

A Tree by Any Other Name
Language Use and Linguistic Responsibility

Karen L. F. Houle

COMMON APPLE (MALUS DOMESTICUS)

September. I am flying to Washington from Toronto for a philosophy conference. Since I'll be at the airport from about 11 a.m. to 2 p.m., I pack a lunch: a tuna sandwich, a juice box, and an apple from the Guelph Farmers' Market. She's a beauty, an "Empire." The apples are sublime this time of year. I eat my sandwich and drink the juice on the way to Toronto Pearson Airport. Passing through U.S. customs, the agent asks me if I have any *agricultural products* to declare? I say "No." Either I forgot that I still had the apple, or the expression "agricultural product" was not the name for what I had in my bag. Then they search my bags. They find an "agricultural product." I am sent into the bad people zone. In the bad people zone, an unsmiling customs agent, guns on both hips, motions me forward and asks to see my passport. He makes some notes on the computer. He then lectures me on the dangers of bringing Canadian fruit into the U.S. of A.: (1) pesticides harm the environment; (2) introduction of foreign species including pests; (3) collapse of trade. He then states, "The offending item is a security threat." I ask him, "Are all apples doomed to be forbidden fruit?" Evading the biblical query, he answers that had the apple originated in the United States it would not be forbidden, but this particular one was an *invasive species.*

My humble, delicious apple: an offending item, a security threat, an invasive species. He hands my lovely apple back to me and then motions for me to dump her into a waste bin filled to overflowing with, well, what looked like perfectly edible food. I say to him, "I don't want to throw it out. I'll just sit down over there and eat it, then go." He startles, puffs himself up to full size, and shouts at me: "Ma'am! I can't let you do that." I giggle as he actually puts his hand on his revolver.

"Ma'am!" he warns, sharply. I stop smiling. "Why can I not eat my apple?" I ask. He replies: "It's a choking hazard. We are security officers and we are not trained to perform medical assistance. Put the item in the bin! Now!" he says, gesturing toward the garbage with his gun.

JUSTICE

Even though the word *justice* is used in many scientific papers and every-day conversations, there is not only one concept of justice. There are many variants, some slight gradations, some radically heterogeneous. These variants have been described theoretically and defended con-cretely in different texts, in different regions of the world, at different scales, at different times. We are often using the same word but not always talking about the same thing.

In everyday parlance, in institutional protocol but also in scientific discourse, a single notion of justice dominates: *distributive justice*. The concept of distributive justice basically conceives of justice as a *what*: as measurable and fungible states, goods, or qualities, and amounts of pos-session or dispossession thereof. It focuses on what someone has or does not have; some*thing* one needs or does not need, receives or fails to receive, what one has a right to or has no right to, what one deserves relative to another on the basis of this or that further amount (contribu-tion, expertise, years, effort, charisma, status, legitimacy, force). Under this conception of justice, violations (and thus their prevention and their remediation) are primarily material, take place in the domain of physi-cal action, with identifiable subjects and measurable objects and units, which can be distributed and redistributed. From this point of view of justice, the crucial questions to ask are: how fast, how much, how far, how long, how many?[1]

Let us look at a range of concrete examples that seem to be very different justice milieus and issues but, in fact, all subscribe to one basic concept of justice: *justice as distribution*. These are variants of a single con-cept rather than different concepts altogether.

We worry (rightly) about the grotesquely uneven distribution of global resources, north to south especially. We worry about the levels of injury and mortality of innocent civilians, animals, entire ecosystems in war zones. We count the number of female lives lost to malaria, Chagas' disease, dysentery, leishmaniasis, and tuberculosis, and compare these to the number of lost male lives, and discuss the disproportion as a matter of injustice. We measure heavy metals in blood, placentas, urine, and hair in human communities living downstream from heavy industry or near power plants, and we compare those to levels in other more affluent communities, upstream. We try to track bribery and corruption in government and corporations: the diversion of public funds to private interests. We calculate differential taxation rates. We conduct happiness studies in all the countries of the world. We plot and compare rates of deforestation, species' loss, Alzheimer's and Parkinson's, infant mortality, health care wait times, poison seepage into water tables, genetic mutation, ozone holes, glacier calving, coral bleaching. Whether neuroscientists, environmental engineers, nurses, development workers, or policy makers, we do this measuring not for measurement's sake but to establish baselines and patterns; to generate metrics. We establish baselines and patterns and generate metrics in order to detect and draw attention to deviations. Ultimately, we want to be able to draw attention to empirically sound, statistically significant deviation in order to be able to argue for or against, or, better, to work for or against, such deviations. All this activity is grounded in a basic sense of justice that all rational beings allegedly have.[2] But the particular conceptual form that our basic sense of justice takes is justice as distribution.

If our concept of justice is distributive, *then* our sense of what it means to remedy injustice or *be responsible* will be that we ought to put our energies into distributive and redistributive interventions, in order to stabilize or adjust those patterns and rates: Northern researchers partnering with scientists in the global South, adjusting flow rates in rivers with an eye to species' rebound, lobbying to remove barriers to health care access, developing green technologies to offset carbon, giving time or money to charities, donating a spare kidney, cloning stem cells to fight neuronal loss, developing labor policies to increase numbers of visible minorities and women in certain occupations, working to install and oversee uncorrupt systems of representative governance, where leaders are not currently installed by highest proportion of popular votes. We spend tax monies arming ballot box stations in order to ensure that voting and then vote counting happens because *counting properly* is such an intelligible and fundamental form of fairness.

Two Concerns

Conceptual Monoculture

Although these are very important issues and profoundly important interventions, they are not the only justice phenomena. There are other, entirely different ways of thinking about justice, where the focus is not on amounts of stuff or patterns of distribution.

One of those is *procedural justice,* where fairness and unfairness, rightness and wrongness, are primarily a matter of a *how* rather than of a *what,* and hence locate responsibility in, and for, better processes, rather than in amounts or patterns of outcomes.[3] Another shares a kinship with the camp of normative philosophy known as "virtue ethics," where fairness and unfairness, rightness and wrongness, are primarily a matter of *quality of character* rather than *what* or *how.* This normative framework locates responsibility in, and for, the cultivation of attributes like courage, fidelity, trustworthiness, kindness, and compassion, whether in a border guard or beekeeper, a human rights policy or a zoo.[4]

There is yet another idea of justice, which I will call *linguistic justice,* where rightness and wrongness—and hence culpability—are located in and around word use, the ways that words, utterances, gestures, statements, and symbols shape realities for better and for worse. Word use is something we ought to pay attention to *as a matter of justice.* Words, phrases, and grammar are not just matters of communication, aesthetics, or anthropology. As I will argue, language use is something we are answerable for as a matter of justice. I take up this concept below. For now, I merely want to name it alongside virtue and procedural justice in order to make the point that, although there are several quite distinct concepts of justice, for the most part we are not aware of that, let alone endeavoring to integrate their unique dimensions into our sensitivities, our imaginaries, and our remedial efforts. Our capacity for a sense of justice has been cultivated in just one direction. There is a kind of conceptual monoculture here, a flat, extensive mental terrain that makes it difficult to catch a glimpse of these other forms of justice (and injustice) in the world, let alone to cultivate them. We tend to only see, hear, and react to politically normative parameters like correct speed, proper ratio, right amount, safe distance, ideal number, good density, and reasonable degree of deviation.

Culpability and Nonculpability

Evildoers exist, but for the most part, most of us are not rapists, thieves, river poisoners, or genocidal dictators. Yet, the dominance of the notion of justice as distribution keeps our collective eyes trained to spot just such evildoers and hence away from another cast of characters who also contribute to the ills of the world, in very different ways. This cast includes everyday people. It includes ourselves.[5] Setting our sights on blockbuster injustices funnels money, major awards like the Nobel Prize, and hence public attention toward urgent, sexy, or alarming "problems," distracting from everyday and banal examples of acts of violence, as well as chronic, mundane, and subclinical unfairness that has a negative impact on lives and quality of life the world over.

Why are so many of us (undergraduate students, volunteers, natural and health scientists) so compelled to work in "faraway places"—Southeast Asia, the Amazon basin, and Canada's North—and less inclined to find local projects as legitimate and worthy of our time, effort, and money?[6] There is the compelling point that wealthier regions have already gotten enough attention. But there certainly is abject poverty, species' vulnerability, and environmental degradation within affluent regions: the west end versus the east end of Montreal, for instance. Does justice, *as if* mostly a question of distribution, not focus our concern far away and hence feed into a view of the self that hides our complicity and continuity with the whole? Thinking with this concept of justice makes it hard for scholars and ordinary citizens of the global North to see ourselves as instigators of injustice and to imagine ourselves as anything but vital agents of redemption, doers of good. The injustices we seem to now be able to see best, and the impulses we now have to remedy them, retrace almost exactly North–South geopolitical lines. This fact forces us to consider that the concept of justice we are working with may be an expression of, and extension of, that geopolitical history, rather than a break with it or its mitigation.

LINGUISTIC INJUSTICE

Let us now turn our "capacity for a sense of justice" toward a different concept of justice: justice involving language. This includes what we do with words and what words do to us, what we do or fail to do as language users, what kinds of harm and healing take place in and through language: words, gesture, utterance, grammar, and tone. Language is a mode of activity in which we constantly partake. It is a mode in which our subjectivity and the possibilities for action are both constituted and tested. Our language is thus a part of the direct and actual shaping of the world, shaping possibilities for subjects/objects. It is as powerful as pulling a gun, putting a needle in an arm, grafting trees, or building nuclear plants. Our world, our everyday lives—whether we spend our time in toxicology labs or immigration lineups, whether in intimate conversation or on billboards—are, and have always been, filled with and shaped by *how* and *what* is said and not said, what can be said, and what is blocked from being said.[7] We are what Heidegger named "linguistic beings."[8] Even a single word written on paper involves beliefs, will, concepts, history, emotional states, mental activity, furniture, graphite, trees, muscles in hands and shoulders, opposable thumb, the optic nerve, scratches, vibrations, images, lines, color, scent, garbage, decay. The meaningfulness of linguistic phenomena exceeds their truth value. In *How to Do Things with Words*,[9] J. L. Austin argues that certain kinds of sentences are best understood as "performative," that is, they do not so much say something about X as do something X in their being said. An example he uses is the statement "I take this man as my lawfully wedded husband." Language is not being used here by a subject to describe or state what she is doing, but is actually an operation being performed by the language being spoken, and the act of speaking—constituting the bride and the husband simultaneously. Significantly and surprisingly, Austin also claims that most of our utterances are like that: language is a kind of operation. This includes the statement "The offending item is a security threat."

And so, contrary to the common view that language is neutral with respect to all values, except epistemic ones, or that the chief work accomplished by language is to state (represent) facts "out there," is the view that language is and does something powerful. Language both reveals the meaning of reality and creates reality through its special capacity for directing a particular meaningfulness through, and toward, objects and events spoken about.[10]

Insofar as language is something done to us, something we do, and something we are, and insofar as we have genuine options around exercising it—what we say, how we say it, and even whether we speak or remain silent in the face of objects and events—then language and its use are part of what we are accountable for. Language itself is ethical and political; it is not just a vehicle for ethics, the political action and value creation taking place elsewhere. Language is, or does, real harm and real good.[11]

Douglas Fir (*Pseudotsuga menziesii*)

June 2. I am walking a needle mattress of a path in Cathedral Grove, on Vancouver Island, the northern edge of the territory where Douglas firs grow. These firs did not fall to chainsaws in the last few centuries. They are so enormous, so gravid and magnificent, I can't even think, let alone speak. I've come with other people from the research team but they have sort of disappeared into the green. Cathedral Grove is a sensate pocket of smell, sound, color. The light through the branches spins shadows upward and outward. I have come to pay my respects to the elders. I love them and my heart breaks with gratitude that they exist.

June 3. I am walking along the highway near Parksville, roughly eighty kilometers south of Cathedral Grove with twelve students and a public health nurse named Karen. We are doing a transect. It's part of the Ecosystem Intensive Health course. Our baffling epidemiological case study is the outbreak of *Cryptococcus gatticii* that claimed the lives of dolphins, cats, and elderly persons in the area.[12] To our right is a narrow strip of forest between the highway and the Pacific Ocean. There are Douglas firs among the trees. Each time we come upon one, Karen says, "Carrier," and points to it. The trees turn out to be the critical vector in the pathology transmission. She says this over and over and over, each time lifting her finger and pointing while she says the word *carrier*.[13] The walk continues. I feel sick. I avert my eyes from her finger. I plug my ears. In my head, I say over and over and over to the trees, "I'm sorry, I'm sorry, I'm sorry."

June 4. I'm driving down the highway toward Victoria. I see the Douglas firs and the word *carrier* keeps coming into my head. Worse than a catchy pop tune. It's overwriting my mental files. I'm filled with sadness. "If I knew as well as a forester what sick trees looked like, I fear I would see them everywhere."[14]

June 5. I'm with my friend, Ze'ev, a geographer, at his study site on Rocky Point, on the Finlayson Peninsula, 120 kilometers south of Parksville. He studies climate change through tree rings, specifically Garry oaks *(Quercus garryana)*. There's a good population of them left untouched (unlogged) on the peninsula, which was a former Canadian Forces military ammunition depot. We needed security clearance to get in. We drive in. It's kind of spooky, like a provincial park that nobody is allowed to tent in. All along the shady road are giant Douglas firs. I didn't think I would see any more of these old ones this far south. The Pacific Ocean hammers away to their left. I feel so happy again. I can feel and see the trees directly: there is no epidemiological filter through which they are passing into me. Maybe because Ze'ev loves trees too. But: the farther we go into the site, we see more and more of these firs ringed with caution tape. But: blue, not yellow. At one point, an entire row of maybe one hundred trees is ringed in tape, like you see along driveways of missing soldiers or along the edge of new housing developments where they are going to slaughter the forest. I feel disoriented. I ask Ze'ev to stop. "What is it?" Written on the blue tape surrounding a tree: "Culturally modified tree." ("Also known as a CMT, a culturally modified tree is defined as a 'tree that has been intentionally altered by Native people participating in the traditional utilization of the forest.'")[15] Ze'ev points out some faded burn marks running up the trunks about six feet from the ground; on others, some wounds where bark was stripped away and the cambium removed. Apparently, these particular trees were used by the Coast Salish Indians who had inhabited the region for at least the past two thousand years for a wide number of things: pitch to fasten arrowheads to shafts or patch canoes, fuel, food, to fashion head and neck bands for winter dancers, shamans, and hunters, purification rites, puberty training, mythological stories, sunburn salve, dye (see Figures 8.1 and 8.2).[16]

Figure 8.1. A Douglas fir *(Pseudotsuga menziesii)* on Rocky Point, British Columbia. Hundreds of these trees on the Finlayson Peninsula are wrapped in blue "caution tape" to indicate that they are valuable "cultural objects." Anthropologists have detected alterations in the bark made by the First Nations inhabitants of the area, the Coast Salish, as they used the trees for various purposes. (Photo credit: Karen Houle, July 2008)

"However, the Indians were always careful not to strip all the bark from a tree, because it would die and a nearby tree would curse a harvester."[17]

June 10. University of Northern British Columbia. I escape the airless classroom where my research team is teaching the Ecosystem Approaches to Health summer short course. I wander into a glassy, spacious atrium. In the atrium there is a small glass box. There is a glass box in a glass box, in a university perched on a hillside right up against the lush B.C. forest. In the box is a two-foot section of a tree. Debarked. Skewered on a display plinth. You can't tell anymore what kind of a tree it was when it was living in that forest. Not if it's a fir or a hemlock. Its treeness has been subtracted from it. In a groove in the tree are markings in Carrier, in lead pencil. Carrier is the written language used by local First Nation (Nadleh) peoples after contact. I stare at the markings, none of which I recognize. Sideways spirals with dots in them. Upside

Figure 8.2. A Douglas fir (*Pseudotsuga menziesii*) on Rocky Point, British Columbia, and author (for scale). This particular tree has evidence of burn scars. (Photo credit: Ze'ev Gedalof, July 2008)

down fan shapes. Hockey sticks. I can see myself in the reflection in the glass. Underneath the skewered and cross-sectioned tree, on a shiny brass plate is written: "This is a culturally modified tree" (Figure 8.3).

And the English translation of the Carrier markings:

> There is a body
> Pierre
> Hello I am saying this
> Antoine (1877)

THERE IS A BODY

These examples serve to express the perpetual vulnerability of living things, of the vegetable world, to the names we give them; to the *kinds* of names we are capable of giving them: litiginous, Latinate, epidemiological, poetic, utilitarian, qualitative, cause–effect, technological, negatively relational (combative, oppositional, as in "the nonhuman"), or

Figure 8.3. "Culturally Modified Tree," in glass display case, in the Atrium, on the University of Northern British Columbia (UNBC) campus, Prince George, British Columbia. In the cambium of this tree, there is a message written in graphite, in the Carrier language and dated 1877. The message was discovered in the forest by a member of the Nadleh First Nations during a logging operation, salvaged, and brought to the university as a cultural object. (Photo credit: Margot W. Parkes, December 2015)

positively relational (friend, beloved). Language is a continuous, total, gestural performativity we enact, continuously shaping the world we name and, in turn, being continually shaped by the particular overall nature of the world that our naming calls forth.[18] Through each of these linguistic choices, no matter how banal or how quickly they are uttered, our own ways of being in the world shift. There is activating and deactivating of our impressively varied capacities to be: with, or apart from, observer of or experiencer in, questioner of or teacher to, receptive to or hostile toward, loving or blocked, from the total situation in which we find ourselves. Each of these *dispositions* performs what Heidegger calls an "enframing," or an "ordering." Not all of what is, is seen or heard or felt at once. Only one aspect of an apple, or a Douglas fir, is reached for with one kind of word, and *that* quality of the thing—its color or its species' name, or what it means for national security—emerges in response to that mode of address. It is ordered and "stands forth" each time we speak, and in a certain way.

That all of our reality is composed of these kinds of relations—a kind of call and then a response to that kind of call—means that language use is something we must take great and continuous care with. The question of justice enters at every syllable. What if our words for a thing or a phenomenon or a situation we care about far away, in the Amazon or in the Canadian North, are wholly scientistic: *mercury, cadmium, poisoning, samples, sample size, bioaccumulation, contamination.* That world, and our relationships in that world with those special collections of entities—spruce trees and fish and clouds and lakes and babies—will more and more be *revealed* as chemicals, vessels, numbers, polluted. That we have options with respect to what words we can use, and where we use particular words in contexts that we then leave, leaving those words there, compounds the justice question enormously.[19]

What we like to call in philosophy "the question of justice" arises not as a matter of how much harm is done to what components, even the harm done to things by the names we use for them. The question of values happens at an even more profound and concerning level. It involves the risk of getting stuck in one particular mode of speaking about, and in the world, and then "continually approaching the brink of the possibility of pursuing and promulgating nothing but what is revealed in ordering, and of deriving all his [our] standards on this basis."[20]

What is meant here is that we can begin to mix up what is of real value with the kinds of things that we see *as valuable on the basis of a certain way of talking and questioning.*

Thinking that justice is about amounts of stuff and patterns of distribution (more versus less, higher versus lower) and perceiving the remedy of injustice as an outcome achievable by the relocation of amounts or patterns are, of course, how "justice" presents itself to our imaginations. This concept is the result of looking at the world predominantly in terms of objects and numbers, of facts and the names of those facts. If you speak about nature as stuff, causes and effects, as countable, fungible objects, then "nature presents itself as a calculable complex . . . of the effects of forces [that] can indeed permit correct determinations."[21]

If this is the only way you write or speak about nature, then it is the only way you will see nature. Eventually, everything except your own position will start to look like generic "stuff out there" (or what Heidegger develops and names in "The Question Concerning Technology" as "standing reserve").

It is not that justice as distribution is dangerous or false or bad on its own. It is that it is only one mode among many. It is a mode of thinking that dominates, which makes it very hard to fight back against and regain the neutrality of perceptual capacity through which any beautiful thing—an apple—can be seen again *as* beautiful and wonder filled. And further, thinking about justice as a metric, an allotment gesture, is a mode of thinking that conceptually extends what it seeks to remedy. What the "capacity for a sense of justice" in us senses and then seeks to remedy is the unbearableness of our situation. But counting and redistributing are only ever that: counting and redistributing. Therefore this form of justice cannot ever constitute a deep overturning of the actual situation from which our sense of justice arises. The real danger, according to Heidegger, is that "where everything . . . exhibits itself in the light of a cause–effect coherence, even God, for representational thinking, can lose all that is exalted and holy, the mysteriousness of his distance."[22] In simple terms, what Heidegger is saying here is that the wonder and beauty of life become flattened and dead under a certain mode of thinking. This extends to a certain automatic mode of speaking. How, exactly, does talking about Douglas firs as "carriers" or beautiful fall apples as "choking hazards" do that? I quote Heidegger again at length:

> When destining reigns in the mode of enframing, it is the supreme danger. This danger attests itself to us in two ways. As soon as what is unconcealed no longer concerns man even as object, but exclusively as standing-reserve, and man in the midst of objectlessness is nothing but the orderer of the standing-reserve, then he comes to the very brink of a precipitous fall; that is, he comes to the point where he himself will have to be taken as standing-reserve. Meanwhile, man, precisely as the one so threatened, exalts himself and postures as lord of the earth. In this way the illusion comes to prevail that everything man encounters exists only insofar as it is his construct. This illusion gives rise in turn to one final delusion: it seems as though man everywhere and always encounters only himself.[23]

167

What this says is that, if we only think about and talk about the world as if numerable objects, we will find it harder and harder to think about reality as anything but stuff to extract, move around, own, sell, throw away. Our power of extracting and making things into neat piles or skewering complex phenomena with a single noun will become the only thing we know about ourselves: that nothing in or about the world can resist our questioning, our naming. Or, perhaps just as deceptively, our fixing. We come to actually believe everything can be known and known by us. But, of course, this means only in a certain way. Everything can be named, and named by us. But only in a certain way.[24]

Consider the way we speak about all the other others. Consider the term *nonhuman animal*. It seems a positive, inclusive term, and indeed has been used by the biggest advocates of animal welfare, like Peter Singer. It is common currency in contemporary animal welfare science. But, putting it this way—*the* nonhuman makes it seem as though nature really is as Aristotle named it: an onion filled with one kind of thing (human subjects) and everything else lined up, equidistant to each another, on the periphery of the next level, *animalia*. Consider how the phrase "*the* nonhuman" makes invisible and thus irrelevant the infinitely complex dimensions of the creaturely world—and that in spite of the infinite space that separates the lizard from the dog, the protozoan from the dolphin, the shark from the lamb, the parrot from the chimpanzee, the camel from the eagle, the squirrel from the tiger, or the elephant from the cat, the ant from the silkworm, or the hedgehog from the echidna.

Using this term *(nonhuman)* reproduces in the imaginary a pathologically distorted frame of how the natural world stands to itself, and hence how we stand in and with it: "Confined within this catch-all concept, within this vast encampment of the animal . . . are all living things that man does not recognize as his fellows, his neighbors or his brothers."[25]

The fundamental wonder that is or passes through life and nature and humanity—"the infinite space" between—and the loving, humble, careful way appropriate to that can be lost or overwritten by the use of a single word. That is the most critical injustice, because it is really an injustice to everything.

This includes ourselves. This modality of injustice is "of concern" to us. Heidegger rightly worries that, in this way of being, we will even think of each other and ourselves that way. As though the *real injury* done to me by being raped can be compensated for by a legal settlement. As though the *true meaning* of the Israeli–Palestine conflict is readable in boundary lines. As though the *loss* of wonder that accompanies tagging Douglas firs as "carriers" can be matched by gains in protection against disease.

But this path only gains more and more distance from the lifeworld, from objects, from our own profound objecthood: our sensuous bodies, the natural rhythms of light and water and shadow we are, our mortality and natality, our creaturely nature. Little by little, this prevents us from being able to be activated, acted upon, or impacted by anything.[26] This, in the end, makes us the living dead, heading out earnestly, whether theoretically or practically, "into nature" to experience or to act on behalf of the "nonhuman"[27] or just be in solidarity "among the downtrodden" but finding only our own echo bouncing back to us, from Cathedral Grove, from the orchard, from the glass box.

NOTES

1. A more extensive critique of the reduction of the sense of justice to distributive justice can be found in the work of feminist political philosopher Iris Marion Young in *Justice and the Politics of Difference* (Princeton, N.J.: Princeton University Press, 2001).

2. One of the most important philosophers of the twentieth century was John Rawls, who, in his two major works *A Theory of Justice* (Cambridge, Mass.: Harvard University Press, 1971) and *Political Liberalism* (New York: Columbia University Press, 1993) argued that a fundamental power all rational subjects have and have in common, whatever their individual station in life or peculiar "conception of the good," is a "moral point of view" and a "capacity for a sense of justice" (*Political Liberalism*, 19).

3. The name most often associated with this view of justice is Jürgen Habermas.

4. The name most often associated with this normative perspective is Aristotle.

5. This claim is related to the analysis that Hannah Arendt leveled in her work *Eichmann in Jerusalem: A Report on the Banality of Evil* (New York: Viking

Press, 1963). There, Arendt was illustrating the ways in which "everyday people" (bureaucrats, neighbors) can become caught up in murder, bribery, extortion and normalize it, psychologically. My point is slightly different. That conceiving/discoursing on ethics and politics *as if* the true and fundamental essence of justice, fairness, wrongness is expressed in acts like murder, dam building, North–South extortion, and rape has the effect of distancing ourselves from considering the ways that what we do (how we eat, what we eat, what we call what we eat, where we toss our waste) also raises justice questions, though different kinds of justice questions from distribution questions. Further, these other kinds of justice concerns are entwined with violence like extortion and genocide.

6. The mandate of Canada's International Development Research Centre to support health research outside Canada is a good example. Many students and researchers have projects with an international component (say, Cuba or Ghana) and see national, local components as providing a good companion study (say, the Baie des Chaleurs or downtown Windsor) but these linkages are not as easily funded as those whose work is entirely "elsewhere."

7. Two examples, one laughable to the not so laughable. In 2008 "a Reno man filed a lawsuit to try to keep U.S. Senator Hillary Clinton off the Nevada ballot with the argument that the U.S. Constitution prohibits a woman from holding the office." The man, Douglas Wallace, "contends that because the U.S. Constitution relies on the pronouns 'he' and 'his' in describing the duties of the president, no woman can hold the office." See Arturo Garcia, "Texas Marketing CEO: Women Can't Be President Because of 'Hormones' and 'Biblical Sound Reasoning," *Rawstory* April 4, 2015, http://www.rawstory.com/2015/04/texas -marketing-ceo-women-cant-be-president-because-of-hormones-and-biblical -sound-reasoning/.

Additionally, in a rape trial in Kansas City, "the judge prohibited her [Tory Bowen, the appellant] from uttering the word 'rape' in front of a jury. The term 'sexual assault' was also taboo and Bowen could not refer to herself as a victim or use the word 'assailant' to describe the man who allegedly raped her. . . . To force a victim to say, 'when the defendant and I had sexual intercourse' is just absurd, said the prosecutor representing Bowen, who filed a lawsuit against the judge as a First-Amendment violation." In cases where the defendant's version of events is pitted against that of the alleged victim, "words are really important." See Tony Rizzo, "Judge's Ban on the Use of the Word 'Rape' at Trial Reflects Trend," *Kansas City Star*, June 7, 2008, http://www.alternet.org/story/87515/ judge_bans_use_of_the_word_%22rape%22_in_rape_trial.

8. This is a central and perpetual claim of Martin Heidegger's. It is a thesis throughout his magnum opus *Being and Time,* trans. John Macquarrie and

Edward Robinson (New York: Harper & Row, 1962), but also found in shorter works such as *On the Way to Language* (New York: Harper & Row, 1971).

9. J. L. Austin, *How to Do Things with Words,* William James Lectures, ed. J. O. Urmson (Cambridge, Mass.: Harvard University Press, 1962). The lectures were delivered at Harvard University in 1955.

10. See Daniel O. Dahlstrom, *Heidegger's Concept of Truth* (Cambridge, UK: Cambridge University Press, 2001).

11. Some of the most important political works in this domain include Andrea Dworkin's on the text and images of pornography constituting real harm and sex-discrimination, especially *Pornography: Men Possessing Women* (New York: Perigee Books, 1981); and Catharine A. MacKinnon, *Only Words* (Cambridge, Mass.: Harvard University Press, 1993).

12. Laura MacDougal et al., "Spread of *Cryptococcus gattii* in British Columbia, Canada, and Detection in the Pacific Northwest, USA," *Emerging Infectious Diseases* 13, no. 1 (2007): 42–50, doi: 10.3201/eid1301.060827; Kausik Datta et al., "Spread of *Cryptococcus gattii* into Pacific Northwest Region of the United States," *Emerging Infectious Diseases* 15, no. 8 (2009): 1185–91.

13. The gesture is a basic part of how children learn language. Notice that it is a performative utterance. The action carried by the nonverbal, which "puts" the sign onto the proper object while simultaneously withdrawing and canceling the arm of the teacher and the word *this,* as in "*This* is a carrier." For an excellent elaboration of the complex production of signification, see Michel Foucault, *This Is Not a Pipe* (Berkeley: University of California Press, 1983).

14. Bill McKibben, *The End of Nature* (New York: Random House, 1989), 211.

15. Morley Eldridge, "The Significance and Management of Culturally Modified Trees: Final Report Prepared for Vancouver Forest Region and CMT Standards Steering Committee," Ministry of Forests, Lands, and Natural Resource Operations, January 13, 1997, http://bcwildfire.ca/ftp/Archaeology/external/!publish/Web/culturally_modified_trees_significance_management.pdf. There are to date more than five thousand CMTs inventoried in British Columbia, many of them distinguishing natural fire scars from cultural scars. And, it is in establishing the latter that these trees are "designated provincial heritage sites." This evidence has subsequently been used in legal cases involving treaty claims and logging blockades.

16. Nancy Chapman Turner and Marcus A. M. Bell, "The Ethnobotany of the Coast Salish Indians of Vancouver Island," *Economic Botany* 25, no. 1 (1971): 63–99.

17. H. G. Barnett, *The Coast Salish of British Columbia* (Eugene: University of Oregon, 1955), quoted in Turner and Bell, "The Ethnobotany," 76.

18. For a reading of the impacts on phenomenal life of technological language use (or "enframing," as he calls it), see Martin Heidegger, "The Question

Concerning Technology," in *The Question Concerning Technology, and Other Essays*, trans. William Lovitt (New York: Harper & Row, 1977), 3–35.

19. Reporting a conversation with an Innu student: "In Innu territory, we can call it 'mercury' or 'assikumanapui.' We can write it as a periodic table shorthand (Hg) or 'a heavy silver metal that is a liquid at room temperature and is found in very small amounts throughout the earth and is unhealthy to fish, animals and humans.' We can define it as 'xenochemical' or the eleven-word Innu definition used in ecotoxicology literature: 'namesh-assikumanapui eka minuat tshekuan eninakuak apu mimishkakut aven mak aveshish.' We can call it 'toxic' or 'contaminant' or call it 'a chemical or bacteria that is either present in a place it does not belong or at levels that might cause harm to animals, plants or people.' Each of these choices calls forth a slightly different world, and a slightly different reaction on our parts, to the thing named. I, for one, recoil from anything called 'contaminated.'"

20. Martin Heidegger, *Basic Writings,* ed. David Farrell Krell (San Francisco: HarperCollins, 1977), 331.

21. Ibid.

22. Ibid.

23. Ibid., 332.

24. Verbs and verb conjugations among the Pacific Northwest tribes were not organized by time (past, present, future) but on where they heard it and the reliability of the source, for example, in a dream, from a friend, from a gossipy person.

25. Jacques Derrida, "The Animal That Therefore I Am (More to Follow)," *Critical Inquiry* 28, no. 2 (2002): 369–418 [402].

26. Thus the insidious persistence of a Cartesian ontology.

27. For an account of how this gesture in visual art backfires, see Karen Houle, "Infinite, Indifferent Kinship," *C-Magazine: International Contemporary Art* 107 (2010): 12.

NINE

What Vegetables Are Saying about Themselves

Timothy Morton

ARTHUR SCHOPENHAUER IS ONE OF THOSE TOO NEGLECTED PHILOS-ophers of the nineteenth century who deserves fresh attention. Schopenhauer is a Kantian who takes Kant's thought and pushes it toward a disturbingly realist vision. In this vision, a gigantic swirling will "represents" itself in objects: a pencil, a galaxy, a lizard, a human. This will is the one and only truly real thing in the universe. Everything else is a phenomenon of will. We cannot access will directly: we can only access will-phenomena such as raindrops and starlight. When you hit me, will is really hitting itself. Will, then, is like the *ouroboros* in Greek mythology, or Jörmungandr in Icelandic mythology: a self-swallowing snake, sucking its own tail. The only way to counter the endless, ferocious machination of will is to suspend it—one cannot disable it, but one can, as it were, put the gears in neutral and allow them simply to spin, without engaging. This neutral state is reached in art, and in particular music, in which will just moves forward without an agenda. In a pop song, for instance, chorus follows verse, but the chorus is not therefore "better" than the verse. There is no actual progress. Note twenty-one is not "worse" than note twenty. The suspension of will is what Schopenhauer believes is aimed at in the forms of Buddhism he studied, and this is roughly accurate: Theravada Buddhism does aim for suspending the machination of will, which it calls the chain of nidanas (causation), between "craving" and "grasping." Between, in other words, my feeling the urge to smoke a cigarette and my actually reaching out for the cigarette packet. In that moment of suspension, I have transcended samsara, which is imagined as an endless cycle of desires chasing their own tail.

Thus the disorienting endlessness, the vastness, and the pessimism of Schopenhauer's vision combines with a turn to Eastern thought that also disorients the normative trajectories and boundaries of Western philosophy. Perhaps it is not surprising, then, that Schopenhauer also disorients the normative anthropocentrism of Western thought. Schopenhauer talks seriously about nonhuman beings—animals and their suffering loom large, for instance, in his prose. And not only animals—plants, and, indeed, minerals, are also explored in a highly nonstandard way that is very pleasing for contemporary thought. We are now beginning to attribute agency to nonhuman beings. Not only are primates allowed to have some kind of consciousness; we are beginning to see how sentience as such is not restricted to certain forms of life at all. Only consider the field of plant sentience. When I was a child, jokes about suffering vegetables were standard—I recall one comedian refusing to serve a lettuce on a TV comedy sketch on the very popular BBC series *A Bit of Fry and Laurie*. The waiter (played by Hugh Laurie) starts to slice a lettuce. A scream is heard off camera. Upon being asked what is going on by the diner (Stephen Fry), the waiter replies, "Never heard a lettuce scream before?" He continues: "Frightening, isn't it? Never occurred to you that a lettuce might have dreams, hopes, ambitions, a family?"[1] But to a large extent, lettuces *do* scream, it turns out!

Schopenhauer makes it terribly easy to appreciate lettuces. I am a manifestation of will. A lettuce is a manifestation of will. We are merely phenomena, not things in themselves. There really is not that much of a difference. Schopenhauer argues that I am a more perfect "representation" of will than a lettuce. Likewise, a lettuce is a more perfect representation of will than a crystal. In this sense, there is some teleological thought in his philosophy, the kind that awards consciousness to beings that he supposes to have been more highly evolved. Lettuces, indeed, *do* have dreams and ambitions, just as Hugh Laurie observes in the 1989 comedy sketch.

It is quite astonishing that a whole region of comedy, to wit the gratuitous mockery of vegetables, has been challenged in one's lifetime. Schopenhauer, who incidentally is uncannily good for a philosopher at explaining jokes, would readily have had something to say about it if he had been alive. Laughter erased, and in a decisive way. We simply cannot laugh at a fellow sufferer. Of course we might be able to laugh *with* one. Perhaps a whole new realm of humor could open up, akin to the humor of the survivor (Jewish humor for instance), in which lettuces and humans bemoan their common lot. And this commonality is uncannily anticipated in the Fry and Laurie sketch itself. Fry is trying to finish a joke for his female partner at the restaurant table. The joke is a classic "Englishman, Scotsman, Irishman" joke where all three are giving reasons for killing themselves. Typically the Irish are mocked for being less than highly civilized—a joke sequence popular during the British demonization of Sinn Fein and the Irish republicanism movement. A joke involving implicit and absurd distinctions between humans, distinctions that, if they were made explicit, would render the joke extremely unfunny. Then along comes the waiter, Hugh Laurie. In turn, Laurie interrupts the joke and the normative situation in which a man is impressing a woman at a dinner table with his command of humor. Laurie interrupts the joke about inferior humans by offering to prepare "Chicken à la Croix" (crucified chicken?) at their table. The preparation involves slaughtering a live chicken right in front of Fry and his friend. Evidently they refuse, and the exasperated Laurie goes back into the kitchen to make a salad, whereupon we hear a scream. The interruption puts the demeaning of humans in a far wider context of speciesist demeaning of nonhumans.

Schopenhauer reckons that humor is based on the exploitation of the obvious gap between the thing in itself and its phenomenon. Since this gap is irreducible, it is potentially funny when someone acts as if it were reducible, or when the gap is pointed out in a situation where it has become unclear. Fry's joke is about humans committing suicide in a number of culturally specific ways—putting an end to their phenomena of will, so to speak. Schopenhauer would argue that the supposed differences between the Englishman, Scotsman, and Irishman cover over the obvious similarity, which is the basis for the joke's humor. The supposedly absurd gap between a joke about suicide and the killing of a live chicken at one's table shows us something else about will. It is no different, in essence, to kill a live chicken at the table than it would be to eat a chicken killed in a factory farm. It is simply that its "representation" differs—we do not see the slaughter of the chicken, so we do not mind eating it. We do not mind exercising our will when it is invisible.

The gap between will and its expression is enacted by Laurie's constant interruption of Fry's joke in ever more absurd yet seriously de-anthropocentric ways. Our will to laugh becomes weird as the scope widens: laughing at the Irish, laughing at people horrified by chickens, laughing at vegetarians, laughing at people who ascribe feelings to lettuce . . . Yet, on another level, the will *really is invisible* in an irreducible way. All we have are will-phenomena, remember: comedians, chickens, Englishmen committing suicide . . . In this respect, the more beings we allow to suffer and to inflict suffering ("Never heard a lettuce scream before?"), the stranger and more obvious the gap grows, the irreducible gap between phenomena and the thing, between humans and chickens on the one hand, and will on the other. When I eat a chicken, will is in fact eating itself. When I make a joke about a lettuce screaming, will is in fact laughing at itself.

Now that I have made jokes about lettuce extremely unfunny by explaining them to death, let us proceed. We shall see as we go on that in all kinds of ways, funny and disturbing by turns, what we are exploring are anthropocentric categories and the ways we can jump between them—and the reasons why we can defy them like that. These categories are not only between humans and nonhumans, but also between animals and plants; between consciousness and unconsciousness; between sentience and nonsentience; between life and nonlife; between being a thing and being a pattern; and even between existing and not existing.

We shall take our cue from Schopenhauer himself, who argues that, although everything is a manifestation of will, there are foxes and cacti, and they are different. We shall be jumping between these categories not to erase them and reduce everything to an unspeakable one, but to explore the irreducible paradoxes involved in a reality in which the following seems to be true: things are what they are, yet never as they seem. In other words, what we are going to end up talking about is not simply botany, not simply biology, and not even epistemology. We are going to be talking about ontology—what it means to be a thing at all: a thought, a bunny rabbit, a quasar, a mathematical formula.

Schopenhauer's thought operates at a geotemporal scale sufficient for thinking ecological awareness: his prose is saturated with the non-human, the vast, the disturbing. He argues that plants are manifestations of will: they just grow. Indeed, he crams his prose about plants with examples of seeds from early agrilogistical times being repotted and sprouting, uncannily alive after all this time, zombie slaves of an undead will.[2] (And, furthermore, this is a genuine phenomenon.[3]) In this sense, plants are like algorithms, since algorithms do not know anything about numbers, they just execute computations. Thus algorithmic models of plants work just like plants, hence the success of the beautiful book *The Algorithmic Beauty of Plants* (Przemyslaw Prusinkiewicz and Aristid Lindenmayer, 2004). A flower is a plot of an algorithm. And a trope is an algorithm, a twist of language that emerges as meaning by simply following a recipe such as "Stick two nouns on either side of the verb *to be*." A trope is a flower of rhetoric (*anthos, anthology*). Milton's Satan curls around like a snake trying to turn into a vine.

Disturbingly, rhetoric and algorithms and plants and Satan exhibit a zero degree of intelligence, or not . . . we cannot know in advance. Plants haunt us with what Lacan says "constitutes pretense": "in the end, you don't know whether it's pretense or not."[4] They might be lying, which in a sense means that they *are* lying. Just as an algorithm could pass a Turing test (I could discern thinking and personhood in its "blind" execution), so plants are posing and passing Turing tests all the time. In looking at a flower, you are doing the flower's job. Bees complete the test all the time, by following the flower's nectar lines. Or, as Schopenhauer puts it, plants want to be known, because they cannot quite know themselves. They have something approaching a rudimentary awareness, for Schopenhauer, but not enough to recognize themselves—they require a mediator.

A plant in this sense is the zero degree of personhood—as Nietzsche said, people are halfway between plants and ghosts.[5] This zero degree is a strange loop that says something like "This is not just a plant," something like "I am not just a vegetable." For Descartes, a fundamental uncertainty is key to reasoning that I exist: "Maybe I'm just a puppet of an all-powerful demon."[6] In this sense, paranoia is the default condition of being aware. Before you get into the theology aspect, whereby a good god would never deceive you like that, you have to traverse a layer of deep uncertainty: "I might be a robot"—to exist is to be paranoid that you might be an algorithm. To be a person is to be worried that you might not be one.

The zero degree of the Cartesian cogito is the paranoia that I might simply be a puppet of some demonic external force. Is this not the creeping sensation that I might just be a vegetable? Since we now know about plant sentience—the capacities of plants to perceive, whether or not that implies consciousness—this must be accepted even more.[7] In this light, T. S. Eliot's line about flowers is perfect, from the plant's own point of view: "The roses / Had the look of flowers that are looked at."[8] A disturbing chiasmus lies at the heart of correlationism. Or, to put it in Schopenhauerian, trees and plants want to be known, because they cannot know themselves as bodies. "Blind willing" requires "the foreign intelligent individual" to be perceived, to "come . . . into the world of the representation."[9] The further point being that this uncertainty applies to my own (mammalian) so-called sense of conscious "awareness."

Let us turn to two paragraphs from Schopenhauer's philosophical father, Immanuel Kant:

> In order to consider something good, I must always know what sort of thing the object is [meant] to be; i.e., I must have a [determinate] concept of it. Flowers, free designs, lines aimlessly intertwined and called foliage: these have no significance, depend on no determinate concept, and yet we like [*gefallen*] them.[10]

> Consider flowers, blossoms, even the shapes of entire plants, or consider the grace we see in the structure of various types of animals, which is unnecessary for their own use but is selected, as it were, for our taste. Consider above all the variety and harmonious combination of colors, so likable and charming to our eyes (as in pheasants, crustaceans, insects, down to the commonest flowers); since these colors have to do merely with the surface, and even there have nothing at all to do with the figure [i.e., (visible) structure] of these creatures—which might be needed for these creatures' inner purposes after all—is seems that their sole purpose is to be beheld from the outside.[11]

Kant is arguing that the decorations and colors of flowers and animals suggest "that their sole purpose is to be beheld from the outside."[12] He says he cannot accept this, since it would violate a law he likes against "multiplication of principles."[13] Moreover, it risks implying that nature, which for Kant mechanically does things that look nice (like crystallization), was trying to look nice. Schopenhauer harps on this in his analysis of plants and ice crystals that *look like* trees and flowers.[14] Indeed, since we know about sexual selection—Darwin argues that aesthetic display goes all the way down at least to beetles—Kant's suggestion is worth pondering, despite his own deletion of it.[15] Indeed, Kant seems to contradict himself a little: we find things good because we know what they are for; yet we are not allowed to know what flowers are for, because then we would be "beholding them from the outside," and this would suggest that we are carrying out some kind of nonhuman intention to be seen, an intention that resides in the plant! Humans get to decide what things are for, and thus to find them good, Kant argues. It is no good if plants are already telling us what they are for. In the first quotation, flowers are disturbingly *meaningless*. In the second, they are disturbingly *meaningful*.

Kant claims that thinking that plants intend us to look at them would add too many principles to the argument. Yet all we need to reach the idea that plants want to be looked at is to *remove* a principle or two. The first is anthropocentrism. The second is necessity. Kantian aesthetics depends on a paradox, a purpose of no purpose. It seems as if a beautiful thing is designed for me to enjoy its having been designed for me to enjoy, and so on. We have a loop: *This is not (just) a sentence.* But what constitutes this loop? Consider the fact of sexual selection, which is why flowers exist. The only reason, argues Darwin, why I have reddish facial hair and white skin is because someone thought it was sexy a few million years ago, and she probably did not have a choice. In other words, she was not performing something like bourgeois self-fashioning through taste.[16] She cleaved closer even than Kant to the nonconceptuality of the aesthetic dimension. There was *even less purpose*. There is no reason for these huge horns or this iridescent wing pattern. It is actually terribly expensive from DNA's point of view. Just as atelic patterns subtend DNA (crystals that self-replicate for no reason are the condition for DNA's existence), so purposive purposeless depends on weaker purposelessness formats. Under the current ultra-utilitarian conditions universities find themselves subjected to, DNA would not be able to apply for state funding for its multimedia project.

What we glimpse in sexual selection is precisely a form of purposive purposelessness. Yet the phenomena on which it depends, such as flowers, are held by Kant to be outside true aesthetic judgment. The only reason for this must be anthropocentrism, because by Kant's own argument, flowers do indeed appear to have no purpose (the first quotation), and yet it also appears as if they might have a purpose (the second quotation). Is this not exactly what he might say about an oil painting or a piece of music? The whole point is that we can never point to the purpose of an oil painting either. Its purposiveness is weirdly virtual, floating somewhere between itself and the viewer, in a loop.

Since there is no good reason why an insect is gorgeously iridescent, apart from the recursive reason that it looks nice, is it not easy to imagine that the conditions of possibility for human beauty are beautiful flowers, which are also just there to look nice, in a sexual display mediated through bees? Like Kant, scientism also tries to contain the de-anthropocentric explosion that comes with thinking through the aesthetic power of nonhumans. Aesthetic–sexual "niceness" for its own sake induces an anxiety that provokes scientism to want to "know more"—that is, to be able to reduce the disturbing excess of display. The "more" in question is inevitably a non- or antiaesthetic reduction to neo-Darwinian utility. Such an anxiety is based on the long history of substance ontologies, where what a thing is underlies how it appears, such that its appearance is superficial and meaningless.

The anxiety about appearance does not have to do with heterosexuality, in particular, but with display as such. Sexual activity itself is by no means just heterosexual or monogamous, as trees prove every day by exploding huge clouds of pollen to be spread by insects and birds.[17] Sexual activity is purposeless in that sense. Is it not possible that the conditions for *that* are to be found *below* plants, in the logical conditions for life-forms as such, namely, self-replicating loops that are both physical and informational at the same time? There is no good *reason* why squiggles of organic chemicals should "mean" things to other squiggles that constitute their environment. The appearance–thing gap goes down at least to DNA and RNA. And one wonders what causes such things to exist in turn. Is it not precisely because there is an appearance–thing gap at all, as a condition of possibility for existing as such?

Bacteria came before viruses chronologically, but logically viruses come before bacteria. Viruses are the condition of possibility for life-forms: nonliving patterned strands, *foreign intelligences,* as Schopenhauer puts it, that force other patterned strands to go into a loop and become ciphers. Because is this not what being a strand of code in a physical format means in the first place? DNA is an inconsistent molecule that is trying to unzip itself into nonexistence. In the very attempt to cease, it ironically reproduces itself, since, as Schopenhauer and his follower Freud argued, the purpose of life is death, but in following that purpose, you make more of yourself and continue life.[18] The useless beauty of a flower is thus not a cynical ruse to make more flowers. It is a viral cipher that serves no purpose, but which, when caught in another system, say a bee's search for nectar, ends up ironically reproducing itself.[19] Thus viruses, flowers, iridescent wings, Kantian beauty, tropes, earworms, and daft ideas that float around in my head all share something. They are symptoms of an irreducible gap between being and appearance that eats away at the metaphysics of presence from the inside. So not only might viruses and tropes and flowers share some kind of family resemblance. They might *actually* be part of the same physical family. When I see a flower, I really am seeing a trope. And when I use a trope, I really am reproducing a virus. And when I get a virus, I am flowering. Do not forget that an old definition of *parasite* concerns not animals but plants. And that plants require parasites, "the foreign intelligent individual" (Schopenhauer again) to reproduce.

As Baudelaire's title *Les fleurs du mal* begins to suggest, flowering is thus indeed a kind of "evil," a necessary evil that comes with existing, since existing means having a gap between what you are and how you appear, even to yourself. Flowers of evil. Is it not the case that the aesthetic dimension has been seen with great disfavor by philosophers who have varying degrees of allergy to the phenomenon–thing gap? So, for instance, Plato argues that the aesthetic is a demonic realm of influence where some other entity can pull me in with its tractor beam, without my say-so, before I even notice. And thus he spends a lot of his time proving why Socratic dialogue is not that: it is giving birth to the idea, rather than having sex for no reason, as it were. And thus we find these beautifully ambiguous lines in the *Phaedrus*:

> SOCRATES. But let me ask you, friend: have we not reached the plane-tree to which you were conducting us?
>
> PHAEDRUS. Yes, this is the tree.
>
> SOCRATES. By Hera, a fair resting-place, full of summer sounds and scents. Here is this lofty and spreading plane-tree, and the agnus cast us high and clustering, in the fullest blossom and the greatest fragrance; and the stream which flows beneath the plane-tree is deliciously cold to the feet. Judging from the ornaments and images, this must be a spot sacred to Achelous and the Nymphs. How delightful

> is the breeze:—so very sweet; and there is a sound in the air shrill and summerlike which makes answer to the chorus of the cicadae. But the greatest charm of all is the grass, like a pillow gently sloping to the head. My dear Phaedrus, you have been an admirable guide.
>
> PHAEDRUS. What an incomprehensible being you are, Socrates: when you are in the country, as you say, you really are like some stranger who is led about by a guide. Do you ever cross the border? I rather think that you never venture even outside the gates.[20]

This is seductive, descriptive, scene-setting language for no reason, the very thing that Plato worries leads us by the nose like a stranger in a strange land. Yet it has something to do with the dialogue that ensues being all about the experience of beauty. Kant shows that this slightly evil dimension is intrinsic to thinking as such. There are flowers in your head. Let us return to the first Kant quotation. There is a gap between appearing and being, which things such as squiggles and plants exemplify—suggestively, Kant does not make much of a distinction between an actual plant, an arabesque, and calligraphy. There is no way to know in advance what they are for. It is the problem of pure decoration, which is the problem of givenness. Which is, of course, the problem of reason—since reason is just given, like a flower that pops up for no reason. So there is always a gap between a thought and the thinking that thinks it, and its inner logical content—an insight that Husserl goes on to elucidate beautifully. This is the same as the gap between the appearance of a flower, and what a flower might be for. The trouble is thoughts and flowers and appearances are not just anything. Foliage is foliage, not spaghetti. We cannot distinguish the appearance of a thing from what it actually is. If only things came with a little dotted line on them and a little picture of a pair of scissors, saying "cut here." It was Plato who said the good philosopher knows where to cut the eidos, but does she? How can she check in advance? As David Byrne says, if this is paradise, I wish I had a lawnmower.[21]

Now we can talk in more detail about narcissism, that much maligned state, maligned principally in philosophy by Hegel and Hegelians in denial about the Kantian explosion. This is because Narcissus was a kind of flower: someone stuck in a loop between what he was and how he appeared, even to himself, a loop in which he ended up metamorphosing into what he already was. All entities are narcissists, insofar as they consist of weird loops of being and appearance, where there is an irreducible gap between being and appearance. These loops are weird because they are tight. Raindrops are raindroppy, and sadly not lemon-droppy or gumdroppy. But raindroppiness is not the same as being a raindrop. Such is Kant's own example from the *Critique of Pure Reason*.[22]

Imagine this insight applying to humans and nonhumans alike. Imagine, in other words, an object-oriented, de-anthropocentrized version of what Derrida says here, a version implied already in the source in so many ways and in so many (other) texts: "There is not narcissism and non-narcissism; there are narcissisms that are more or less comprehensive, generous, open, extended. What is called non-narcissism is in general but the economy of a much more welcoming, hospitable narcissism . . . without a movement of narcissistic reappropriation, the relation to the other would be absolutely destroyed, it would be destroyed in advance."[23] Narcissism and coexistence intertwine. We want coexistence to mean the end of narcissism, but it is this very thought that would destroy in advance the relation to the other.[24] Yet this fact is difficult to think in the face of our emerging awareness that we are a hyperobject (species) inhabiting another hyperobject (planet Earth); a hyperobject being an entity that we can think and compute, but that we cannot see or touch directly.[25] A being, in other words, that exemplifies what it means to be a thing, withdrawn yet vivid at the same time. While huge and hard to discern, these large finitudes are such that it becomes obvious how, as the philosopher Emmanuel Levinas puts it, referring to Pascal, "'My place in the sun' is the beginning of all usurpation."[26] We know that other humans and other life-forms are suffering, and we also know that their suffering is in part a determinant of our own existence and, at any given moment, our lesser suffering (because there is one of us and billions of them). Does narcissism, in the face of this intuition, not seem really, really disingenuous?[27]

Yes, if we think that existence means solid, constant, present existence. That would be the ultimate disaster. It is based on the fantasy that all the parts of me are me: that if you scoop out a piece of me, it has *Tim Morton* inscribed all over it and within it, just as sticks of English Brighton rock contain a word in pink all the way through their deliciously peppermint sugar tubes. This is not the case. All entities just are what they are, which means that they are never quite as they seem. The first part of that sentence gives us Hegel's dreaded A=A, the night in which all cows are black. For Hegel, this is not even logic yet, it is prior to logic, like a plant or a hallucination. The second part of the sentence shows how being what you are is also a kind of loop—even a night in which all cows are black has cows in it. There are cows in the darkness, cows of darkness. "Equals A" does something to "A" even if that something is hardly different from A. A goes into a loop. It has the look of a flower that is looked at.

What is the sin of A=A? It is the sin of a virus. A=A is a parasitical piece of code that cannot survive on its own. But it is also a maddening little loop that reveals the necessary incompletion of all linear logical sequences. A=A is a virus, which is to say it is a trope, which is to say it is a flower. Which is to say that it is a paradoxical loop.

We are likely to call someone a narcissist who is in fact suffering from *wounded narcissism,* that is, a kink or two in his or her loop. In a sense, a wounded narcissist *is not narcissistic enough.* Therefore he or she attacks or withdraws from the other—from the loop that *is* the other— precisely because the other loop is provocatively efficient and closed. Is it possible, then, that Hegelianism and other forms of Western anti-Buddhism, so quick to use the term *narcissist* as a term of abuse, are symptoms of a philosophical form of wounded narcissism?

With his anxiety about A=A, what Hegel is warding off most of all is the possibility that this loop thinks all by itself, that it is a kind of artificial intelligence. That it was unnecessary to have undergone the linear sequence of logical reasoning, which begins to look like an imperialist leveling of a foreign territory.[28] A logical imperialism that reasserts a rigid boundary between the Neolithic and the Paleolithic, for instance; between the indigenous and the "inevitably" tragic "civilized man." Just as imperialists would argue that Africa is outside of history and needs to have history imposed upon it, or just as Asia is perversely historical, stuck in narcissistic loops of feminine laziness, and needs a good electrical jolt of the negative to wake it up. What this logical imperialism disavows is precisely the ouroboric, abyssal swirl of A=A.

Yet the mathematician Kurt Gödel showed that any complete logical system must be able to say things that it cannot prove as a condition of possibility for its truth—as a condition of possibility, that is, for existing at all. In other words, you can find loops in a coherent logical system—you can make it say A=A, you can make its metalanguage into an object language, to the precise extent that it is coherent and true.[29] So the more you try to eliminate the virus of narcissus, as Russell and Whitehead tried to do in the *Principia Mathematica* (they called it self-reference), the more you are able to make logical systems sprout flowers, as if things were like the guns the Blue Meanies fire at the end of *Yellow Submarine*: flowers come out of them every time they pull the trigger.[30] Every attempt to reduce a system to simplicity (by firing a gun at it, for instance, or by trying to unzip oneself, like DNA) ends up with the system reproducing itself, flowering into contradiction.

For every logical system there is a Gödel sentence. For every cell there is a virus. For every stem there is a flower. For every life-form there is the possibility of death. Try to eliminate the virus, and you get a much more virulent one. Autoimmunity is hardwired into the structure of a thing. Or: a thing is saturated with *nothingness*. Entities are so incredibly . . . themselves. Yet in this selfsameness they are weird, self-transcending. The contrary motion of what things are and how they appear makes a mockery of presence. Things emit uniqueness. They *bristle* with specificity. Purple, pale violet, light blue, their soft and sharp spines and flower spines bristle forth. Bristle forth despite me, despite my subject–object scissions. This flickering between a thing and its appearance is the reason why coexistence cannot be holistic. Something is always missing. My self-awareness is a sense of incompletion.

From a certain speculative realist point of view, this looks like tragedy, and let us wallow in it. The horror of doom. But as even the ancient Greeks already knew, there is a comedy level embedded within the doom. For every three tragedies in the City Dionysia, there was a comedy. It is as if one needs to notice that the very attempt to escape the web of fate *is* the web of fate, and one needs to notice it about three times—but then you notice that, hilariously, you are caught in the loop, and you are the loop. Every moment of speculative realist horror prose is a farcically sincere reaction to the necessary evil of flowers, viruses, tropes, and self-reference. The horror at the void just is what it is, performing itself endlessly, stuck in a loop, and as we know from Bergson, what is funny is when you see someone totally caught in their phenomenological style.

And since, as Gödel showed, even a logical system is caught in its phenomenological style, the active ingredient of reality is not tragedy but comedy. Yet this comedy is not to be found outside of tragedy, but rather inside it, just as the tree of reality, Yggdrasil, is found inside the closed loop of Jörmungandr, the universe serpent.

The aesthetic equivalent of this A=A is the corny, cheesy—nay, narcissistic—art of Keats and the Pre-Raphaelites and Salvador Dalí, condemned as feminine, kitschy poster art before poster art—flowers, as it were, that know that they need to be looked at to be completed. This kind of kitsch has been roundly condemned by avant-gardism—indeed, that is what avant-gardism has been since 1790, a culture of disgust at kitsch, which is the closed enjoyment loop of the other and their self-pleasuring abyss of plastic flowers. Dalí's poem about the painting *Metamorphosis of Narcissus* is about how the human and his brain give way, leaving an idea, which is a flower:

> when that head bursts,
> it will be the flower,
> the new Narcissus,
> Gala—my narcissus.[31]

The true enemy of modernity is *kitsch that makes you think,* the kitsch that contains irony, not as a way to go meta, but as a way to realize you cannot go meta.[32] For every nature poem, there is an artificial flower poem, because flowers are just artifice, as I have been showing. Warning: this is not the same as saying that kitsch is the next big thing or the new avant-garde. To say so would be for the avant-gardist approach to win, which in the end is a win for cynical reason.

Horror is what does not know this yet, a masculine style of scaring oneself precisely by the possibility that one might be in a loop—because, heaven knows, loops are scary. Speculative horror prose is thus also saying, quite firmly, "I am not a self-fellating snake." Which is to say, *I am not a flower.* Which unfortunately is what flowers are saying, so you *are* a flower. Remember that homosexuality, in medieval theology, is called self-love, which is what it is also called by Kellogg, the inventor of Corn Flakes, that homophobic cereal that was designed to stop boys from going into a self-pleasuring loop, a kind of medicine that by ingestion would work better than just chaining boys to the bed, just as for Plato, maieutics (midwifery) is the medicine of art. Cereals in general are metonymies for crops on which I have minimized the flowers and maximized the seeds. No wasted parts. Nothing added, nothing taken away, as Shredded Wheat says. It might be interesting to see agriculture, which is the law of noncontradiction applied to the earth's surface, as a war on flowers, which would make it a war on self-reference. Flowers, as D. H. Lawrence pointed out, are paths to the underworld, where the daughter of Ceres was dragged.[33] Flowers are evil, because evil is self-reference, because self-reference is art, because art is a virus, because viruses are the logical precursors of cells, because A=A is a precursor of logic, because appearance cannot be peeled off of a thing. Here is a quotation from an evil narcissist, otherwise known as a Buddhist:

> Since things neither exist nor don't exist,
> Are neither real nor unreal,
> Are utterly beyond adopting and rejecting—
> One might as well burst out laughing![34]

Horror still wishes that it could electrocute itself awake from its self-induced nightmare, the realization that Hegel covered his ears to, the Kantian gap between phenomenon and thing. Laughter, on the other hand, simply erupts because the insight arises that the very attempt to electrocute oneself awake *is* the nightmare. Which brings us back to the question of comedy, back to where we began, like a snake biting its tail. When I first read Schopenhauer, I thought he was cosmically depressing. But now I read him again—and he does instruct you to do so in the introduction to *The World as Will and Representation,* a point that you miss the first time around, and even if you do not miss it, you do not understand it—he is freaking hilarious, with the floral pun on "freak" well intended.[35]

Perhaps the best way to close is to quote from this poem by Robert Frost, "Spring Pools":

> The trees that have it in their pent-up buds
> To darken nature and be summer woods—
> Let them think twice before they use their powers
> To blot out and drink up and sweep away
> These flowery waters and these watery flowers
> From snow that melted only yesterday.[36]

What we have discovered is that in some strange but not totally figurative sense, flowers *do* communicate. In so doing, flowers tell us something about the capacities of appearance as such, which is not just the decoration of a substance, but an active causal power. The power of appearance has nothing to do with how it is used toward some aim, and, in particular, the power is not activated by a human or even by something we consider sentient—or, indeed, even by something we consider alive. The power of appearance resides *within itself,* operating in the form of a loop. The seeming superficiality of flowers and our attempt to reduce them to utility are two aspects of two related things: an anthropocentric restriction of meaning, intelligence, and agency to the human; and an equally anthropocentric anxiety about the loop-like intertwining of being and appearance at levels of reality we still consider "below" us. In turn, this means that what we consider to be agency, intelligence, sentience, or consciousness is not exclusively human. It has to do with processes we take to be automated—processes that happen all by themselves.

NOTES

1. Stephen Fry and Hugh Laurie, "Chicken," *A Bit of Fry and Laurie,* season 1, episode 6 (BBC, 1989).

2. Arthur Schopenhauer, *The World as Will and Representation,* trans. E. F. J. Payne, 2 vols. (New York: Dover Publications, 1969), 1:137 n. 13.

3. Stephen Messenger, "Extinct Tree Grows from Ancient Jar of Seeds Unearthed by Archaeologists," *Treehugger,* October 5, 2013, http://www .treehugger.com/natural-sciences/extinct-tree-grows-anew-after-archaeologists -dig-ancient-seed-stockpile.html.

4. Jacques Lacan, *Le séminaire, Livre III: Les psychoses* (Paris: Editions de Seuil, 1981), 48, my translation.

5. Friedrich Nietzsche, *Thus Spoke Zarathustra* (Cambridge, UK: Cambridge University Press, 2006), 6.

6. René Descartes, *Meditations on First Philosophy* (Indianapolis: Hackett, 1993), 16–17.

7. See Daniel Chamovitz, *What a Plant Knows: A Field Guide to the Senses* (New York: Scientific American/Farrar, Straus and Giroux, 2012). Depending on which scientist is talking, these capacities are called "sentience," "signaling," "perception," and even "neurobiology."

8. T. S. Eliot, "Burnt Norton," in *Four Quartets* (Orlando, Fla.: Mariner, 1968), lines 28–29.

9. Schopenhauer, *The World as Will and Representation,* 1:201.

10. Immanuel Kant, *Critique of Judgment,* trans. Werner Pluhar (Indianapolis: Hackett, 1987), §4, p. 49.

11. Ibid., §58, pp. 221–22.

12. Ibid., p. 222.

13. Ibid.

14. Schopenhauer, *The World as Will and Representation,* 1:182.

15. Charles Darwin, *The Descent of Man* (London: Penguin, 2004), 114–16.

16. Ibid., 241–49.

17. Joan Roughgarden, *Evolution's Rainbow: Diversity, Gender, and Sexuality in Nature and People* (Berkeley: University of California Press, 2009), 30.

18. Sigmund Freud, *Beyond the Pleasure Principle,* trans. and ed. James Strachey (New York: Liveright, 1950).

19. I use Judith Roof's invaluable concept of DNA as *cipher* rather than as *code*. Roof, *The Poetics of DNA* (Minneapolis: University of Minnesota Press, 2007), 78, 81–82.

20. Plato, *Phaedrus,* trans. Benjamin Jowett, http://classics.mit.edu/Plato/phaedrus.html.

21. Talking Heads, "(Nothing but) Flowers," *Naked* (Sire Records, 1988).

22. Immanuel Kant, *Critique of Pure Reason,* trans. Norman Kemp Smith (New York: St. Martin's Press, 1965), §8, pp. 84–85.

23. Jacques Derrida, "There Is No *One* Narcissism," in *Points . . . : Interviews, 1974–1994,* ed. Elisabeth Weber, trans. Peggy Kamuf et al. (Stanford, Calif.: Stanford University Press, 1995), 199.

24. Attempts to banish narcissism fall under the category of what I am elsewhere calling *agrilogistics*. See, for instance, Timothy Morton, "She Stood in Tears amid the Alien Corn: Thinking through Agrilogistics," *diacritics* 41, no. 3 (2014): 90–113.

25. See Timothy Morton, *Hyperobjects: Philosophy and Ecology after the End of the World* (Minneapolis: University of Minnesota Press, 2013).

26. The most profound discussion of this is found in Emmanuel Levinas, *Totality and Infinity: An Essay on Exteriority,* trans. Alphonso Lingis (Pittsburgh, Pa.: Duquesne University Press, 1969), 37–38. See also Emmanuel Levinas, interview with François Poirié, in *Is It Righteous to Be? Interviews with Emmanuel Levinas,* ed. Jill Robbins (Stanford, Calif.: Stanford University Press, 2001), 23–83 [53]. The Pascal quotation forms one of the epigraphs to Emmanuel Levinas, *Otherwise than Being: Or Beyond Essence,* trans. Alphonso Lingis (Pittsburgh, Pa.: Duquesne University Press, 1998), vii.

27. Peter Atterton, "Do I Have the Right to Be?" *New York Times,* July 5, 2014, http://opinionator.blogs.nytimes.com/2014/07/05/do-i-have-the-right-to-be/.

28. See G. W. F. Hegel, *The Philosophy of History,* trans. J. Sibree (Mineola, N.Y.: Dover Publications, 2004), 8–79.

29. Kurt Gödel, "On Formally Undecidable Propositions of *Principia Mathematica* and Related Systems" (1931), trans. Martin Hirzel, November 27, 2000, http://www.research.ibm.com/people/h/hirzel/papers/canon00-goedel.pdf.

30. Alfred North Whitehead and Bertrand Russell, Principia Mathematica, 3 vols. (Cambridge, UK: Cambridge University Press, 1925–1927); *Yellow Submarine,* dir. George Dunning (1968; Apple and United, DVD).

31. Salvador Dali, *The Collected Writings of Salvador Dali,* ed. Haim Finkelstein (Cambridge, UK: Cambridge University Press, 1998).

32. The best existing exploration of the vitality of kitsch within and against modernity is Daniel Tiffany, *My Silver Planet: A Secret History of Poetry and Kitsch* (Baltimore: Johns Hopkins University Press, 2014).

33. See D. H. Lawrence, "Bavarian Gentians," in *The Complete Poems,* ed. Vivian de Sola Pinto and F. Warren Roberts (Harmondsworth, UK: Penguin, 1994).

34. Longchen Rabjam, *The Practice of Dzogchen,* trans. Tulku Thondup (Ithaca, N.Y.: Snow Lion, 2002), 316.

35. One sense of *freak* is a streak of color, for instance on a flower. *Oxford English Dictionary,* s.v. "freak," http://www.oed.com.

36. Robert Frost, "Spring Pools," in *The Poetry of Robert Frost: The Collected Poems,* ed. Edward Connery Lathem (New York: Henry Holt, 2002), 245, lines 7–12.

III. Literature

TEN

The Language of Flowers in Popular Culture and Botany

Isabel Kranz

ONE OF THE MAIN CHARACTERS IN VANESSA DIFFENBAUGH'S BEST-selling novel *The Language of Flowers* (2011), Elizabeth, explains the concept of a secret floral code to her foster daughter: "It's from the Victorian era, like your name. If a man gave a young lady a bouquet of flowers, she would race home and try to decode it like a secret message. Red roses mean *love*; yellow roses *infidelity*. So a man would have to choose his flowers carefully."[1]

According to a dictionary from the nineteenth century that Elizabeth uses, each flower stands in for a specific concept, and by combining several blooms, one can even compose an entire statement. In this way, it is possible to convey a secret message to a loved one by sending him or her an arrangement of flowers. The protagonists in Diffenbaugh's novel are attracted to this idea of a language without words since it promises them a means of communication free of all the problems that a natural language entails, such as misunderstandings, untruthfulness, lies, and so forth. Needless to say, this supposedly easy way of avoiding communication difficulties is doomed to fail in the end, albeit in a different way than the novel wants to make its readers believe.

Both within the story and its paratexts, Diffenbaugh goes to great lengths to recover the presumably lost language of flowers around which her novel's central intrigue revolves. Besides having her characters explain and employ this floral code, she includes a short dictionary with the most important expressions at the end of her book. Along with the novel, the publishing house Ballantine released *A Victorian Flower Dictionary: The Language of Flowers Companion,* written by Mandy Kirkby with an introduction by Vanessa Diffenbaugh. The novel and its accompanying handbook were extremely popular with readers worldwide. *The Language of Flowers* entered the New York Times best-seller list in the fall of 2011, and translations into German, French, and other languages were published almost concurrently with the original American edition, turning the novel into a global book club favorite.[2]

The impressive commercial success Diffenbaugh enjoyed comes as no surprise to someone interested in the history of the language of flowers. One need only take a look at the extensive bibliography compiled by the historian Beverly Seaton in her seminal study on the language of flowers to realize that, around 1840 in particular, books on the subject were very popular in Continental Europe and England as well as in the United States. However, it remains questionable whether the language of flowers ever translated into everyday practice, as is claimed in Diffenbaugh's book, or whether it was merely a publishing phenomenon, consisting mainly of keepsakes and coffee-table volumes.[3]

Diffenbaugh's novel from the early twenty-first century and the language of flowers books from the mid-nineteenth century display several common features that extend beyond the mere fact of a two-hundred-year-old idea being reintroduced in a contemporary novel. Both partake of a commercial publishing culture that is associated with sentimentality, femininity, and flowers. On a more fundamental level, they exhibit a similar understanding of the possibilities and limits of language that I will disentangle in what follows. I propose to do so by reading the popular idea of a sentimental flower language alongside a considerably more influential "language of flowers" that was established in the middle of the eighteenth century, namely, Carl Linnaeus's botanical systematics.

Even if in the popular and the scientific code flowers are central to the production of knowledge, these two "languages" seem very much at odds with one another at first glance. By establishing a new method of ordering the plant kingdom (taxonomy) as well as a new naming procedure (nomenclature), Linnaeus's aim was to systematize the study of nature. He chose flowers, or, more precisely, the reproductive organs of plants as the structural element by which comparison between different plant genera was made possible and a hierarchy of vegetative beings established. The binomial plant names that accompany his new system function as a scientific code enabling the exchange of botanical knowledge beyond national languages. By contrast, the sentimental flower language of the nineteenth century is based on the idea that one can communicate feelings by encoding them in floral signs. This language employing flowers, instead of words, is intended for intimate conversations between lovers. Hence, it aims to restrict the circulation of information, not to enlarge it.

Nevertheless, these two systems of meaning production are interrelated. My claim is that it is precisely by assigning flowers the role of ordering the plant kingdom, thus making them the key indicators for systematic and structured knowledge, that an esoteric undercurrent emerges in which flowers convey a surplus of meaning that cannot be contained. In particular, they become signs of affection, supposedly transmitting secret messages of intense emotions. From a historical constellation around 1750 that still has repercussions today, flowers emerge as meaningful signs that are claimed in the service both of science and of sentimentality, capable of encoding an exoteric as well as an esoteric knowledge that can be deciphered.

While almost all handbooks published in the nineteenth century claim that the language of flowers is a universal language that has been known throughout history, many among them evoke its Oriental origins. For the first time, a secret language using (not only) flowers is mentioned in a French book about the Ottoman Empire at the end of the seventeenth century that was immediately translated into English as *The Turkish Secretary, containing the Art of Expressing One's Thoughts, without Seeing, Speaking, or Writing to One Another; With the Circumstances of a Turkish Adventure as Also a Most Curious Relation of Several Particulars of the Serrail That Have Not Before Now Even Been Made Publick.*[4] The author of the treatise calls himself "Du Vignau" and has just recently been identified as the French writer and former secretary in Istanbul Édouard de La Croix.[5] As indicated by the convoluted title, *The Turkish Secretary* contains three parts: the depiction of the so-called *selam*, which consists of sending messages without the use of written words (Figure 10.1); a love story in which this secret code plays a decisive role; and a description of life in the Turkish harem.[6]

In Du Vignau's book, the *selam* is related to the expression of erotic feelings in Turkey, where the contact zones between men and women are severely limited, and it is thus difficult for lovers to communicate. In addition to these strict social norms, the majority of Ottoman subjects cannot read and write, according to Du Vignau, so in most cases, epistolary communication is difficult, if not impossible. Whereas the

Figure 10.1. Young woman in Constantinople handing over a bouquet of flowers that encodes a secret message to her lover outside her window. Frontispiece in *Le Sélam, morceaux choisis, inédits, de littérature contemporaine, orné de dix vignettes anglaises* (Paris: F. Astoin/A. Levavasseur, 1834), after a painting by Paul-Émile Destouches, engraved by William Ensom.

Turks lack the means of written expression, Du Vignau asserts that they are endowed with an excess of feelings: "they seem to be more susceptible of Love, than any other of the *Eastern* Nations, where the Commerce of Women is equally difficult."[7] This discrepancy has supposedly led the Turks to invent a secret love code that is not based on writing but on the arrangement of objects: "all these Things which the *Turks* call *Selam* in this use, that is to say Safety, or desire of Peace, have their Naturall or Allegoricall Signification and Worth, insomuch that a little Packet of about an Inch bigg, if you have a regard to what it contains, composes a very Expressive Discourse, which is understood by the Interpretation of the name of Each thing they send."[8] The objects used are everyday items such as flowers, fabrics, and herbs. Whole objects or parts of them (such as a half-eaten almond and a piece of silk) are wrapped in a little bundle, forming a letter without the use of writing. The meanings ascribed to these objects are based on words rhyming with their Turkish names. So, for example, a piece of sugar, which in Turkish is called *şeker,* will then stand in for the sentence "Seni madem tcheker," meaning "My heart is longing for you."[9] In order to decipher a *selam,* one therefore first of all needs to know the name of the object one receives. The name then forms the basis for the rhyming word, which, in turn, will stand at the end of the sentence that is the message of the respective object, a poetic principle called *bouts-rimez* (rhyming ends).[10] Several objects can be combined to form a longer message. As announced in the book's title, this secret code thus presents a means of circumventing both vocal and written expressions. What is not lost, however, is the link to verbal language (i.e., the close connection to Turkish).

It has been claimed that the *selam* depicted in Du Vignau's little treatise makes use of many different objects and can therefore not be considered a precursor of the language of flowers popular in the nineteenth century.[11] It is true that of the 191 objects listed in Du Vignau's dictionary, 26 belong to the category of fruits and vegetables (including nuts), 17 are dried herbs and spices, and 39 are different sorts of fabric and clothing items (among them, 16 kinds of silk, a finely grained register of fabrics for which we lack the precise vocabulary today). One can even find wood and marble. Strictly speaking, only 16 of the items are flowers: [12] "Ambrette-flower," "Aneimony," "Orange-flowers," "Crows-toe" (or "Jacint"), "Jasmin," "Lilly," "Dazy-flower," "Liricumfancy" (lily of the valley), "Narcissus," "Pink-flower" (carnation), "Paunsy-flower," "Rose," "Marygold," "Tubereuse-flower," "Tulip," and "Violet-flower."[13] The signs used in the Turkish love code are thus obviously not limited to flowers and blooms.

Nevertheless, it is worthwhile to include the *selam* in a genealogy of the language of flowers for (at least) two reasons: first, because the nineteenth-century dictionaries explicitly name it as their precursor; second, and more important, because the code's restriction to flowers around 1800 is not an argument against considering similarities to the *selam* but rather an incentive to find out just why such limitation took place.

In order to answer this question, we need to trace how the idea of a language of flowers developed. Throughout the eighteenth century, references to the Turkish language of love are scarce. After 1688, the secret code is mentioned prominently in Mary Wortley Montagu's account of her experiences in Constantinople, an early document of women's lives in the Ottoman Empire as seen through the eyes of a Western woman. In a letter dated March 16, 1718, she makes a reference to the *selam* (without, however, naming it thus): "I have got for you, as you desire, a Turkish Love-letter, which I have put in a little Box, and order'd the Captain of the Smyrniote to deliver it to you with this letter."[14] Almost thirty years after Du Vignau's *Secretary,* the concept of a "Turkish love letter" was already well known to Montagu's English correspondent— apparently, information about it had spread.[15] Montagu mentions the heterogeneous nature of the materials that can be used: "There is no colour, no flower, no weed, no fruit, herb, pebble or feather that has not a verse belonging to it"—but broadens the code's scope in terms of content: "and you may quarrel, reproach, or send Letters of passion, friendship, or Civillity, or even of news, without even inking your fingers."[16] In all later references, however, the semantic range of the *selam* stays limited to emotions.

After Montagu, the *selam* is briefly cited in Jean-Jacques Rousseau's *Essay on the Origin of Languages,* begun in 1755 and published posthumously in 1781,[17] and there are several entries about it in French dictionaries from the 1770s onward.[18] Almost one hundred years after Montagu's depiction of the *selam,* the Viennese Orientalist Joseph von Hammer-Purgstall responds to it, claiming that this secret code was only used in

Figure 10.2. The three flowers depicted each represent a symbolic concept (the rose stands in for *beauty*, ivy for *constant friendship*, and myrtle for *love*), thus forming a bouquet that conveys the message "To beauty, friendship, and love." Frontispiece and title of Madame Charlotte de Latour's *Le langage des fleurs* (Paris: Audot, ca. 1820). (Photo credit: Deutsche Nationalbibliothek, Leipzig)

the Turkish harems among the women for their *"lesbian* declarations."[19] In terms of plot, the *selam* then figures prominently in a melodrama by the French playwright René-Charles Guilbert de Pixérécourt, *Les ruines de Babylone* from 1810.[20] And in the notes on Johann Wolfgang von Goethe's collection of poems inspired by the Persian poet Hafiz, "West-östlicher Divan" (1819), one can find a whole section devoted to the exchange of flowers and signs ("Exchanging Flowers and Symbols").[21]

In 1819, however, a book was published in Paris that is generally considered the model for most of the dictionaries on the language of flowers popular in the nineteenth century: *Le langage des fleurs*, written by a Madame Charlotte de Latour (Figure 10.2).[22] As with Diffenbaugh's best-selling novel,[23] Latour's flower dictionary was quickly translated into many languages. A German translation appeared in 1820; Italian and Spanish versions, as well as an early English adaptation, soon followed.[24] But translations were only part of Latour's success story. Her book initiated a fashion for the language of flowers that finds its expression in the publication of many supposedly original versions of the secret floral code.

This abundance of dictionaries in Latour's wake is central to Diffenbaugh's aforementioned novel. When the protagonist Victoria falls in love with the flower seller Grant, she learns that the language of flowers is not, as her foster mother has put it, "nonnegotiable." Quite the contrary: it relies solely on a prior convention. One day, Victoria receives a flower that she cannot find in her handbook. At her local library, she finds out that her dictionary is not the only one available but that there are many handbooks for the language of flowers with often contradictory associations between flowers and their meaning. The security once promised by the floral code is gone. Given the great number of dictionaries available, misunderstandings are therefore not only possible but quite probable.[25]

Convention also regulated the meaning of the *selam* messages in Du Vignau, from which Latour, however, departs in two ways: first, the secret language of love that she proposes relies solely on flowers as signs and not on other kinds of material; and second, the attribution of meaning to the floral signs is not regulated by rhymes but follows a symbolic structure. Both changes can be related to a major shift in what Michel Foucault in *The Order of Things* calls the *episteme* of the Enlightenment and, in particular, a change in how language was perceived, as we shall see later in this chapter. But let us first examine Latour's version of the language of flowers in more detail.

LE LANGAGE DES FLEURS

Le langage des fleurs consists of several parts.[26] It starts with a preface providing some historical background on the language of flowers, followed by a narrative section that is ordered according to the seasons and presents plants during their time of bloom, and ends with its centerpiece: the dictionary of flowers and their significations. In addition to these three central components, we find a verbalized version of Linnaeus's flower clock. In her introduction, Latour situates the language of flowers in a longer tradition of sign languages. She mentions, among other things, that "the Chinese" have an alphabet that consists entirely of flowers and plants,[27] and also evokes plant motifs in Egyptian hieroglyphs as a similar way of attributing meaning to signs instead of words.[28] Du Vignau is not cited explicitly, but Latour claims that she has borrowed most of the flower meanings from "the Orientals."[29] These and other historical references are employed to substantiate the claim that the language of flowers is as "old as the world," therefore rendering Latour's version not a new invention but merely an actualization.

Where Du Vignau, after explaining its basic rules, presents the *selam* in action by telling the story of two lovers communicating through object messages, Latour offers nothing of the kind. The second part of her book is devoted to single plants, starting with grass in spring and ending with the violet (the sign for modesty). The paragraphs about the respective plants are heterogeneous in style. They consist of elaborate depictions of bucolic scenes, as well as information about the history of certain plants or other plant-lore-related issues, and are interspersed with citations from poems and other literary texts.

The centerpiece of Latour's volume is a bidirectional translation dictionary in two parts comprising slightly more than three hundred entries. The first section lists the meanings one wants to convey and the flowers used to express them, providing some background on why a certain flower stands in for a concept.[30] The second list gives their respective meanings in alphabetical order. With a few exemptions, the messages consist of nouns such as "absence," "desire," or "frugality," some of which are further qualified by adjectives (such as different kinds of love from "pure and lively" to "platonic" to "voluptuous"); very few are further explained (such as "egotism": "You only love yourself"). Only a short list of phrases is included, half of which begin with the pronoun "I" and express feelings ranging from "I am burning" to "I declare war on you" to "I have lost it all." The other half starts with the addressed "you" accordingly and presents statements from "your charms are traced in my heart" to "you give me death."

Obviously, the language of flowers is no universal language but rather a very special code. It is restricted in both the semantic range of meanings conveyed (emotions, most of them referring to interpersonal relations) and the signs used to express them (mostly flowers with some flowering plants and herbs).[31] The choice of the latter is related to Latour's audience. As the ideal reader of her encyclopedic elaborations, the French writer depicts a young girl, uninitiated into the ways of the world, whose sole interest lies in the study of nature.[32] The language of flowers is particularly apt for her since she is interested in botany, which, by the end of the nineteenth century, "had become known as the feminine science *par excellence.*"[33] Besides being interested in the world of plants, the girl Latour envisions is experiencing the first budding feelings of love. Since, from Du Vignau onward, the language of flowers had been characterized as particularly useful for the exchange of secret messages between lovers, this makes it the ideal code for expressing a young girl's longings.

It is important to note that the social rules governing the commerce between the sexes, even if they were almost as strict in early nineteenth-century France as in Du Vignau's Constantinople, are not mentioned as the most important reason for choosing flowers as signs in Latour. Rather, there is some innate quality of flowers themselves that is credited as the reason why they are used in a secret code. A later handbook explains that flowers are considered to be particularly close to female subjectivity in general and to the sentiments of young girls in particular: "As soon as a delicate thought has started budding within the soul of a young girl, before she even dared to confide in the one who made her dream in such a way, the young girl has had to take a flower as her confidant. In this moment of her life, she herself is as innocent as the flower. The flower is therefore a companion who can understand her. Thus, the delicate desire finds its emblem."[34] Flowers, those aesthetically pleasing objects that the botanist John Ray at the end of the seventeenth century described as "the most delicate part of the plant, fleeting, conspicuous in colour and shape or both, coming before the fruit,"[35] are here parallelized to the sentimental thoughts of a young girl. It is suggested that the young girl and the flower are alike in their undiscerning innocence. Whereas the flowers are first personified as the confidants of the girl's thoughts,[36] they are then turned into the *means* by which the girl communicates.[37] In the context of the language of flowers, flowers are therefore clearly gendered as feminine signs. The supposed femininity of the flowers—which contradicts the botanical findings that flowers contain both the female and the male reproductive organs of plants—is related to their frailty and beauty: "Putting to action beings as fresh and graceful as flowers in order to represent words and ideas seems to me something better than the mere alignment of words in a dictionary. . . . It means turning nature herself into an enormous book whose lightest creatures are the living words."[38] Whereas the Oriental *selam* started out as a way of circumventing writing, the language of flowers in its nineteenth-century version is indicative of what Jacques Derrida in *Of Grammatology* has termed the *phonocentrism* of Western philosophy, namely, the primacy of the spoken word over writing. By symbolically standing in for a whole concept and by virtue of being living things, flowers can supposedly achieve more than mere writing. They are "better than writing," according to Latour, and therefore preferable to written signs.[39] This falls in line with the widespread idea that writing (*écriture*) is a secondary notation for a prior meaning that is only afterward laid down in letters.

By supplanting the cold letters of the alphabet with living flowers, the old metaphor of nature as a book to be read is given an interesting twist. However appealing this variation may be, such reasoning has very obvious limits. Each spring, Latour underlines, the signs used in the secret floral code, are rejuvenated by nature. This is a euphemistic description of the fact that in nature (and most certainly at the beginning of the eighteenth century, when greenhouses were not as common as they are today) flowers only bloom at a certain period of the year, even though a dictionary like Latour's presupposes their availability at all times. The feature that qualifies flowers as perfect signs for love—namely, their "freshness"—is also that which renders them impracticable as signs. It is therefore implied from the beginning that one cannot use real flowers as signifiers but has to resort to either dried specimens (which also lack the quality of aliveness) or to some representation, be it drawings or, more likely, writing (i.e., spelling out the name of the respective flower instead of presenting the real object).[40] Writing is, therefore, not avoided in the language of flowers but once again proves to be the basis for this—as for any other—system of communication.

Besides its limited set of signs, the language of flowers is severely restricted in terms of its subject matter. Just like Du Vignau's *selam* at the end of the seventeenth century, the nineteenth-century version is marked as a language of emotions only.[41] Even though Latour allows for its use in matters of friendship and motherly love (among others), its main area of application remains the expression of feelings between the sexes that Latour refers to as "this pure and chaste love."[42] What is striking is the complete elimination of sexuality and reproduction from this code, since the decisive shift that occurs between the late seventeenth-century idea of a Turkish love code and the sentimental *langage des fleurs* of the early nineteenth century is precisely the recoding of flowers by Carl Linnaeus. Some eighty years before Latour, Linnaeus in his *Systema Naturae* first proposed that flowers be considered the most important part of a plant when it comes to order. Starting with him, flowers occupied a central position in what would be called *taxonomy,* the ordering of the plant kingdom in different groups, at the beginning of the nineteenth century.[43] The establishment of classes, orders, families, genera, and species relied on the respective number, proportions, and positions of the male and female parts of the plants' flowers.[44] This way of ordering the plant kingdom has been described as the Linnaean sexual system.

In the language of flowers following Latour, the reproductive organs of plants are stripped of their botanical function and recoded. Sexuality is replaced by sentiment and issues of reproduction are reserved for a later time, when feelings are successfully integrated into societal expectations (i.e., after marriage).[45] Thus, the language of flowers reinserts a vocabulary of love into the botanical register that since Linnaeus seemed only to be concerned with sexuality: "Why should botany not have its graceful language, its idiom of pleasures and pains of the heart, in a word, its vocabulary of love?"[46] From this point of view, the sentimental language of flowers and botany complement each other, encompassing both the mechanical, reproductive aspect of sexual relations and the realm of emotions and feelings related to them. However, seen from today, this attribution of spheres cannot be upheld in such a clear-cut way. Even if Linnaeus's proposition to take the reproductive organs of plants as the marker by which to establish different plant genera was met with the reproach of sexualizing nature in an inappropriate way, Linnaeus hardly transcended the bourgeois gender concepts of the time himself, placing "plant sexuality . . . almost exclusively within the bonds of marriage," as the excellent studies of Londa Schiebinger and others have shown.[47] This can be clearly seen by examining the passage where

Linnaeus explains the parts of the flower in the chapter on sex in his *Philosophia Botanica*:

> 146. Therefore the CALYX is the *bedroom*, the COROLLA is the *curtain*, the FILAMENTS are the *spermatic vessels*, the ANTHERS are the *testicles*, the POL-LEN is the *sperm*, the STIGMA is the *vulva*, the STYLE is the *vagina*, the [VEG-ETABLE] OVARY is the [*animal*] *ovary*, the PERICARP is the *fertilized ovary*, and the SEED is the *egg*.[48]

So both the sentimental language of flowers and the new botanical systematics played into stereotypical gender notions as well as into bourgeois ideals about sexuality prevalent in the early nineteenth century.[49]

But taxonomy was only half of Linnaeus's program. The question of how to arrive at the correct name of a plant is the other central problem he treats explicitly later in his *Philosophia Botanica* (1751):

> 151. The FOUNDATION of botany is two-fold, *arrangement* and *nomencla-ture*. . . . Arrangement is the foundation of nomenclature. Knowledge of botany bears on these two hinges; thus all plants become known in a single year, at first sight, with no instructor and without pictures or descriptions by means of a stable recollection. Therefore anyone who knows this is a botanist, and no one else is.[50]

Linnaeus suggests to learn as much as possible about plants by closely observing them throughout the year, a directive echoed in Latour's presentation of flowers according to their blooming periods. Time and experience form the basis upon which nomenclature can be built, since being able to identify a plant allows the researcher to position it within the arrangement of other plants. In order to make any kind of exchange about this knowledge possible, however, it is indispensible to apply a precise name to the respective plant. Without the names, all is in vain, as Linnaeus makes clear in a later passage: "If you do not know the names of things, the knowledge of them is lost too."[51]

Linnaeus's main achievement lies in having established a new binomial system for the naming of plants that consists of the designation for the plant genus accompanied by a modifier detailing the species. In pre-Linnaean times, plant names were not necessarily less precise but usually much longer since the modifying part could consist of complete phrases. Brevity is, therefore, one of the advantages of the new system, as Linnaeus himself states in the preface to his *Genera Plantarum* published in 1737, the only text in which he explains his taxonomic method.[52] Post-Linnaean botanical names are units of condensed knowledge, allowing for the concise placing of a respective plant within the whole system of flora. In a different way than Linnaean taxa, plant names in the sentimental language of flowers also condense information. In Du Vignau, knowing the name of the object is the first step in de- or encoding a message; in Latour and her followers, those using the code can only do so if they can properly identify and name the respective flowers (if they have some basic botanical knowledge). In the sentimental flower language, plant names encapsulate the meaning one attempts to convey, which can consist of only one word, but also of a phrase or a whole sentence.

Yet, although it is necessary to know one's plant names in order to use the sentimental language of flowers, Linnaean taxonomy is not used in Latour's handbook.[53] Instead, we find the French names of the respective flowers (such as *marguerite des prés, gui commun,* or *anemone*). Most likely, Linnaean binomial names (which were the standard in botany by the early nineteenth century) are deemed too technical for this kind of popular writing. Relying on the vernacular language certainly has the advantage of appealing to all kinds of readers, not only the ones versed in botanical nomenclature. The French names that are used, however, also impact the meaning these floral signs come to stand in for, as later dictionaries remark:[54] "Just as the flora varies according to region and its ensuing nomenclature (the *souci* [French for marigold, also meaning "sorrow"], for example, does not evoke the same idea in German), it follows that each country has a particular *language of flowers.*"[55] By relying on vernacular names instead of botanical taxa, the language of flowers cannot attain the status of a universal language to which it aspires. Rather, it reveals itself as a code that is deeply entrenched in the respective national language. This fact is duly noted by Georges Bataille in his seminal essay "The Language of Flowers" (French original in 1929) when he remarks that if "pissenlit" (the vernacular French name for "dandelion," literally, "piss in the bed") signifies "extension," the reason is not hard to find.[56] The relation between signifier and signified in the language of flowers may therefore seem arbitrary in some cases; in most, however, the names of the plants already prefigure the meaning they are meant to convey.

However, as anyone interested in botany knows, Linnaean taxonomy and the systematics that followed after it, although intended to replace several imprecise national naming systems with a universal one, have also not done away with national or even regional plant names. Popular plant names continue to be in use alongside the universal botanical taxa. The relations between these two systems are rather complex, since designations in one language cannot always be easily translated into the other. For one, the botanical system is much more elaborate than the naming practices of laypeople, as it assigns a name and place to every single plant variety that exists on earth. Therefore, there are many more taxa than vernacular names. Popular names are not measured against the same standards of precision as botanical names. Neither does one name always only refer to what is considered a single botanical genus or species, nor is the existence of several names for one and the same plant considered a problem.

The two nomenclatures of botanical taxa and vernacular plant names are not always compatible, but the distinction between them is also not as clear-cut as one may surmise at first sight. Both ways of plant naming are informed by popular knowledge. Morphological features, known medicinal powers, preferred habitats, as well as the names of mythical figures, have been fundamental in establishing botanical taxa despite Linnaeus's effort to establish strict rules, such as that only botanists can apply names to plants since "private individuals have applied absurd names."[57] His directive that "generic names that are identical with the terms used by anatomists, pathologists, healers and artisans, should be dropped"[58] was not followed in many cases. The sentimental language of flowers and botanical nomenclature are, therefore, built upon a similar basis of popular knowledge about plants and their respective names. This repertoire, however, functions differently in the respective floral codes. Whereas, in the sentimental flower language, it directly influences how the code works by prefiguring meaning in the respective languages, at least in quite a few instances, in botany it points toward the history of the discipline and, in particular, to its roots in medicine, as well as to the fact that despite its universal claim after Linnaeus, until then botany had always been grounded in local circumstances. National languages and a universal code, be it in the form of the sentimental register of love or the systematic naming procedures of botany, are therefore inextricably linked.

Both flower codes' claim to universality, however, has its foundation in a similar understanding of language. For Michel Foucault, in his seminal study *The Order of Things*, Linnaeus's meticulous strategies for purifying the study of nature from earlier nonsystematic approaches are indicative of a deeper shift in the Western *episteme* that is characterized by a change in the status of language. Whereas language before the eighteenth century was understood to be on the same level as the things it designated, it then detached itself from the objects it refers to in what Foucault calls the "Classical age" (from the middle of the seventeenth century until 1800). Language was then understood to hold the power "of providing adequate signs for all representations, whatever they may be, and of establishing possible links between them."[59] Language was capable of representing something outside of it (such as "nature"), and, in order to be able to do so, it had to become transparent.

This understanding of language as something external to the objects it designates made the grammar of nature that is Linnaeus's ordering of the realm of flora possible in the first place. It is echoed in a somewhat less obviously scientific, but nevertheless also systematic, way in the language of flowers books, which registered the different kinds of emotions one could express through floral signs. Linnaeus's grammar of nature is here countered with a grammar of feelings supposedly absent from his botanical endeavors, which therefore needed to be supplemented. Both codes share a belief in the possibility of classification—of nature's beings and humankind's emotions, respectively.

These two languages of flowers are of rather different historical relevance. Whereas Linnaeus's reformulation of botanical standards was nothing short of a revolution in science—even if his sexual system is no longer in use—the sentimental flower language remains until today a popular phenomenon with little to no importance beyond the sphere of entertainment. Within this realm, however, it is still successful in the marketplace. If, contrary to his rival, the French naturalist Buffon, Linnaeus was never a commercially successful author, probably due to his rather dry and aphoristic style of writing,[60] writers of language of flowers books in Latour's tradition are as successful today as their French predecessor was in the early nineteenth century.

NOTES

1. Vanessa Diffenbaugh, *The Language of Flowers* (New York: Ballantine, 2011), 29, italics in original.

2. The novel appeared at number 13 on the *New York Times* best-seller list (hardcover fiction category). See "Best Sellers," *New York Times*, September 18, 2011, http://www.nytimes.com/best-sellers-books/2011–09–18/hardcover-fiction/list.html.

3. This is Beverly Seaton's claim in her comprehensive study *The Language of Flowers: A History* (Charlottesville: University Press of Virginia, 1995).

4. Du Vignau, sieur de Joanots, *The Turkish Secretary, containing The Art of Expressing ones Thoughts, without Seeing, Speaking, or Writing to one another . . .* (London: printed by J. B. and sold by Jo. Hindmarsh at the Golden Ball over against the Royal Exchange and Randal Taylor at Stationer's Hall, 1688). All subsequent quotes are taken from this edition. The original French title is *Le Secrétaire turc, contenant l'art d'exprimer ses pensées sans se voir, sans se parler & sans s'écrire, avec les circonstances d'une avanture turque, & une relation très-curieuse de plusieurs particularitez du Serrail qui n'avoient point encore esté sceuës* (Paris: Michel Guerout, 1688).

5. John-Paul Ghobrial, *The Whispers of Cities: Information Flows in Istanbul, London, and Paris in the Age of William Trumbull* (Oxford: Oxford University Press, 2013), 4–5. I will keep referring to the pseudonymous Du Vignau as the author.

6. The first study on Du Vignau's *selam* to my knowledge was Gerhard F. Strasser, "'Lettres muettes, ou la maniere de faire l'amour en Turquie sans sca voir nÿ lire nÿ escrire:' Manuskript und Druck einer türkisch-französischen 'Liebes-Chiffre' an der Pforte," in *Opitz und seine Welt: Festschrift für George Schulz-Behrend zum 12. Februar 1988*, ed. Barbara Becker-Cantarino (Amsterdam: Rodopi, 1990), 505–23.

7. Du Vignau, *The Turkish Secretary*, 1–2, italics in original.

8. Ibid., 3.

9. Alain Grosrichard rightly observes that the reliance on names and rhyming words puts the *selam* closer to the rebus than to the hieroglyphs as is often claimed in the language of flowers books; see, for example, Alain Grosrichard, *The Sultan's Court: European Fantasies of the East* (London: Verso, 1998), 174.

10. See, for example, Erika Greber, *Textile Texte: Poetologische Metaphorik und Literaturtheorie. Studien zur Tradition des Wortflechtens und der Kombinatorik* (Köln:

Böhlau, 2002); and Erika Greber, "Metonymy in Improvisation: Pasternak, Mayakovsky, Jakobson and their 1919 Bouts Rimés," in *Eternity's Hostage: Selected Papers from the Stanford International Conference on Boris Pasternak, May, 2004, Part I*, ed. Lazar Fleishman (Stanford, CA: Department of Slavic Languages and Literatures, Stanford University, 2006), 193–209.

11. See, for example, Seaton, *Language*, 37–38; and Brent Elliott, "Le langage des fleurs au XIXe siècle," in *L'Empire de Flore: Histoire et représentation des fleurs en Europe du XVIe au XIXe siècle*, ed. Sabine van Sprang (Brussels: La Renaissance du Livre, 1996), 307–17 [311].

12. I have left out plants that are regularly found in language of flowers dictionaries such as basil and other herbal plants that are not usually called flowers.

13. Du Vignau, *Turkish Secretary*, 47–50.

14. *The Complete Letters of Lady Mary Wortley Montagu*, vol. 1, *1708–1720*, ed. Robert Halsband (Oxford: Clarendon Press, 1965), 388. This instance is also cited in Seaton, *Language*; Jack Goody, *The Culture of Flowers* (Cambridge, UK: Cambridge University Press, 1993), 233; and Srinivas Aravamudan, *Tropicopolitans: Colonialism and Agency, 1688–1804* (Durham, N.C.: Duke University Press, 1999), 168–69.

15. Even if the letters were in fact not addressed to real people but formed a fictional epistolary exchange instead, as the editor Robert Halsband suggests, the *selam* is here introduced as a known phenomenon.

16. Montagu, *Complete Letters*, 389.

17. "Salaams are any number of the most common objects, such as an orange, a ribbon, a piece of coal, etc., the sending of which conveys a meaning known to all lovers in the country where this language has currency." Jean-Jacques Rousseau, *Essay on the Origin of Languages*, in *The Discourses and Other Early Political Writings*, ed. Victor Gourevitch (Cambridge, UK: Cambridge University Press, 1997), 247–99 [251].

18. The earliest citation I could find is in *Dictionnaire universel françois et latin, vulgairement appelé Dictionnaire de Trévoux contenant la signification et la définition des mots de l'une et de l'autre langue* (Paris: Compagnie des Libraires Associés, 1771), 629.

19. M. Hammer [Joseph von Hammer-Purgstall], "Sur le langage des fleurs," in *Annales des voyages, de la géographie et de l'histoire . . .* (1809), 346–58 [349]; italics in original, my translation.

20. See, for example, Isabel Kranz, "Die stumme Sprache der Blumen: Selamographie in Pixerécourts *Les Ruines de Babylone, ou le Massacre des Barmécides* (1810)," in *Das Melodram: Ein Medienbastard*, ed. Bettine Menke, Armin Schäfer, and Daniel Eschkötter (Berlin: Verlag Theater der Zeit, 2013), 75–95.

21. Johann Wolfgang von Goethe, "Exchanging Flowers and Symbols," in *West-East Divan: The Poems, with "Notes and Essays"; Goethe's Intercultural Dialogues*, trans. Martin Bidney and Peter Anton von Arnim (Albany: State University of New York Press, 2010), 229–31.

22. The author's name is most likely a nom de plume for Louise Cortambert, the wife of a French geographer, as revealed in Joseph-Marie Quérard's *Les supercheries littéraires dévoilées: Galerie des écrivains français de toute l'Europe qui se sont déguisés sous des anagrammes, des astéronymes, des cryptonymes, des initialismes, des noms littéraires, des pseudonymes facétieux ou bizarres, etc., découverts ou non* . . . (Paris: Chez l'auteur, 1865), 674.

23. Diffenbaugh's reference book is not Latour but Henrietta Dumont, *The Floral Offering: A Token of Affection and Esteem; Comprising the Language and Poetry of Flowers* (Philadelphia: H. C. Peck & Theo. Bliss., 1851), an American version of the language of flowers that was quite popular in the United States in the 1850s and 1860s, according to Seaton.

24. Fredric Shoberl's translation, the standard English edition of Latour, is, while true to the original in many ways, not a word-for-word translation, as the author himself concedes: "though this work is founded on a small French volume, yet, from the alterations which have been introduced, it cannot, strictly speaking, be called a translation." Frederic Shoberl, *The Language of Flowers, with Illustrative Poetry* (Philadelphia: Lea & Blanchard, 1839), 8. The preface is heavily edited, for example.

25. In the novel, the two young lovers solve their communication problems by the only way reasonable. They establish their own version of the language of flowers with a private dictionary so that both parties involved in the exchange of messages use the same medium of reference. From a universal language of love with fixed meanings, the language of flowers is therefore transferred into the realm of a private language, open only to the two people sentimentally involved.

26. The oldest edition I was able to consult dates from around 1820 and is held at the German National Library (Deutsche Nationalbibliothek) in Leipzig. Since there is no publication date included, this edition is most likely to be identical to the original one from 1819. In content and pagination it complies with the digital edition I will be citing from, published under yet another pseudonym, Aimé Martin [Madame Charlotte de Latour (Louise Cortambert)], *Le langage des fleurs* (Brussels, 1830).

27. The idea of a floral alphabet—i.e., replacing each letter with a certain flower—is also mentioned in Louis-Antoine Caraccioli, *Le livre à la mode* (Verte-Feuille [Paris]: Impr. du Printems [Duchesne], 1759), xvi–xvii.

28. Latour, *Langage,* 7.

29. Ibid., 11.

30. Such justification for the coupling of flower and meaning cannot usually be found in other dictionaries. See, for example, Beverly Seaton, "A Nineteenth-Century Metalanguage: *Le Langage des Fleurs,*" *Semiotica* 57, no. 1/2 (1985): 73–86 [75].

31. And yet Latour does not address this and the many other paradoxes that her sentimental flower language entails (such as the contradiction between being both a universal and a secret language at the same time).

32. Latour, *Langage,* 5. Amy M. King, *Bloom: The Botanical Vernacular in the English Novel* (Oxford: Oxford University Press, 2003), focuses on this figure of "the female whose social and sexual maturation is expressed, rhetorically managed, and even forecast by the use of a word (bloom) whose genealogy can be traced back to the function of the bloom, or flower, in Linnaeus's botanical system" (4) without, however, engaging in depth with the language of flowers or spending sufficient time on problematizing Linnaeus's rigid gender concepts.

33. Londa Schiebinger, "The Private Life of Plants," in *Nature's Body: Gender in the Making of Modern Science* (New Brunswick, N.J.: Rutgers University Press, 2004), 11–39 [36], italics in original.

34. *Sélam, almanach fashionable* (Paris: A. Royer, 1843), 37–38; my translation.

35. As cited in *Linnaeus' Philosophia Botanica,* trans. Stephen Freer (Oxford: Oxford University Press, 2003), 68.

36. On the topic of flowers and personification, see, for example, Dorri Beam, *Style, Gender, and Fantasy in Nineteenth-Century American Women's Writing* (Cambridge, UK: Cambridge University Press, 2010), chapter 1, "Florid Fantasies: Fuller, Stephens, and the 'Other' Language of Flowers"; and Beverly Seaton, "Towards a Historical Semiotics of Literary Flower Personification," in *Poetics Today* 10, no. 4 (1989): 679–701.

37. The concepts of allegory, metaphor, emblem, and so forth, meticulously scrutinized by literary criticism, are used interchangeably in the nineteenth-century flower dictionaries.

38. *Sélam,* 40–41; my translation.

39. Latour, *Langage,* 6–7; my translation. In a telling passage, Latour calls flowers "interpreters of sweet feelings" (6), thus turning them into beings endowed with life.

40. This solution is proposed in a language of flowers book from the end of the nineteenth century: "Since it is not always easy to procure the flowers one needs to express the feelings that one is animated by and that one wants to convey through allegory, people have been seen to supplant them by writing the name of the flowers they would have used to form their bouquet if they had

had them at their disposal." *Almanach du langage des fleurs* (Paris: Delarue, 1870), 6; my translation.

41. Alexander Schwan, "'Blumen müssen oft bezeigen, was die Lippen gern verschweigen': Floriographie als Sprache der Emotionen,"in *Gefühle. Sprechen. Emotionen an den Anfängen und Grenzen der Sprache,* ed. Viktoria Räuchle and Maria Römer (Würzburg: Königshausen & Neumann, 2014), 199–221.

42. Latour, *Langage,* 6; my translation.

43. The first use of the term is recorded in French in Augustin-Pyrame de Candolle, *Théorie élémentaire de la botanique* (Paris: Déterville, 1813), 19.

44. Schiebinger, "The Private Life," 14–17.

45. It is quite telling when a later dictionary uses the word *passion* in this context: "flowers can only express matters of sentiment or passion; all that is purely material or insensible has not translation in this language" (*Sélam,* 63; my translation). According to Niklas Luhmann, the discourse of passionate love "absorb[ed] the revaluation of sexuality that occurred in the eighteenth century." Niklas Luhmann, *Love as Passion: The Codification of Intimacy,* trans. Jeremy Gaines and Doris L. Jones (Cambridge, Mass.: Harvard University Press, 1986), 10. Whereas Luhmann bases his theory mostly on epistolary novels, the language of flowers, this popular and sometimes contradictory phenomenon, leads us to a similar conclusion.

46. *Sélam,* 60–61; my translation.

47. Schiebinger, "The Private Life," 25; Alan Bewell, "'On the Banks of the South Sea': Botany and Sexual Controversy in the Late Eighteenth Century," in *Visions of Empire: Voyages, Botany, and the Representations of Nature,* ed. David Philip Miller and Peter Hanns Reill (Cambridge, UK: Cambridge University Press, 2011), 173–93. Erasmus Darwin's playfully erotic poem "The Botanic Garden" (1791) shows that Linnaeus's rather limited view of both plant and human sexuality was noticed in his time already; see, for example, Dahlia Porter, "Scientific Analogy and Literary Taxonomy in Darwin's *Lores of the Plants,*" in *European Romantic Review* 18, no. 2 (2007): 213–21.

48. Linnaeus, *Philosophia Botanica,* 105. However, the explanations following these definitions alternatively propose to call the calyx "the *lips of the cunt* or the *foreskin,*" so there some kind of hesitation on Linnaeus's part seems to have been at work.

49. Conversely, it has been argued, the sexualization of the plant kingdom made it possible for women to talk about sexuality by speaking about botanical matters. Cf. Bewell, "On the Banks," 175–76.

50. Linnaeus, *Philosophia Botanica,* 111.

51. Ibid., 169.

52. Staffan Müller-Wille and Karen Reeds, "A Translation of Carl Linnaeus' Introduction to *Genera plantarum* (1737)," *Studies in History and Philosophy of Biology and Biomedical Sciences* 38, no. 3 (2007): 563–72. Linnaeus's preference for concise expressions is, however, not limited to the admitted aim of his new method—the names—but also characterizes his own writing, which he makes clear is one of efficiency instead of rhetorical eloquence: "I have expressed my ideas with as few words as possible, caring more for weighty words than pompous and eloquent Latin phrases" (570).

53. This nonreception of Linnaean nomenclature does not hold true for more grammar-oriented language of flowers books such as B. Delachénaye's *Abécédaire de Flore, ou langage des fleurs, méthode nouvelle de figurer avec des fleurs les lettres, les syllabes, et les mots, suivie de quelques observations sur les emblêmes et les devises, et de la signification emblématique d'un grand nombre de fleurs* (Paris: Didot l'Aîné, 1811), in which the binominal names are cited. The absence of Linnaean names in the most important German handbook on the language of flowers, Daniel Şymanski's *Selam, oder die Sprachen der Blumen* (Berlin: Christiani, 1820), was even criticized in a contemporary review in the *Allgemeine Literatur-Zeitung* 322 (1820), no pagination.

54. This claim could be further substantiated by comparing handbooks in different languages as Beverly Seaton has done in a "Combined Vocabulary"; see, for example, Seaton, *Language,* 168–97.

55. Artaud de Montor, *Encyclopédie des gens du monde, répertoire universel des sciences, des lettres et des arts; avec des notices sur les principales familles historiques et sur les personnages célèbres, morts et vivans, par une société de savans, de littérateurs et d'artistes, français et étrangers* (Paris: Treuttel et Würtz, 1833–44), 196–97; my translation.

56. Georges Bataille, "The Language of Flowers," in *Visions of Excess: Selected Writings, 1927–1939,* ed. Allan Stoekl (Minneapolis: University of Minnesota Press, 1985), 10–14 [10].

57. Linnaeus, *Philosophia Botanica,* 169.

58. Ibid., 178.

59. Michel Foucault, *The Order of Things: An Archaeology of the Human Sciences* (London: Routledge, 2002), 94.

60. Müller-Wille and Reeds, "A Translation," 564.

ELEVEN

Phytographia
Literature as Plant Writing

Patrícia Vieira

Can the Plant Speak?

Humans have always been fascinated by the possibility that plants could share their stories. If they could converse, what would they tell us?[1] What language would they use and how would they describe their wordless existence? It is now well established that plants communicate, for instance, through biochemical signals, both among themselves and with other living beings, notably insects, in order to warn of danger, to attract pollinators, to repel potential predators, and so on.[2] But the plant tales that appeal to humans the most are not the ones that testify to the pragmatics of survival. We want to learn flora's innermost secrets that appear so hermetic to us, and to penetrate the core of plant being. What would plants say about themselves, about their environment, and, especially, what would they say about us?

Writers and artists have been at the forefront of attempts to translate plant stories into a language humans would understand. From the talking trees in J. R. R. Tolkien's fiction and plant narrators in more recent novels,[3] to installations that capture human–plant interactions, we have long endeavored to learn what vegetal beings convey. A revealing example is the art of Christa Sommerer and Laurent Mignonneau, whose works *Interactive Plant Growing* (1992) and *Data Tree* (2009) use a computer program to "translate" a plant's electrical signals into intelligible language.[4] Beyond the artistic realm, Cleve Backster's experiments of connecting plants to a polygraph to determine their reaction to various stimuli or, more recently, the "Midori-san" blog, "written" by a plant sitting in a Japanese coffee shop and linked to a computer program that interprets its sensations, show how spellbound we are by what plants have to tell us.[5] Perhaps our desire to hear what vegetal beings convey still harks back to the all-encompassing ideals of the Enlightenment. The light of rationality that was to illuminate the darkest recesses of a person's soul should now be extended not only to animals but also to plants. In the Enlightenment's push toward total visibility, which goes hand in hand with the dream of complete translatability, plant tales were simply awaiting their turn to be rendered fully available and intelligible to the human mind.[6]

But what would it mean to shed light into the soul of plants, so as to ferret out their accounts of themselves? Unlike humans, plants do not possess a hidden core buried deep into their psyche. As Goethe perceptively noted, the leaf, completely exposed to sunlight and to the elements, is the archetypal form of a plant, the rest of it being nothing more than a metamorphosis of this basic organ.[7] The intersection of Enlightened reason with the exposure of plant life that offers itself to the exterior world and reveals its riddles on the surface of its skin, both above and below ground, holds the promise to unravel key binaries that continue to plague contemporary approaches to nonhuman living beings. For how can we distinguish matter from form, action from thought, nature from culture, if we adopt the perspective of a plant?[8]

The Enlightenment aspiration toward full visibility conceals a dark underbelly. Does the human yearning to know the stories of plants not express a burning wish to dominate and possess the vegetal world? In our push to render everything and every being completely transparent, are we not obliterating what we are trying to know? It would serve us well to heed the warning of philosopher Emmanuel Levinas, according to whom knowing the other is tantamount to destroying her, him, or it (although the other was always a person for Levinas, I add "it" to the list of possible others, a pronoun that encompasses both plants and animals). Contrary to the Enlightenment's insatiable craving for knowledge, Levinas advocates respect for the other and her/his/its stories, which will always retain an irreducible aura of mystery.[9]

A more charitable understanding of human attraction to plant tales would take its cue not from the Enlightenment's fantasy of complete visibility but from more recent insights originating in postcolonial studies. One might ask, as a rejoinder to Gayatri Spivak's famous question about the subaltern, "Can the plant speak?" What would be the parameters of such an utterance? Would we be prepared to listen to flora's paradoxically silent speech? Or would we rather, as Spivak warned in the case of the subaltern, superimpose our thoughts, reasoning, and preconceived ideas, perhaps even in a well-intentioned manner, onto the plant?[10]

The analogy between the plant and the subaltern is clearly not a seamless one. After all, the subaltern is a human being endowed with an intelligible form of language and a worldview of his or her own. The problem is the failure to recognize the validity of the subaltern's claims to a specific mode of existence that is erroneously regarded as inferior. This predicament resulted in extreme brutality in colonial contexts, where colonizers, deaf to the stories of their subjects, insisted on foisting their own master narratives onto the different lands they occupied. Yet, the similarities between the subaltern and the plant are also striking. Relegated to the margins of Western thought, both categories have been posited as negative images of modernity's triumphant ideals.[11] At least since the time of the first European voyages around the coast of Africa, to India and to America in the fifteenth century, modernity has been charted as a crusade of civilization against barbaric customs and, at the same time, as an effort to tame a wild and unruly nature. Tropical flora that defied the domesticated seemliness of European landscapes was particularly singled out as a peril to be overcome as it stood in the way of the West's march of progress.[12]

I suggest that, following in the footsteps of postcolonial studies, we make an effort to interpret the stories of plants. Even more than in the case of the subaltern, however, this is a challenging endeavor. The tales of plants "as such" will always elude us, given that our relationship to flora is necessarily mediated by human sense perception, scientific knowledge, and an extensive cultural history that includes, among others, georgic and pastoral literature, as well as a large number of utopian and, more recently, dystopian writings. Still, our inability to fully abandon a human standpoint does not spell out the compartmentalization of humans and plants in spheres destined to remain apart. But how can we decipher the mute language of plants and immerse ourselves in their fables?

My tentative answer is that we avail ourselves of the notion of inscription as a possible bridge over the abyss separating humans from the plant world. All beings inscribe themselves in their environment and in the existence of those who surround them. The plants' inscription depends primarily on their physical configurations that shape both the contours of a landscape, as in the case of a tall cloud forest, in contrast to a savanna, and of their relation to animals, determined, for instance, by the color of a flower that attracts a given pollinator or a human being who admires the pleasing combination of shapes and hues in a bouquet. The study of plant modes of inscription in the biosphere is the domain of scientific research that strives to understand how they interact with other living and nonliving entities. Vegetal inscription in human lives, in turn, takes place, at a very basic level, through the food we eat, the spaces we inhabit, and the oxygen we breathe. In this chapter, however, I will work with a narrower notion of inscription. I will foreground the specific modes in which the vegetal word is embedded in human cultural productions, which I call *phytographia,* using literature—in this case, literature about the Amazon—as an example of the porous boundary between artistic portrayals of flora and the imprints left in texts by the plants themselves. The goal is not to argue for a radical break between the heterogeneous composite we tend, for simplicity's sake, to designate as "nature," and the equally vague concept of "culture." On the contrary, this chapter rests upon the premise that a continuum extends from plant to human forms of inscription, which necessarily interact and get entangled in one another.

Plant inscription is not synonymous with the cognate notion of plant agency. While plants are clearly not inert, unresponsive entities, positing agency in flora veers dangerously close to anthropomorphizing its behavior by using a model derived from human action to describe it. Such a move is here deemed unnecessary, for agency is nothing but a longing for inscription, which is tantamount to saying, a desire for being.[13] Rather than framing it as agency, inscription can be understood in terms of the Spinozan *conatus essendi,* the wish of all things to persevere in existence, a yearning that leaves traces in and through other entities. In what follows, I will turn to the traces of flora in literature, the remnants or remainders of plants' ongoing process of inscription, that is to say, of their very lives.

From *Signatura Rerum* to *Phytographia*

The understanding of the world as a complex chain of significations has deep roots in Western thought. Perhaps the most cogent prescientific enunciation of this idea is the doctrine of *signatura rerum,* widely espoused by alchemists and physicians and codified by German mystic Jakob Böhme in the book *The Signature of All Things* (1621). According to Böhme, all entities bear the mark of God, mediated by the different properties the Creator attributed to them. Each inner characteristic or essence of a thing is expressed in its outward shape, form, or signature: "Therefore the greatest understanding lies in the signature, wherein man . . . may learn to know the essence of all essences; for by the external form of all creatures, . . . the hidden spirit is known; for nature has given to everything its language according to its essence and form."[14] If "this is the language of nature, whence everything speaks out of its property, and continually manifests, declares, and sets forth itself for what it is good or profitable,"[15] humanity should simply learn how to understand correctly the signature of each being.

Böhme's doctrine was particularly pertinent as a guide to the relationship between humans and plants. Given that plant life bore the signs of its qualities in its signature, or outward form, human beings could easily uncover the best use of each tree, bush, herb, or flower by attending to the shape of a given plant: "therefore the physician, who understands the signature, may best of all gather the herbs himself."[16] The notion that certain plants were appropriate to treat an illness related to a given part of the body because they resembled this particular organ had been around at least since classical antiquity.[17] Böhme revealed the theological underpinnings of this ancient belief by positing the Maker as the ultimate guarantor of the veracity of the signature. God, like a proud artist, had left his indelible imprint upon even the smallest being of his creation, and humans placed their trust in these signatures precisely because they could be traced back to the will of a benevolent deity. The system of *signatura rerum* was therefore hierarchically organized, mirroring the late medieval view of a pyramidal creation. God stood at the apex—both as the origin and the crowning—of a string of properties that manifested themselves through their signatures onto the bodies of each being, while humans occupied the ambiguous position of both bearing God's signature and being the decoders of his marks upon the world.

One of Böhme's most compelling arguments was his insistence on the correspondence between inside and outside: "as the property of each thing is internally, so it has externally its signature, both in animals and vegetables."[18] The bodies of animals and plants (and also of humans) expressed who they were, in an ongoing commerce between inner and outer realities that resulted in the undoing of this very distinction. For if the shape of an entity expresses its essence, then that essence is already determined by the form. The division between inside and outside is particularly meaningless in the case of plants, which open themselves to the exterior world in their efforts to maximize the surface of their bodies exposed to sunlight. For flora, the signature *is* clearly the essence.

Another name for the plants' signature is their inscription in the world through their bodily manifestations. In fact, the Latin term *signatura rerum* can be read, following Böhme, as God's signature on things, or, alternatively, as the signature *of the things themselves*. If we were to remove the figure of the Creator as the root of all signatures, the system's hierarchical structure would collapse and we would be left with a multiplicity of signatures that express the mode of being of each thing. The complex of immanent signatures and their interrelations amounts to the language of things, of which the materially inscribed language of plants is a subset.

Walter Benjamin expounded his version of the language of things in an essay titled "On Language as Such and on the Language of Man," written in 1916. Similar to Böhme, he maintains that all entities are pregnant with signification: "There is no event or thing in either animate or inanimate nature that does not in some way partake of language. . . . This use of the word 'language' is in no way metaphorical."[19] The means through which things express themselves is their "more or less material community," which, in the case of plants, stands for their physical inscription in their environment. This material community, writes Benjamin, "is immediate and infinite, like every linguistic communication; it is magical (for there is also a magic of matter)."[20] He paints the portrait of an enchanted world, alive not for being populated by the spirits of animism but due to its unfolding in material inscriptions that are equivalent to the language of every being.[21]

It is therefore disconcerting to read in the same Benjaminian essay that "the languages of things are imperfect, and they are dumb" (67).[22] This is not only because things "are denied the pure formal principle of language—namely, sound" (67), but, more decisively, because they lack the ability to name both themselves and other entities. Benjamin, retracing Böhme's hierarchical view of the world, believed that human beings occupied a special position within language, since they alone possess the capacity of naming. To give each entity a name confers upon humankind the enormous power of ruling over the entire creation: "All nature, insofar as it communicates itself, communicates itself in language, and so finally in man. Hence, he is the lord of nature and can give names to things" (65).

Benjamin highlights the redemptive features of naming that liberate things from their enforced muteness and allow them to come into their own by letting the divine breath of creation trapped in them shine forth through human language. Yet, he also recognizes that to shackle nature to the vagaries of humanity is to do it a disservice. "It is a metaphysical truth," he writes, "that all nature would begin to lament if it were endowed with [verbal] language" (72). Nature silently laments—through its "sensuous breath" or through "a rustling of plants"—its yoke to humankind and mourns its powerlessness to name itself that enslaves it to human language. "To be named . . . perhaps always remains an intimation of mourning," he acknowledges, "but how much more melancholy it is to be named not from the one blessed paradisiacal language of names, but from the hundred languages of man, in which name has already withered" (73). If the helplessness of nonhuman entities was already manifest when they were subjected to naming in Eden, how much more despondency do they experience when named in the post-Babelic confusion of languages?

Benjamin hints at a fissure in his seemingly flawless edifice when he notes that the deepest reason for the things' "melancholy" and "deliberate muteness" is their "overnaming" by humans (73). After the Fall, humans lost touch with the sacredness of naming, and language turned into "empty prattle." This was the precondition for the subsequent "turning away" from and "enslavement of things," whose different modes of being are trampled upon when humanity considers them as simple tools or raw materials, a means to an external end (72). Nature's muteness is thus not only the cause of its servitude but also a form of resistance against its subjection to humans. However, while Benjamin admits to the shortcomings of humanity's relation to its others, he fails to take the next step of disentangling the language of things from its dependence on humanity.

Benjamin posits the move from the language of things to the language of man as the infinite task of translation, his version of the interpretation of each thing's *signatura*. While, in a prelapsarian world, this translation would be univocal, things can only be named imperfectly after our expulsion from paradise. Translation has a metaphysical import, namely, that of permanently striving toward the exact rendering of one language into another, an ideal that forever eludes fallen humanity. Still, for Benjamin as for Böhme, the relationship between humans and nonhumans only becomes possible at all because God ensures the adequateness of all significations: "The objectivity of this translation is, however, guaranteed by God" (70). At the origin of both humans and the rest of the world, brought into being by the same creative word, God bears responsibility for the correspondence between the name and the named.

There seems to be an alternative in Benjamin's text to the hierarchical model of translation, whereby an "imperfect language" of things, gives way to "a more perfect one" of human beings (70). Artworks do not aim to translate the language of things into that of humans, but, rather, to stage an encounter between the two: "art as a whole, including poetry, rests not on the ultimate essence of the spirit of language but on the spirit of language in things, even in its consummate beauty" (67). Artists, the demiurges of Romanticism, point in the direction of a non-hierarchical world, where all languages are equally valid and translation moves horizontally, rendering the language of one nonhuman or human being into that of another. Benjamin is fully aware of the utopian undertones of his take on art: "here we should recall the material community of things in their communication. Moreover, the communication of things is certainly communal in a way that grasps the world as such as an undivided whole" (73). Art partakes of the things' communitarian nature in its endeavor to bring together nonhumans and humans and make the broken, post-Edenic world whole again.

Benjamin singles out the fine arts as the ones more attuned to the language of things: "it is very conceivable that the language of sculpture or painting is founded on certain kinds of thing-languages. . . . We are concerned here with nameless, nonacoustic languages, languages issuing from matter" (73). There is, however, no compelling reason to exclude other art forms as potential propitiators of a communion between the languages of different entities. Literature can also set the stage for this encounter, open not only to the Bakhtinian heteroglossy of various human discourses but also to the convergence of nonhuman and human languages. As we shall see, *phytographia,* or plant writing, denotes one such encounter: the coming together of the wordless, physically inscribed language of plants with an aesthetically mediated form of human language in literature.

How to conceive of Böhme's *signatura rerum* without its religious trappings? Could Benjamin's language of things ever be on par with that of humans? What would be the contours of a language of plants and of plant writing? The path to addressing these quandaries takes us to the notions of arche-writing, trace, and *différance* developed by Jacques Derrida. The French philosopher positions himself against Western thought's bias in favor of spoken language, or phonocentrism, which creates the illusion that each utterance can be traced back to its origin by following the voice to its source in the body of a given human being. Moving away from this metaphysical fixation on presence, Derrida suggests, instead, that only a generalized inscription of entities and events in the world (and, in fact, of the world itself) creates the conditions of possibility for any form of language—be it nonhuman or human, spoken or written—to thrive. *Arche-writing* is Derrida's term for this generalized inscription and inscribableness. Drawing on the particular characteristics of the written word, which presupposes a spatial and temporal lag between the moment of enunciation and the time of reading, arche-writing opens "the possibility of the spoken word, then of the *'graphie'* in the narrow sense"[23] and, I would add, of the language of things and of *phytographia.*

Derrida demarcates arche-writing from previous understandings of the world and the beings in it as laden with meaning in that he posits an original dissemination of all inscriptions, or traces, which cannot be ascribed to a unified cause such as the Unmoved Mover, God, or Spirit in the tradition of Western philosophy. For Derrida "this trace is the opening of the first exteriority in general, the enigmatic relationship of the living to its other and of an inside to an outside: spacing."[24] *Différance* is the term he coined to express, simultaneously, the deferral of identity and the difference that always contaminates sameness, underlies all language and life itself, and opens the possibility of spacing and of time. He continues: "The outside, 'spatial' and 'objective' exteriority . . . would not appear without the grammè, without differance as temporalization, without the nonpresence of the other inscribed within the sense of the present, without the relationship with death as the concrete structure of the living present."[25] The ties binding the sayer to the said, the I to her actions, and even the I to herself, are always discontinuous, and otherness, failure, and the prospect of death inform all inscriptions.

Arche-writing is inherently violent, given that it stands for the breach, or tearing apart, of all unity and innerness, a breaking asunder and contamination by alterity that, as Derrida puts it, has *always already* happened before any beginning, birth, or even dream of inception could take place. "To think the unique within the system, to inscribe it there," he writes, "such is the gesture of the arche-writing: archeviolence, loss of the proper, of absolute proximity, of self-presence, in truth the loss of what has never taken place, of a self-presence which has never been given but only dreamed of and always already split, repeated, incapable of appearing to itself except in its own disappearance."[26] The violence of arche-writing is thus a creative one, since beings in the world are nothing more than their inscriptions and traces conceivable only through *différance*.

What is the place of *phytographia* within arche-writing? A Derridean answer to this interrogation would point out that, sessile entities par excellence, always tethered to a given place by their roots, plants are nevertheless the most widespread of beings, not only because they populate a large portion of our planet but also given that, through photosynthesis, they make life on earth possible. The inscriptions of all living beings in the world are, in a very literal sense, a kind of *phytographia,* enabled by the incessant work of plant life. Furthermore, plants not only deliver their seeds to chance, counting on the elements and on other animals for their dissemination in the same way as a piece of writing is often dispersed through circuitous channels, but they also share another central characteristic of written texts, namely, their iterability. Plants endlessly repeat parts of themselves by producing multiple leaves, flowers, and fruits, all sharing similar traits but also displaying minuscule differences. Vegetal life and inscription are thus eminently *graphic* and could be understood as the paradigmatic example of arche-writing.

The more restricted understanding of the term I am proposing here considers *phytographia* to be one of the modes of plant inscription, which, in turn, is embedded within the context of a broader arche-writing. *Phytographia* is the appellation of an encounter between writings on plants and the writing *of* plants, which inscribe themselves in human texts. At its most basic, this inscription has, throughout history, relied on papyrus, pencil, ink, paper, and countless other writing instruments. But the *phytographia* that will occupy us in the rest of this chapter, though beholden to the material substratum of writing, has a still more specific meaning. It does not depend exclusively on a writer and her sovereign authority to define plant being, which would amount to the resurrection of a naive form of realism, whereby the author purports objectively to depict the world. Nor does it rest upon a belief in a mystical communion with vegetal life that would take possession of the writer's soul and determine her prose. Rather, it stands for the literary portrayal of plants that is indebted both to the ingenuity of the author who crafts the text and to the inscription of plants in that very process of creation.

Phytographia can perhaps best be grasped by analogy with *photography*, the writing of light. In a photograph, the materiality of the things themselves interacts with light to create an imprint of reality, filtered through and molded by the artistic vision of the photographer. In their inscription in the environment, made possible by photosynthesis, plants already perform a proto-*photographia*. They use sunlight to create their material articulations in the world and, in doing so, imprint themselves in the biosphere, enabling the inscription of all other living beings in the process. Similar to their physical, *photo*graphic inscription in their surroundings, plants also leave impressions of themselves in human cultural creations, such as literature. *Phytographia* designates this communion between the *photo*graphic language of plants and the *logo*graphic language of literature. I see literature, the realm of the imagination, as a mediator in the aesthetic encounter with plants, knowing full well that the medium always colors the message and that the mediators themselves are all but evanescent.

In the last section of this chapter, I will discuss literary portrayals of flora as possible sites of *phytographia*. While any text on plants can potentially be *phytographic,* my focus will be on literature about the Amazon. By far the world's largest rainforest, covering an area that spans nine nations and housing a treasure trove of biodiversity, much of which still remains to be identified, the Amazon epitomizes the exuberance of flora in the global imaginary and, therefore, provides fertile ground for *phytographic* experiments.

AMAZONIAN LITERATURE AS PLANT WRITING

From the first wave of Spanish and Portuguese explorers, through the late eighteenth- and nineteenth-century naturalist adventurers, to more contemporary visitors, all of those who have traveled in the region marveled at the lushness, variety, and sheer immensity of Amazonian plants. Spanish Jesuit priest Cristóbal de Acuña, who crossed the Amazon basin from Quito to Belém in 1639, highlighted in his *New Discovery of the Great River of the Amazons* (1641) the fertility of the jungle and the abundance of food, a natural wealth that reminded him of the biblical paradise on earth.[27] Centuries later, the British naturalist Henry Walter Bates called the region a "naturalist's Paradise" due to the variety of its vegetation.[28] "Fancy if you can," wrote Bates in his *The Naturalist on the River Amazons* (1863), "*two millions of square miles of forest.* . . . You will hence be prepared to learn that nearly every natural order of plants has here *trees* among its representatives."[29] Bates's associate Alfred Russel Wallace described "the beauty of the vegetation, which surpassed anything I had seen before,"[30] and Richard Spruce, another British botanist, marveled at the "enormous trees" in the region, "crowned with magnificent foliage."[31] Already in the twentieth-century, Theodore Roosevelt, who traveled in the Amazon in 1913–14 after leaving office as president of the United States, was struck by the "immensely rich and fertile Amazon valley" and by its "magnificent," "splendid," and "impenetrable" forest.[32]

Despite consensus about the awe-inspiring plant life of the Amazon, however, portrayals of the region's flora have, for the most part, fallen back upon tired clichés. The most pervasive of these is the dichotomous depiction of vegetation either as reminiscent of earthly paradise or as a green hell. While Acuña and the botanists quoted above espoused, for the most part, an Edenic view of the forest, filled with natural wonders, others have emphasized the dangers of Amazonian nature. Brazilian writer Euclides da Cunha considered humans to be "impertinent intruders"[33] in the region, facing a "dangerous adversary," a "sovereign and brutal nature."[34] His friend Alberto Rangel characterized the Amazon as a "green hell" in a collection of short stories from 1908. The last narrative of the book describes the agony of an engineer from the South of Brazil who goes to the region hoping to acquire wealth and dies of malaria in the hellish jungle.[35]

Perhaps the sheer otherness of the Amazon intimidated travelers, who resorted to religious metaphors as a means to pigeonhole a foreign and potentially threatening vegetation and reduce it to familiar tropes. This is not to say that the paradisiacal/infernal view of the region is a mere figment of human imagination, completely detached from local plant life. On the contrary, this imagery testifies to (mostly) Western engagement with the area's vegetal existence and to our attempts at translating what the forest tries to tell us, in its own cyphered language. Still, beholden to a hierarchical worldview, the writings describing Amazonian plants either as heavenly or hellish overlook the complexity of vegetal existence, since they approach it through the lens of simplistic categories, such as benign or dangerous, useful or useless, beautiful or unattractive, and so forth. Such writings fail to encounter plants fully and distort their physical inscription in the environment by clouding human relationships to vegetation with preconceived notions of what the forest should be like. They neglect listening to the plants' own tales and, therefore, fall short of *phytographia*.

Even though literary texts on the Amazon often reproduce hackneyed representations of the region, literature has managed, in some instances, to move beyond trite descriptions of local flora and has allowed a genuine *phytographia* to emerge. In the final pages of this chapter, I will focus on the so-called "novel of the jungle" that flourished, roughly, in the first half of the twentieth century. Still indebted to previous frameworks for depicting the Amazon, some of these narratives have nevertheless broken new ground in their portrayal of an active, often sentient forest that, more than any of the human protagonists, is the main character in the texts.

True, many "novels of the jungle" inherited elements of the "green hell" narrative: most of them dwell on the misfortunes of travelers and, especially, of workers lured to the Amazon during the rubber boom in the late nineteenth and early twentieth centuries. However, the "green hell" depicted in these texts is mainly human-made. Life in the forest is hellish due to the exploitative labor conditions that reduce workers, many of them migrants from other regions, to de facto slaves, who give their lives for the enrichment of rubber lords. Rather than victims of the forest, many characters fall prey to a ruthless version of capitalism uninhibited by state rules and protections. These texts can therefore be read in terms of a critique of modernization that at times anticipates environmental, conservationist discourses from a later period.

Another reason for discussing the "novel of the jungle" in the context of *phytographia* is its representation of the Amazon as a "frontier," an area of anomie where anything can happen, similar to other liminal spaces such as the American "Wild West" of the 1800s. Unfettered by the limitations imposed by social norms and by a strong political authority, this borderline region, a meeting point between (Western) human society and the forest, allows for an encounter where the prejudices and preconceptions that govern the relations between humans and nonhumans still have not taken root.

The notion of the Amazon as a "frontier" ties in with another trope of the "novel of the jungle," namely, the desire for a return to nature. In many of these texts the protagonists leave a large city and penetrate deep into the jungle, a trajectory reminiscent of the plot of Joseph Conrad's *Heart of Darkness* (1899), which could be considered a precursor of the genre. Even though Amazonian vegetation does not always offer the main characters the idyllic communion with the natural world they were hoping for, they do find themselves face-to-face with an environment that is completely foreign to them, an experience that prompts the inscription of the vegetal world at the core of the protagonists' existence.

The Vortex (*La vorágine*, published in 1924) by Colombian novelist José Eustasio Rivera is a quintessential jungle novel. Its storyline begins with the protagonist and first-person narrator Arturo forced to leave his native city of Bogotá to find himself, first, in the Colombian plains and, in the second part of the text, in the depths of the Amazon, following the city-to-jungle trajectory typical of these narratives. But the plot progressively evolves from a fairly conventional clash of man against nature to a deeper appreciation of the Amazonian forest that, at points, appears to speak in its own voice, *phytographically* mediated by the text.

As he and his friends penetrate into the jungle's mysterious environment, Arturo feels both entrapped (he writes: "Oh jungle, wedded to silence, mother of solitude and mists! What malignant fate imprisoned me within your green walls?")[36] and seduced by the forest. The most telling moments of communion between the men and the vegetal life surrounding them take place when the members of the group come down with fever. In a delirious state, when the constraints of reason and logic loosen, they approach the forest anew and report their extraordinary visions: "He [Pipa] spoke of the trees of the forest as paralyzed giants that at night called to each other and made gestures. . . . They complained of the hand that scored them, the ax that felled them. They were condemned to flourish, flower, grow, perpetuate their formidable species unfructified, unfecundated, uncomprehended by man."[37] Pipa, one of Arturo's companions, reproduces in his monologue a long-standing trope of religion and of literature, that of talking trees who complain about the destruction brought to the forest by greed and about their inability to communicate their aspirations to humankind.[38] His hallucinations continue, offering a glimpse into a humanless future on earth: "Pipa understood their [the trees'] bitter voices, heard that some day they were to cover fields, plains, and cities, until the last trace of man was wiped from the earth, until over all waved only a mass of close-grown foliage, as in the millennia of Genesis when God still floated in space in a nebulous cloud of tears."[39] The feverish man interweaves visions of the indomitable proliferation of vegetation in the Amazon with human fears of being outlived by plants that will rule over our cities long after humanity's extinction. He conjures up a neo-paradisiacal flora freed from the havoc wrought in forests by human destructiveness, a *phytographic* vision of a future golden age of plant life that is simultaneously alluring and frightening.

The encounter between the Amazon's towering presence and the human ability to interpret the forest's imposing inscriptions so as to express them artistically also comes through when the protagonist Arturo addresses the jungle directly to describe the profound impression it made on him: "Unknown gods speak in hushed voices, whispering promises of long life to your [the jungle's] majestic trees, trees that were the contemporaries of paradise. . . . Your vegetation is a family that never betrays itself."[40] Each plant is addressed by a divine voice, akin to the Socratic daemon, that articulates its specific *signatura* and its entanglement with its family members, the other vegetal beings who inhabit the forest. But this quasi-animistic language of things, magical as it may seem, goes back to the trees' concrete existence and their rootedness in the earth: "You [the jungle] share even in the pain of the leaf that falls. Your multisonous voices rise like a chorus bewailing the giants that crash to earth; and in every breach that is made new germ cells hasten their gestation. You possess the austerity of a cosmic force. You embody the mysteries of creation."[41] The plants react to their physical transformations— the falling of a leaf or the felling of a tree; their voices are nothing but a response to their evolving inscriptions in the environment that the novel's narrator brings to light in his prose.

Arturo returns to the "multisonous voices" of the forest when he ponders plant sensation later in the novel. "Vegetal life," the narrator writes, "is a sensitive thing, the psychology of which we ignore. In these desolate places, only our presentiments understand the language it speaks."[42] Not only does the protagonist acknowledge the sentience of plants but he also realizes that they share a language, which humans are only able to recognize imperfectly. He acknowledges that plants tell their own tales and strives to put these into writing. At another point in the text, Arturo, here clearly enunciating his speculations as an alter ego of the author, mentions how the language of the forest inspired his literary pursuits: "What cities? Perhaps the source of all my poetry was in the secrets of the virgin forests, in the cares of gentle breezes, in the unknown language of all things."[43] At its most thought-provoking, *The Vortex* succeeds in weaving this "unknown language of all things," the "secrets of the virgin forest," into tales that humans beings can relate to, a *phytographia* that articulates plant inscription in literature.

It is time to ask, by way of a conclusion, "Can the Amazon write?" The answer entails reading literary works as spaces of inscription, where we find traces of vegetal language. This does not mean that we will fully abandon our human perspective, an endeavor that would, in any case, be doomed to failure. It does, however, in the wake of Rivera's *The Vortex*, require us to broaden our human horizons, and to make them capacious enough to accommodate our animal and vegetal others. Literature offers us a hint of what Amazonia's tales would be like. It behooves us to learn how to listen to and interpret this *phytographia*.

NOTES

1. An earlier version of this chapter was published in *Environmental Philosophy* 12, no.2 (Fall 2015): 205–20. I thank the journal for granting permission to reprint here a revised version of the article.

2. See, for instance, Richard Karban's and Robert Raguso, as well as André Kessler's chapters in this book.

3. See Erin James's chapter in this collection.

4. See Christa Sommerer, Laurent Mignonneau, and Florian Weil, "The Art of Human to Plant Interaction," in *The Green Thread: Dialogues with the Vegetal World*, ed. Patrícia Vieira, Monica Gagliano, and John Ryan (New York: Lexington Books, 2015), 233–54.

5. For an analysis of the "Midori-san" blog, see Michael Marder's chapter in this collection.

6. For a discussion of the Enlightenment's dream of full visibility, see Patrícia Vieira, *Seeing Politics Otherwise: Vision in Latin American and Iberian Fiction* (Toronto: University of Toronto Press, 2011), 23–26.

7. Johann Wolfgang von Goethe, *The Metamorphosis of Plants* (Cambridge, Mass.: MIT Press, 2009).

8. Michael Marder's *Plant-Thinking: A Philosophy of Vegetal Life* (New York: Columbia University Press, 2013) has precisely highlighted the potential of plant life to undo some of the most enduring metaphysical biases, such as the preference for depth over surface, the preponderance of the whole over its parts, or the denigration of materiality in favor of the loftiness of spirit.

9. Emmanuel Levinas, *Totality and Infinity: An Essay on Exteriority*, trans. Alphonso Lingis (Pittsburgh, Pa.: Duquesne University Press, 2001).

10. Gayatri Spivak, "Can the Subaltern Speak?" in *Marxism and the Interpretation of Culture*, ed. Cary Nelson and Lawrence Grossberg (Urbana: University of Illinois Press, 1988), 271–313.

11. Marder, *Plant-Thinking*, 2, 186–87.

12. Nancy Leys Stepan, *Picturing Tropical Nature* (Ithaca, N.Y.: Cornell University Press, 2001), 53–54; David Arnold, "'Illusory Riches': Representations of the Tropical World, 1840–1950," *Singaporean Journal of Tropical Geography* 21, no.1 (2000): 7–11.

13. Human agency is one of the forms of human inscription. We share many other forms of human inscription, such as bodily expression, involuntary movements, and so on, with animals and plants.

14. Jakob Böhme, *The Signature of All Things* (London: J. M. Dent & Sons; and New York: E. P. Dutton, 1912), http://jacobboehmeonline.com/.

15. Ibid.

16. Ibid.

17. For a detailed analysis of role of resemblance and correspondence in early modernity, see Michel Foucault, *The Order of Things* (London: Routledge, 2002), 19ff.

18. Böhme, *The Signature*.

19. Walter Benjamin, "On Language as Such and on the Language of Man," in *Selected Writings*. vol. 1, *1913–1926,* ed. Marcus Bullock and Michael W. Jennings (Cambridge, Mass.: Belknap Press of Harvard University Press, 1997), 62–74 [72]. Subsequent parenthetical page citations refer to this edition.

20. Ibid., 67. Benjamin expresses the same idea later in the text when he mentions the "communication of matter in magic communion" (70).

21. Benjamin might be implicitly dialoguing in his essay with sociologist Max Weber's famous verdict that the scientific, Western mind-set has led to a disenchantment *(Entzauberung)* of the world. Positing a language of things would go a long way toward "reenchanting" our existence.

22. For Benjamin, things only come into their own when they communicate their wordless language to humans: "To whom does the lamp communicate itself? The mountain? The fox?—But here the answer is: to man. This is not anthropomorphism. The truth of this answer is shown in human knowledge [*Erkenntnis*] and perhaps also in art" (64).

23. Jacques Derrida, *Of Grammatology,* trans. Gayatri Spivak (Baltimore: Johns Hopkins University Press, 1997), 70.

24. Ibid.

25. Ibid., 70–71.

26. Ibid., 113.

27. See Jorge Marcone, "Nuevos descubrimientos del gran río de las Amazonas: La 'novela de la selva' y la crítica al imaginario de la Amazonía," *Estudios. Revista de Investigaciones Literarias y Culturales* 8, no. 16 (2000): 132. Acuña accompanied the Portuguese Pedro Teixeira on his return trip from Quito to Belém in 1639. Teixeira had been the head of the first expedition to ever navigate the Amazon and the Napo rivers counter-current; his expedition left Belém, on the Atlantic coast, in 1637 and arrived in Quito, in present-day Ecuador, ten months later.

28. Henry Walter Bates, quoted in John Hemming, *Tree of Rivers: The Story of the Amazon* (London: Thames & Hudson, 2008), Kindle ed., chapter 5.

29. Ibid.

30. Alfred Russel Wallace, *A Narrative of Travels on the Amazon and Rio Negro,* quoted in Hemming, *Tree of Rivers,* chapter 5.

31. Richard Spruce, *Notes of a Botanist on the Amazon and Andes,* quoted in Hemming, *Tree of Rivers,* chapter 5.

32. Theodore Roosevelt, *Through the Brazilian Wilderness* (1914), Project Gutenberg e-book.

33. Euclides da Cunha, *Um Paraíso Perdido: Reunião de Ensaios Amazônicos,* ed. Hildon Rocha (Brasília: Senado Federal, Conselho Editorial, 2000), 116. This and all other citations from a Portuguese original have been translated into English by the author.

34. Ibid., 125.

35. Alberto Rangel, *Inferno Verde: Scenas e Scenarios do Amazonas* (Tours, France: Typographia Arrault, 1927).

36. José Eustasio Rivera, *The Vortex,* trans. Earle K. James (Bogotá: Panamericana Editorial, 2001), 155.

37. Ibid., 179.

38. See Marder's chapter in this collection for more information on talking plants in literature and religion.

39. Rivera, *The Vortex,* 179.

40. Ibid., 155–56.

41. Ibid., 156.

42. Ibid., 273.

43. Ibid., 124.

TWELVE

Insinuations

Thinking Plant Politics with The Day of the Triffids

Joni Adamson and Catriona Sandilands

> It was . . . quite a while before anyone drew attention to the
> uncanny accuracy with which [the triffids] aimed their stings,
> and that they almost invariably struck for the head. Nor did
> anyone at first take notice of their habit of lurking near their
> fallen victims. . . . There was no great interest, either, in the
> three little leafless sticks at the base of the stem. . . . It was
> assumed . . . that their characteristic of suddenly losing their
> immobility and rattling a rapid tattoo against the main stem was
> some strange form of triffidian amatory exuberance.
>
> —*The Day of the Triffids*

JOHN WYNDHAM'S CLASSIC POSTAPOCALYPTIC NOVEL *THE DAY OF*
the Triffids (1951) is generally read as a work of Cold War science fiction
that also raises provocative questions about altruism, self-sufficiency,
and gender politics. In the novel, Earth's human population is almost
entirely blinded by a meteor event (implied later to be the result of the
detonation of an orbiting satellite weapon). Told from the perspective of
Londoner Bill Masen, one of the world's few sighted survivors, the story
includes his and others' ethical deliberations about which of the blinded
masses to help (if any), how to form a productive new society that moves
beyond scavenging in the remains of the old, and the need for women
to be able to tackle previously "masculine" tasks such as the driving of
trucks and the running of engines (not surprisingly, Wyndham does
not pay equal attention to men's needs to perform "feminine" labors).[1]
The eponymous triffids—a fictional species of large, biologically enhanced
plants that happen to be able to walk, communicate with one another,
sense motion, and kill people with a single blow from their strong
stingers (in order, we discover, to eat the rotting flesh)—tend to be
lumped in the general category of "alien antagonist." Although Wynd-

ham ultimately blames the triffids on Soviet agrobotany (the plants are understood to be the result of highly secret experiments, and the global distribution of their seeds the undisputable result of Cold War espionage), and although he also explores the specific ways in which the plants gain purchase in the West because of industrial exploitation of their oil-producing capacities (both of which we will discuss below), critics tend simply to note that the triffids may represent an "environmental parable" and leave it at that: triffids are allegories of a natural order mutated by Cold War desire, and the fact that they happen to be plants is beside the point.

In contrast, in this chapter we consider the vegetality of the triffids to be highly significant. Specifically, we argue that Wyndham had an excellent understanding of the specific abilities of plants to *insinuate* themselves into human relations in ways in which their capacities—for movement, for communication, and most of all for agency independent of human desire—are routinely ignored in both individual and industrial instrumentalizations of plant life. By paying close attention to Wyndham's detailed characterization of the triffids, in addition to the political context of this characterization—including pointed critiques of the agrobotanical politics of the Stalinist Soviet Union and also of the blindness and hubris of profit-driven Western extractive industries—we argue that the novel is not only an allegory of Cold War corruptions of nature, but also a specific depiction of the multiagential biopolitics of plant–human relations in a post–World War II, globalizing world. We thus take *The Day of the Triffids* as both a relatively sophisticated rendering of plant capacities, especially those of movement, communication, and distributed intelligence, and as an invitation to a larger conversation about the biopolitics of (other) so-called invasive plant species. After a consideration of the novel's triffid biopolitics, then, we turn our attention to other "triffids" located in our own, early twenty-first-century ecosystems: the Sonoran Desert of the U.S. Southwest (tumbleweeds, *Kali tragus*) and the Great Lakes Basin (dog-strangling vine, DSV, *Cynanchum rossicum*). Where Wyndham's novel imagines and predicts a plant-oriented dystopia premised on an always already antagonistic conception of the mobility, communicativity, and intelligence of plants, we take these dystopian understandings as a reflection on and attenuation of current plant/human capacities in the context of neoliberal global ecologies:

triffids share significant family resemblances with tumbleweeds and DSV, and thinking with these plants offers a fruitful avenue for exchanges about plant politics. The first, the tumbleweed, quietly insinuates itself through highly mobile, yet largely invisible distribution methods. The second, dog-strangling vine, is, by contrast, a much louder species of concern in the sense that its insinuations are enabled by a more visible intertwining of its mobility with the mobility of humans (e.g., rail and road corridors).

Triffid Insinuations

According to Masen's retrospective narration, the triffids first come to the attention of the Western world when a mysterious man named Palanguez shows up at the Arctic and European Fish Oil Company with a light pink, highly nutritious vegetable oil originating from Russia. Western scientists quickly realize the value of this oil, and Palanguez offers to return to Russia and smuggle some seeds from which the oil is pressed back to the West. Palanguez fails to reappear, and Masen speculates that he may have been the victim of foul play: perhaps the plane carrying the triffid seeds may have been shot down, scattering the seeds at an altitude that allows them to drift around the globe, prospering first in East Asia, and then quickly spreading around the planet. When it is discovered that triffids are able to lash out with a long stinger that can inject a deadly poison, that they eat the flesh of their victims, and that they can move on three "leg-like" appendages, they are almost exterminated. However, when they are identified as the source of valuable oil, humans learn how to dock their stingers, a practice that renders them harmless and turns them into a popular garden plant. At the same time, they are also cultivated and bioengineered in laboratories and grown on a large scale to produce the highly profitable oil.

Masen is a biologist employed by Arctic and European Oil (after it moved into triffid production, it dropped the "fish"); we see the triffids through his eyes both as he recounts their emergence as key subjects of biotechnological intervention and, post-apocalypse, as he devises and deploys a multitude of ways of dispensing them (there is never any consideration of living *with* the triffids, a subject to which we will also return later). In the narrative, he is *aware* of triffids in ways that almost nobody else is: one of the first triffids growing in the locality was at the bottom of his parents' garden, and his childhood curiosity was rewarded with a nasty, if nonfatal, sting. More important, he is one of the first individuals to make a profession of triffid production, alongside a prescient individual named Walter, who is the first to consider that the plants' rattling of their sticks may be a form of communication, that they exhibit something like intelligence in their modes of attack on humans, and that they could be very dangerous as a result. By and large, however, the triffids are ignored: once the public gets used to their oddness, especially their mobility, they lose interest. Even after the supposed meteor event has blinded most of the world's population and the survivors, sighted and unsighted, are beginning to organize modes of continued subsistence, Masen's concern about the triffids—including his insistence on scavenging a large arsenal of anti-triffid weaponry—is ridiculed and minimized: as he later remarks, "there's a kind of conspiracy not to believe things about triffids" (158).

The triffids remain in the background, inert and instrumentalized plants safely chained on oil farms and docked in front gardens, until it becomes clear to all of the survivors, not just Masen, that the plants are actually the greatest threat to human survival on the postapocalyptic (including post-plague)[2] planet. Although Wyndham depicts the triffids as having several important capacities that give them a considerable advantage over the largely blind and scattered human population (which Walter also foresees)—their lethal, poison-filled stingers, their ability to hear and sense movement, their uncanny patience while waiting for human prey—it is their qualities of *movement, communication,* and *collective intelligence* that most clearly define their situational evolutionary superiority. These biosemiotic capabilities allow them to insinuate themselves into the world almost unnoticed: humans have ignored or downplayed these capabilities literally at their peril. In addition, as we discover as the years go by and the survivors enter into longer-term modes of social organization, these are also the capabilities against which humans must mobilize huge resources if they are to survive.

The triffids are described as moving "rather like a man on crutches. Two of the blunt 'legs' slid forward, then the whole thing lurched as the rear one drew almost level with them, then the two in front slid forward again" (30). Although their movement is awkward by a bipedal mammalian standard, it is both intentional and effective: the plants respond to the sounds of prey by picking up their legs/roots from the soil, walking both short and long distances, and rerooting themselves in a new position in order to wait for prey. They are also capable of the sudden movement of "sending [their stings] whistling" in the direction of any human prey within reach, including into the open car, over the garden wall, and "through the broken window" (121). Throughout the novel, triffids lurch, sway, amble, lurk, skulk, plunge, pull up their chains, break loose, and break in (even as they do not like to walk on asphalt); they also ambush, lash out, slash, whip, smack, and let fly their stingers at any passing human. Their movement is originally the stuff of television light entertainment: as one newscaster quips, "Vegetables on vacation! . . . Maybe if we can educate our potatoes right we can fix it so they'll walk right into the pot. How'd that be, Momma" (29)? But by the end of the novel, it is a source of terror, a monstrous perversion of a supposed natural order in which purposeful ambulatory movement is a capacity of animals and not plants.

In addition, although Walter's conjecture about the "pattering and clattering" (38) of sticks against stem as a form of triffid communication marks the inauguration of Bill's dawning understanding of their mode of talking, it is hardly the last occasion in which their capacity for communication plays a significant role in their eventual dominance over people. Unlike human beings in the post-technological world of the novel, triffids are not only able to communicate across large distances, but also appear to be able to broadcast something like concern even after they have been decapitated: "I was about to assure her that it was [dead] when it began to rattle the little sticks against its stem" (179). The novel suggests, here and elsewhere, that Wyndham is aware that plants' communicative ability is distributed throughout the plant and not located in a brain: Masen's naive belief, upon blowing off the triffid's top, that the decapitation of the plant renders it *mute* turns out to be a potentially fatal anthropomorphism. The novel also clearly considers that these communications are meaningful: the dead triffid's final tapping calls its colleagues to investigate; when one triffid hears the anthropogenic noises of generators, tractors, and guns, it broadcasts the location of humans to thousands of its kin either directly or in some form of triffid communicative web. Where humans could not really ignore the obvious fact of triffid mobility, it took Masen and the others a very long time indeed to understand the fact of meaningful triffid conversation: there is always more to the triffids than people think.

Perhaps most important, it also gradually dawns on Masen and others that triffids have an "apparent ability to learn, in at least a limited way, from experience" (205). More specifically, where Walter had early on pointed to signs of individual triffid intelligence in the deadly accuracy of their stings (always aimed at the human head), it becomes clear through the ways in which the plants lay siege in the thousands to any human inhabitation that some sort of collective intelligence is also at work. The triffids learn and remember the times of day their kin are given zaps of electricity in order to keep them off a protective fence; initially effective luring, trapping, and eradication techniques become less effective even by their third iteration because the triffids come to understand the ruse. As Masen explains, drawing on his biological training:

> They used to say that man's really serious rivals were the insects. It seems to me that the triffids have something in common with some kinds of insects. Oh, I know that biologically they're plants. What I mean is they don't bother about their individuals, and the individuals don't bother about themselves. Separately they have something which looks slightly like intelligence; collectively it looks a great deal more like it. They sort of work together for a purpose the way ants or bees do—yet you could say that not one of them is aware of any purpose or scheme although he's part of it. . . . It may be that no single individual knows why it keeps hanging around our fence, but the whole lot together knows that its purpose is to get us—and that sooner or later it will. (208–9)

That people are unwilling or unable to recognize this intelligence in the midst of their instrumentalization and exploitation of the plants for oil and profit allows the triffid collective brain to become massive; once the competitive scale is finally weighted in favor of the triffids, "they, more than anything else, seemed able to profit and flourish on [human] disaster" (167).

TRIFFID POLITICS

Wyndham's depiction of the triffids was not merely a fantastical projection. That plants have capacities for movement, communication, and intelligence is, of course, now the subject of considerable research and lively public discussion, including by other contributors to this volume.[3] Although it is clear that Wyndham's choice to have the triffids *walk* and communicate by *sound and hearing* was tinged with more than a little bit of anthropomorphism, many elements of his depiction resonate with biological understandings of the period and also almost prophetically with current research into plant mobility, signaling, and neurobiology. Wyndham briefly studied farming in the 1920s right after he left school;[4] the story of the triffids has at least partial origin in this expertise, part of a larger engagement with biological/technological politics apparent also to a greater or lesser extent in some of his other novels, including *The Chrysalids* (1955), *The Midwich Cuckoos* (1957), and *Trouble with Lichen* (1960). We can thus make conjectures about some of the novel's key biopolitical engagements.

First, many elements of the novel suggest that Wyndham was not only familiar with Charles Darwin's research on plants, but also animated by some of the same questions, circulating by mid-twentieth century, about plant cognition. As botanist Rainier Stahlberg writes, Darwin was among the first to record extracellular "action potentials" (APs), or the role of electric currents in plant movement. Later, in 1873, the physiologist John Burdon-Sanderson picked up on this research and performed a series of experiments on the leaves of the Venus flytrap (*Dionea muscipula* Ellis).[5] These kinds of experiments led scientists, as early as 1907, to refer to plants as having nerves.[6] Over the course of the first half of the twentieth century, and more definitively by the 1960s, research into plant cognition and intelligence also indicated that so-called "normal" plants (as opposed to "sensitive" plants like the Venus flytrap), such as pumpkins with their trailing vines, were moving in response to stimuli.

Unlike later scientists who wrongly dismissed Darwin's and Burdon-Sanderson's research, Wyndham clearly takes plant mobility seriously. *The Day of the Triffids* indicates that he may have also been aware of research from the 1930s that sought to establish further similarities between plants and animals. This electrophysiological research on "long-distance signals in plants and animals," writes Stahlberg, established the similar and different capacities of vegetal and mammal species "to respond to environmental signals."[7] The triffids, with their carnivorous appetites, their ability to move toward stimuli, and tendency to strike out with a poisonous, tentacle-like appendage, also suggest that Wyndham may even have been familiar with 1930s experiments with an underwater plant called stoneworts *(Nitella)*, a branched multicellular algae, and a giant squid. This research, writes Stahlberg, "elegantly demonstrated" the "mechanistic similarity" of electrical excitations when stoneworts and squid were compared.[8]

The triffids combine the qualities of both "normal" and "sensitive" plants and, like both plants and animals, seem to have an electrophysiology. Moreover, Masen's inability to decapitate a triffid, and the triffid's ability to engage in a kind of communication across a distributed web, suggest that Wyndham may have also been influenced by Darwin's hypothesis about plants having a control center for behavior dispersed across their root tips, something like what scientists today are terming a "root-brain."[9] The triffids anticipate work on concepts of "mind" suggested by Gregory Bateson (1972), who wrote of a "recursive communicative order" that he called an "ecology of mind" and that could not be described as a subjective power operating exclusively *inside* human brains.[10] Wyndham seems fascinated with the idea that plants may possess a mind that, as philosopher Alva Noë expresses it, accounts for the fact that plants orient and react appropriately in response not only to light, but also to wind, water, predators, quality of soil, and the volume of available soil, among many other factors.[11] Today, writes Noë, it is entirely appropriate to ask: Do plants "*forage* for light"? Do they "*decide* where to send out their roots?" Is the "plasticity and growth of plants to be compared with the free movement and action of animals?"[12]

When Masen muses that the individual triffids have something "which looks . . . like intelligence" and "collectively . . . a great deal more," Wyndham anticipates the collective intelligence and shared support systems that scientists who study the root and fungal systems have playfully dubbed the "wood wide web." Trees, writes Manuela Giovannetti et al., collect sunlight in their canopies, turn it into energy, and send it to mutually beneficial arrangements of fungus on the forest floor, which employ a chemical vocabulary of nutrients to feed seedlings and nourish the root systems of other entangled root systems. Arbuscular mycorrhizal (AM) fungi, she and her colleagues write, have a wide host range and they are able to colonize and interconnect contiguous plants by means of hyphae extending from one root system to another as they move nutrients through spreading pathways that exhibit an ability to "discriminate self from non-self."[13] These forms of "self-discrimination" point to the reasons why many scientists, anthropologists, and critical plant studies scholars argue that humans and trees, roots, and fungi might all be considered "selves" that "have legibly biographical and political lives."[14]

Perhaps most interesting, however, Wyndham not only depicted triffid capacities in a decidedly plant-sensitive way but also responded to the contemporary plant *politics* of his time quite directly. In the early days after the meteor shower, Masen recounts: "In the days when information was still exchanged Russia had reported some successes. Later, however, a cleavage of methods and views had caused biology there, under a man called Lysenko, to take a different course" (23). In addition, the triffids are developed and industrially produced, originally by the Soviet Union, for their *oil*. Both elements indicate that Wyndham was referring to the struggle over food scarcity in Russia and competing approaches to science represented by Stalin's anti-Mendelian favorite Trofim Lysenko and world-renowned founder of modern crop breeding Nikolay Vavilov, who "organized 115 research expeditions through some sixty-four countries to find novel ways that humanity could feed itself."[15] Vavilov had been considered the world's leading plant geneticist before his tragic incarceration by Lysenko and Stalin in 1941 and his death from starvation in 1943; Lysenko's state-sanctioned triumph set Soviet efforts to achieve food security back by decades.

Earlier, in the 1930s, Vavilov played a small but critical role in the genetic development of the "Mammoth Russian sunflower," which bears more than one resemblance to the enormous triffids, their large bobbing seed heads and their valuable pink oil. First domesticated by Native American Pueblo groups in North America, sunflowers were taken to Europe by Spanish explorers in about 1500. The plant became widespread throughout Western Europe mainly as an ornamental, but, in Russia, Peter the Great saw the value of the plant as a source of staple food and commercially marketable vegetable oil. By 1830, the oil had increased dramatically in popularity because the restrictive Russian Orthodox Church allowed the consumption of sunflower oil during Lent.[16]

The sunflower-resembling triffid was, for Wyndham, a metonym of Soviet biology à la Lysenko, whose very name came to stand for the deadly notion, linked to mass starvation in the Soviet Union, that "quick fix" ideologies were superior to the "slower, Darwinian plant selection processes" modeled by Vavilov.[17] Here again, *The Day of the Triffids* is linked to Darwin's ideas about plant breeding and neurophysiology through the history of the sunflower. During the early Stalin period, in 1930, Vavilov took his last research expedition to North America and returned with wild sunflower seeds from West Texas *(Helianthus lenticularis)*. He gave them to Russian plant breeder V. S. Pustovoit, who began crossing them with common sunflower, *Helianthus annuus*. Over the next several decades, Pustovoit produced a "stable hybrid sunflower with a level of polyunsaturated oils that dwarfed that of other sunflowers" and by the 1970s, it became widely grown in the Soviet Union.[18] In 1934, however, Lysenko, competing for Stalin's favor, claimed that Vavilov's theories about crossbreeding plants, such as the sunflowers, to increase their yield and nutritional value was proof that Vavilov and other "elitist academicians" were not interested in feeding the people. Lysenko translated his broadly Lamarckian views into what is now largely considered a pseudoscientific program of research and implementation that had more to do with ideological control than scientific rigor. Today, *Lysenkoism* has come to mean the manipulation or distortion of the scientific process as a way to reach a predetermined conclusion as dictated by an ideological bias or political objective. Wyndham is, then, in his references to Lysenko and the sunflowers, making a pointed comment in *The Day of the Triffids*: politics, science, and plants are a deadly combination.

In addition, however, Wyndham is almost as critical of the West as he is of the Soviet Union. As underscored by Masen's deep complicity with the British triffid industry, the global proliferation of triffids is a product of greed on both sides of the Iron Curtain, and the willful ignorance of triffid capacities a product of precisely the blinders put in place by a narrow and instrumental focus on plant production for extraction and profit. It was Arctic and European Oils that put in place the large network of triffid nurseries across the country, plant factories that ended up rendering almost all parts of the postapocalyptic rural Britain just as deadly as the plague-infested cities (and for longer). It was Walter who, through experimentation for the company, discovered "that the quality

of the extracts was improved if the plants retained their stings" (39), leading to the discontinuation of docking that left an immediate army of
tens of thousands of triffids at the ready for institutional collapse and subsequent revolt. In part reflecting on his own involvement in proliferating the creatures that had now become his nemesis, Masen notes: "I saw
them now with a disgust that they had never aroused in me before. Horrible alien things which some of us had somehow created, and which
the rest of us, in our careless greed, had cultured all over the world. One
could not even blame nature for them. Somehow they had been bred—
just as we bred ourselves beautiful flowers or grotesque parodies of
dogs" (167). And later he considers the ways in which capitalism and
communism alike are built on a very fragile and hubristic understanding of control over nature, one in which triffids—or any other wild plant
or human-created cultivar, creation, mutation, or discovery—are never
quite domesticated: "we coped with them all right in normal conditions.
We benefited quite a lot from them, as long as the conditions were to
their disadvantage" (210).[19] Here, conditions of disadvantage allowed the
triffids to insinuate themselves into everyday human life without much
notice: the subtleties of their communication and intelligence (if not their
mobility) were completely obscured under layers of human greed.

The Day of the Triffids ends up on closer reading to be far more than
a work of Cold War science fiction. For both the Soviet Union and in the
West, Wyndham—along with plant breeders and scientists—could see
an impending disaster of "horrible alien things" created by humans in
their "careless greed" (167). The novel is an interesting critique of both
Stalinist and Western scientific and agricultural politics: it is a powerful
condemnation of both communist and capitalist relations to the natural
world and especially to plants whose agential capabilities are both largely
ignored (in their expansive capacities) and greatly exploited (in their
instrumentalized particularities) in industrial biotechnological rationali-
ties; it is a thoughtful and botanically sophisticated musing on the ways
in which plants' capacities for movement, communication, and intelligence place them in complex and changing relations to human capacities; it is a call, in our view at least, to think more carefully about the
biopolitics of plant–human relations in the present moment; and it is a
caution to pay more attention to the ways in which plants are currently
insinuated—and insinuate themselves—into human social relations in

a globalizing biotechnological world because of their communicative and other abilities. Especially in our so-called Anthropocene era, Masen's reflection on the triffids, which he sees now only with disgust, raises all kinds of new questions about plant insinuations and biopolitics. It is with these questions that this chapter will conclude.

CONCLUSION: "REAL-LIFE" TRIFFIDS, TUMBLEWEEDS, AND DOG-STRANGLING VINES

A 2013 article in *Mirror Online* bearing the title "Slay of the Triffids" reports on a new U.K. Department for Environment, Food, and Rural Affairs ban on the sale of five species of "rampant [exotic invasive] pond plants to avoid a *Day of the Triffids*–style rampage through Britain."[20] A 2014 article in the *Ecologist* on a potential biological control for the "difficult to control" Japanese knotweed *(Fallopia japonica)* asks: "could [this] oriental triffid be tamed following the . . . introduction of a specialist pest from Japan's volcanic uplands?"[21] And in South Africa, the highly invasive shrub *Chromolaena odorata* (Asteraceae) is commonly known by English speakers as "triffid weed"; it is considered "the most problematic of the alien invader plants threatening nature conservation areas in Natal,"[22] and its eradication is highly recommended throughout the country.

In the same way that triffids are *always already killable* in Masen's postapocalyptic world, so too are so-called problem invasive plants in early twenty-first-century management discourses almost always *about to be put to death*. In this context, the invocation of triffidian metaphor adds apocalyptic urgency to calls for policy, behavioral, chemical, physical, and biological controls: once a plant becomes a "real-life triffid," the full arsenal of anti–real-life triffid weaponry can be unleashed without much further thought about the context in which certain plants begin to behave in manners we deem invasive in the first place, the potential benefits of some exotic and/or invasive species to the future of biodiversity in some contexts (especially in degraded landscapes), or the ethical and ecological minefields of such highly anthropocentric selective botanical death dealing. Although others have amply discussed the deeply problematic notion of "invasion" that animates much conservation and management discourse in this area,[23] we would like to conclude our consideration of triffid insinuations by considering two so-called invasive species that resemble triffids in interesting ways, the tumbleweed and the dog-strangling vine (DSV).

Like the triffids, fictional plants that became stars of film, television, and radio,[24] the tumbleweed is counted among the most famous onscreen plant characters. Having come to symbolize locations that are desolate, dry, and often peopled with few or no occupants, the plant is also famous for its mobility and can often be seen rolling across the sets of Western films and television programs. So common is this plant in the desert regions of the American West that most modern residents do not know that the "Russian thistle" is said to have arrived in the United States accidently in shipments of flax seeds from Russia delivered to South Dakota in the 1870s, after which it rolled along train tracks and dirt roads, spreading its seeds, and proliferating dramatically in places where cattle were allowed to overgraze.[25] This is one of the ways that the iconic tumbleweed became associated with cowboys and ranching.

Today, this highly mobile plant has spread across one hundred million acres in the arid West, displacing crops and native plants. In wildfires, it bursts into spectacular fireballs, spreading mayhem. On highways, it endangers life by smashing into vehicles in high winds.[26] It has even been seen "taking over towns."[27] In March of 2014, with a similarity that has bizarre parallels to lurking triffids, tumbleweeds took over whole neighborhoods in Colorado.[28] After a particularly strong wind, large numbers lodged themselves against the outer walls of homes. Residents had to literally fight their way out of their doors and garages with rakes and snow plows.

The tumbleweed is a member of the "cosmopolitan family" of amaranth *(Amaranthaceae),* so known because this group of plants can be found in almost every place in the world. *Kali tragus,* or tumbleweed, is a member of the amaranth subfamily *Salsoloideae,* known for its ability to live in salty, dry soils. It is a "diaspora," a plant that, after maturing into large prickly clumps, breaks from its root system after the first freeze and rolls in the wind, dispersing over two hundred thousand seeds.[29] Because it outcompetes native grasses, thrives during drought, and sprouts earlier in the spring than most commercial crops, it is considered a pernicious invasive. As a result, tumbleweeds have become the subject of "vigorous attempts to cut, poison, burn or dig them up."[30]

Despite this nasty reputation, tumbleweeds are coming to be recognized as an important food and forage plant for both indigenous and domesticated animals, including prairie dogs, sheep, cattle, and deer.[31] Also, like its North American cousin amaranth (from the subfamily *Amaranthoideae*), which was long the staple crop that made the Aztecs wealthy (a fascinating history of its own),[32] tumbleweed is becoming identified as a potential "famine food" for humans because it thrives in dry conditions. The young shoots and tips of *Kali tragus* are edible raw or can be cooked like greens. They are a rich source of vitamin A and phosphorus, with a protein balanced with a high dose of the essential amino acid lysine.[33]

Unlike the triffids, the tumbleweed has not been shown to have any sort of collective intelligence, but, interestingly, it can feed the "distributed intelligence" of arbuscular mycorrhizal (AM) fungi, discussed above, which is found in 80 percent of woody and grassland species.[34] Tumbleweed does not host mycorrhizal fungi. Yet, this lack can have an ironically positive effect on soils degraded by overgrazing.[35] If any mycorrhizae survive in the topsoil of degraded land, it will invade *Kali* roots, stunting, or even killing them. When the plant dies, the mycorrhizae consume what remains, then spread out and infect surrounding tumbleweeds.[36] The dead tumbleweed also enriches, mulches, and aerates the new grasses and provides shade for other plants' seedlings. This promotes growth of indigenous plants adapted to forming advantageous symbiotic associations with mycorrhizae. Rangeland scientist Janet Howard warns that ecological interrelationships should not be regarded too simplistically, but, in some cases, this complex process tends to repopulate the soil better and faster than killing *Kali* with herbicides or fire.[37] In short, *Kali*'s ability to provide food and forage and its potential to support revegetation suggest that humans might beneficially explore new relationships to invasives, since, in the case of tumbleweed, fire, chemicals, and drought only encourage them to take over towns and explode spectacularly on impact with speeding cars.

In contrast, dog-strangling vine has no apparent nutritionally or ecologically redeeming features. A milkweed relative, it was likely brought to North America as an ornamental in the nineteenth century, and it has turned out to be stubbornly nonuseful on all fronts since then:

long since escaped from the garden, it is toxic to livestock and not a good source of raw material for industry (although its latexy sap was tested as a rubber substitute during World War II); it has very effectively taken advantage of anthropogenic infrastructure such as roads, railway tracks, and utility corridors for the distribution of its prolific, tiny, wind-borne pollinaria; and it has grown very exuberantly both in disturbed areas throughout the Great Lakes Basin and, increasingly, in more established ecosystems. DSV literally "crept up" on us: although it was first recorded in Ontario in 1899, it did not really register in the public imagination until it began to appear in large numbers in the same spaces that human inhabitants had designs on for other botanical goals such as the preservation of indigenous species. Our current cohabitation is conflictual not only because of the plant's creepiness but also because we did not notice it until it impinged on our interests. Like tumbleweed (and like triffids), DSV originates in Russia/Ukraine; unlike tumbleweed, its vine-like climbing and twining ubiquity in the eastern part of North America has not (yet) become the stuff of Hollywood legend.[38] Instead, the species has become public enemy number one: the Ontario Invading Species Awareness Program has a hotline for reporting DSV sightings; its website links to a "Landowners' Guide" that helps gardeners identify the culprit, followed by a list of the relative merits of various potential ways of destroying the plant(s), including mowing, cutting, pulling, suffocating, and spraying with Roundup; and it emphasizes the plant's criminal tendency to use other species as scaffolding for its enthusiastic tendrilling, effectively "strangling" the trees and shrubs in whose space it spreads.[39]

But there are many interesting things about DSV that indicate that we need to learn to live *with* the plants rather than immediately move to destroy them. For one thing, dog-strangling vine behaves a lot like *Homo sapiens* in North America:[40] prolific, opportunistic, urban, thriving in disturbance, and possessing, in numbers, the ability to significantly remake our habitat in our own interests (including a process called allelopathy in plants involving the chemical alteration of the soil). DSV thus tells us something about our own behavior, including the fact that we tend to create habitats that only support other opportunists. The vine is, in a profound sense, our companion species, enthusiastically inhabiting the disturbed spaces that we create even if it does not yet appear to give us very much in return. For another, DSV is a remarkably mobile, communicative, and intelligent plant: not only are its pollinaria capable of traveling great distances, but its capacities for *advancing, creeping,* and *twining* are what make it so criminally suspect; not only are its pronounced allelopathic tendencies an important form of vegetative communication with surrounding plant communities, but it is also clear that DSV operates like a collective intelligence in its demonstration of the so-called Allee effect, in which the insinuating plant remains fairly unobtrusive in its new landscape until it has reached a critical population density, which appears to then trigger an explosion.[41] Indeed, current experiments with the biological control of DSV using a species of Ukrainian moth, *Hypena opulenta,* seem to recognize that cohabitation with the plant is going to prove necessary for the conceivable future: researchers acknowledge that the moth will not eradicate the vine, and the emerging cosmopolitan ecology of southern Ontario will thus necessarily include new modes of moth–DSV–human relationship that transcend the immediate imperative to kill.[42]

Tumbleweeds and DSV thus powerfully show the limitations of triffid *thinking* even as they show us what we might learn from triffids themselves. Although Wyndham's novel quite brilliantly depicts plant capacities and politics, it ultimately displays a lack of imagination. Rather than imagine mobile, communicative, and intelligent triffids as necessarily our enemies and evolutionary competitors, as Wyndham does, perhaps we can see these capacities as resources for *solidarity.* Even when certain plants frustrate biopolitical insistences on utility, we can and must learn to live with them; to move out of the botanical cold war in which we find ourselves, we need to see plants, with their multiple capabilities, as potential allies rather than as simply resources, nuisances, and antagonists.

NOTES

1. The novel also includes a significant subplot about the productive and reproductive politics of the "new" society, contrasting Masen's desire for an autonomous, heterosexual, monogamous nuclear family with the polygamous rationales of one Michael Beadley, and with the socialist desires of one Wilfrid Coker. John Wyndham, *The Day of the Triffids* (London: Penguin Books, 1999

[1951]). Parenthetical page references are to this edition. Some passages cited are not included in other editions.

2. The novel also alludes to the idea that the plague is a result of Cold War biological warfare; it decimates the urban population and drives survivors into the country, where they are further plagued by huge populations of triffids that have literally pulled up stakes and escaped from farms.

3. See also Daniel Chamovitz, *What a Plant Knows: A Field Guide to the Senses* (New York: Scientific American/Farrar, Straus and Giroux, 2012); Monica Gagliano et al., "Experience Teaches Plants to Learn Faster and Forget Slower in Environments Where It Matters," *Oecologia* 175 (2014): 63–72; Stefano Mancuso, "The Roots of Plant Intelligence," *TEDGlobal,* July 2010, http://www.ted.com/talks/stefano_mancuso_the_roots_of_plant_intelligence/transcript?language=en; Michael Marder, "Plant Intelligence and Attention," *Plant Signaling & Behavior* 8, no. 5 (2013): e239092; Michael Pollan, "The Intelligent Plant: Scientists Debate a New Way of Understanding Flora," *New Yorker,* December 23 and 30, 2013, 92–105; and Anthony Trewavas, "Plant Intelligence: Mindless Mastery," *Nature* 415 (2002): 841.

4. See the John Wyndham papers, housed at the University of Liverpool archives: Andy Sawyer, ed., "'The Return of the Triffids . . .': The John Wyndham Archive," University of Liverpool Library, accessed March 11, 2016, http://www.liv.ac.uk/~asawyer/trifcat.html.

5. Rainier Stahlberg, "Historical Overview on Plant Neurobiology," *Plant Signaling & Behavior* 1, no. 1 (2006): 6–8 [6].

6. Ibid., 7.

7. Ibid.

8. Ibid.

9. See also Paco Calvo Garzón and Fred Keijzer, "Plants: Adaptive Behavior, Root-Brains, and Minimal Cognition," *Adaptive Behavior* 19, no. 3 (2011): 155–71.

10. Gregory Bateson, *Steps to an Ecology of Mind* (Chicago: University of Chicago Press, 1972).

11. Alva Noë, "Do Plants Have Minds?," *13.7: Cosmos & Culture,* December 2, 2011, http://www.npr.org/sections/13.7/2011/12/02/143041917/do-plants-have-minds.

12. Ibid.

13. Manuela Giovannetti et al., "At the Root of the Wood Wide Web: Self Recognition and Non-Self Incompatibility in Mycorrhizal Networks," *Plant Signaling & Behavior* 1, no. 1 (2006): 1–5 [2].

14. S. Eben Kirksey and Stefan Helmreich, "The Emergence of Multispecies Ethnography," *Cultural Anthropology* 25, no. 4 (2010): 545–76 [545].

15. Gary Paul Nabhan, *Where Our Food Comes From: Retracing Nikolay Vavilov's Quest to End Famine* (Washington, D.C.: Island Press, 2009), 9.

16. National Sunflower Association, "All About Sunflower: History," accessed March 11, 2016, http://www.sunflowernsa.com/all-about/history/.

17. Nabhan, *Where Our Food Comes From,* 178.

18. Ibid., 177.

19. Here, Wyndham's suggestion that humans have not lived with triffids long enough to understand how to stabilize them may allude to the daunting challenge of creating stable or F_1 hybrids, which are largely annual and vegetable cultivars produced by crossing two stable seed lines (called *inbred lines*) that give rise to especially uniform progeny that possess good vigor, yield, and other properties. This was the challenge Pustovoit pursued as he crossed *H. annuus* with *H. lenticularis*. It took decades of careful "hand crossing" to produce the stable hybrid that came to be known as the Mammoth sunflower (Nabhan, *Where Our Food Comes From,* 177).

20. Ruki Sayid, "Slay of the Triffids: Five Invasive Plant Species Rampaging across UK to Be Banned from Sale," *Mirror Online,* January 29, 2013, http://www.mirror.co.uk/news/uk-news/invasive-species-to-be-banned-from-sale-water-1562345.

21. Kate Constantine, "Japanese Knotweed—Could a Tiny Insect Tame the Monster?," *Ecologist,* October 17, 2014, http://www.theecologist.org/blogs_and_comments/Blogs/2574293/japanese_knotweed_could_a_tiny_insect_tame_the_monster.html.

22. R. L. Kluge, "Biological Control of Triffid Weed, *Chromolaena odorata* (Asteraceae), in South Africa," *Agriculture, Ecosystems, and Environment* 37, nos. 1–3 (1991): 193–97.

23. Brendon M. H. Larson, "Reweaving Narratives about Humans and Invasive Species," *Études Rurales* 185 (2010): 25–38; and Banu Subramaniam, "The Aliens Have Landed! Reflections on the Rhetoric of Biological Invasions," *Meridians: Feminism, Race, Transnationalism* 2, no. 1 (2001): 26–40.

24. *The Day of the Triffids* was made into a feature film in 1962 and has also been adapted to radio and television.

25. Anne Orth Epple and Lewis Epple, "Tumbleweed, " in *A Field Guide to the Plants of Arizona* (Helena, Mont.: Falcon, 1995), 56–57; Green Deane, "Russian Thistle, Tumbleweed," *Eat the Weeds,* accessed March 11, 2016, http://www.eattheweeds.com/salsola-kali-noxious-weed-nibble-green-2/; and Janet L. Howard, "Salsola kali," in *Fire Effects Information System,* U.S. Department of Agriculture, Forest Service, Rocky Mountain Research Station, Fire Sciences Laboratory, 1992, http://www.fs.fed.us/database/feis/plants/forb/salkal/all.html.

26. Emiline Ostlind, "It May Be High Noon for Tumbleweed," *High Country News,* May 20, 2011, http://www.hcn.org/wotr/it-may-be-high-noon-for-tumbleweed.

27. Jareen Imam, "Tumbleweeds Take over Colorado Neighborhoods," CNN.com, March 24, 2014, http://edition.cnn.com/2014/03/23/us/irpt-tumbleweed-colorado/.

28. Ibid.

29. Hossein Akhani, Gerald Edwards, and Eric Roalson, "Diversification of the Old World Salsoleae s.l. (Chenopodiaceae): Molecular Phylogenetic Analysis of Nuclear and Chloroplast Data Sets and a Revised Classification," *International Journal of Plant Sciences* 168, no.6 (2007): 931–56.

30. Ostlind, "It May Be."

31. Howard, "Salsola."

32. See Joni Adamson, "Medicine Food: Critical Environmental Justice Studies, Native North American Literature, and the Movement for Food Sovereignty," *Environmental Justice* 4, no. 4 (2011): 213–19.

33. Howard, "Salsola."

34. Giovannetti et al., "At the Root," 1.

35. Howard, "Salsola."

36. Ibid.

37. Ibid.

38. Catriona Sandilands, "Dog Stranglers in the Park? National and Vegetal Politics in Ontario's Rouge Valley," *Journal of Canadian Studies/Revue d'études canadiennes* 47, no. 3 (2013): 93–122.

39. Ontario Invading Species Awareness Program, "Dog-Strangling Vine," accessed March 11, 2016, http://www.invadingspecies.com/invaders/plants-terrestrial/dog-strangling-vine/.

40. DSV is rare and patchy in Russia and Ukraine, causing some doubt as to whether one can make conclusions about the species' "essential tendencies," given its radically different behavior in each context. Its common name, interestingly, emerges from Linnaeus: *Cynanchum = kyon =* dog + *anchein =* to strangle or poison.

41. Sandilands, "Dog Stranglers."

42. Carys Mills, "The Tale of the Moth and the Dog-Strangling Vine," *Ottawa Citizen,* August 8, 2014, http://ottawacitizen.com/news/local-news/the-tale-of-the-moth-and-the-dog-strangling-vine.

THIRTEEN

What the Plant Says

Plant Narrators and the Ecosocial Imaginary

Erin James

STEPHEN WRIGHT'S 1983 NOVEL ABOUT THE VIETNAM WAR, *MEDITATIONS in Green,* begins with an unexpected voice: "Here I am up in the window, that indistinguishable head you see listing toward the sun and waiting to be watered. Through a pair of strong field glasses you might be able to make out the color of my leaf (milky green), my flower (purple-white), and the poor profile of my stunted growth. In open country with stem and root room I could top four feet. Want a true botanical friend? Guess my species and you can take me home."[1] The first words of Wright's novel, spoken from the point of view of a houseplant, serve as a surprising introduction for his readers. Opening with the houseplant's voice allows Wright to foreground what will become a dominant plant motif in the text—a motif that appears in the incessant and unrelenting green of the Vietnam jungle, the marijuana and opium plants that provide the drugs upon which the novel's protagonist, Griffin, and fellow soldiers are hooked, and the houseplants that Griffin's therapist encourages him to care for as a means of recovering his life after the war. The houseplant's voice in the novel's opening lines immediately emphasizes this motif and calls attention to the dominant role that plants play in the text.

As readers progress through the novel they discover that these lines technically are not spoken by a houseplant. The novel suggests implicitly that Griffin thinks these words as he pretends to be a plant—as he meditates, per his doctor's advice, on *what it is like* to experience the world as a plant as a means of rooting himself in his postwar reality. The speaking houseplant thus kicks off the novel with an important bit of symbolism. Griffin and the plant are equally stunted—he by his wartime experience and drug addiction and the plant by a lack of care and nourishment—and both are equally eager for someone to take them home. Previously defined by the jungle plants of his war setting and the opiate plants that now dominate his body, here Griffin imagines himself as another type of plant as a means of communicating his troubled sense of self.

But beyond this symbolism, Wright's opening passage also is notable because it serves as an important piece of metalepsis, or a transgression of narrative levels in which, in this case, the character and the reader mirror each other. Readers beginning Wright's text must go through the same process that Griffin does. The metalepsis invites readers to imagine what it is like to be a houseplant. They must model and imaginatively inhabit a story world informed by the houseplant's point of view, and experience time and space according to the houseplant's perspective. Indeed, Wright's novel begins with a powerful statement about the potential of narrative to push readers into unfamiliar imaginative terrain. Wright's houseplant mandates that readers engage in a vivid thought experiment: what is it like to function in the world as a plant?

This chapter builds on Wright's thought experiment by exploring other examples of plant narrators. On the surface, this phrase—plant narrator—seems oxymoronic. Plants endure a reputation for being unmoving, unfeeling, unthinking, unspeaking. Indeed, perhaps the sheer lack of plant narrators, compared to a plethora of narratives told by dogs and other animals, is indication enough of how plants tend to be represented in Western cultures. English speakers do use the term *vegetable* to refer to someone in an unmoving, unfeeling, unthinking, unspeaking state, after all. Yet recent work in botanical science and ethics suggests this view of plants is outdated at best and dangerous at worst. In *Plant-Thinking* (2013), Michael Marder argues that the reason that one in five plant species on the planet is currently on the brink of extinction is that human thinking about plants is too simplistic.[2] For Marder and botanists who use plant behavior to query standard definitions of intelligence, this simplicity is largely based upon a human inability or refusal to imagine space and time from the logic of vegetal life that acknowledges plant intelligence and agency. Marder draws on recent work by botanists sometimes controversially referred to as "plant neurobiology" that tracks the ways in which plants sense and respond beneficially to environmental variables such as water, gravity, temperature, nutrients, toxins, and chemical signals from other plants. Some botanists, such as

Daniel Chamovitz, even go as far as to discuss "what a plant knows," the title of his recent book (2012), and map human modes of perception (seeing, smelling, hearing, feeling, and remembering) onto plants and vegetal systems.[3]

This work on plant intelligence finds a sympathetic project in the recent material turn in ecocriticism. Material ecocritics Serenella Iovino and Serpil Oppermann claim that all matter—including but not limited to vegetable matter—is agentic and capable of producing its own meanings. They argue that all matter is a "site of narrativity" and thus can be the object of critical analysis aimed at discovering its own stories.[4] In addition to exploring the meanings that agentic nonhuman matter produces, Iovino and Oppermann suggest that one way of interpreting the agency of matter is to explore how nonhuman agentic capacities are described and represented in narrative texts.

Iovino and Oppermann suggest that narratives stand to be important sites of reconceptualization, in which readers can grapple imaginatively with perspectives and experiences that do not necessarily correspond to a human experience of the world. Botanists interested in the viability and limitations of plant intelligence make a similar point in their appeal for new language to describe plant ontology. As Richard Firn states, "human language guides and greatly limits our thoughts when trying to appreciate the functioning of plants."[5] Questioning the appropriateness of such imaginative metaphors as "plant neurobiology," Firn calls upon scholars to develop new words to describe plant intelligence more accurately. Leading plant intelligence researcher Anthony Trewavas similarly recognizes the need for new language when he argues that metaphors help "stimulate the investigative imagination of good scientists," and thus will play a crucial role in new work on plants.[6]

Inspired by the material ecocritical interest in stories of agentic matter, I take seriously these calls for new language and apply them more broadly to narratives to explore the contributions that these can make to vegetal reimagining.[7] Stories featuring plant narrators such as Wright's stand to help both expose cultural presuppositions about plants *and* reshape readers' understanding of what it is like to experience the world as a plant. They also stand to help us reimagine plants at agentic beings, capable of producing their own meanings. Such narratives thus help us explore the "language of plants" in two key ways: they offer us new imaginative tools to rethink vegetal life, and thereby point out the special role that literary language—narratives in particular—can play in shifting human understandings of plants and our relationship to them. I do not claim that such narratives can tell us what it *really* is like to experience the world from a plant's perspective—they are, after all, anthropomorphic in that they are representations of sequences of events that originate in and are interpreted by human brains. Yet I am interested in tracking and categorizing the various narrative strategies employed in stories with plant narrators to question patterns of *human* perceptions of plants. In other words, I want to ask: How might narratives with plant narrators help readers better recognize how they imagine the plants that surround them? How might the language of such narratives help readers reassess their understanding of plants and the potentials of plant agency by tasking them with imaginatively inhabiting a plant's point of view? And finally, how might such narratives encourage readers to contemplate the material language that real-life plants *do* speak? How might the language of these texts help readers recognize what Michael Marder calls the "phenomenology of vegetal self-expression"—a mode of communication that he argues is best described as "an articulation without saying" that normally goes unregistered to humans?[8]

THE SPEAKING TREE

Fifteen "Meditations in Green" pepper Wright's novel, and as the narrative progresses their symbolism becomes increasingly clear. At times the plant's life appears to slip into Griffin's own experiences, such as in Meditation 4, in which the recitation of the chemical compound for a common broadleaf herbicide drifts into a common soldier's chant:

> 2,4-dichlorophenoxyacetic acid
> 2,4,5-tricholorophenoxyacetic acid
> 2,4,6-start the engines, pull the sticks
> 2,4,6,8-everyone evacuate[9]

The close union here between the soldier and the plant emphasizes a similar union in the novel's opening meditation that I discuss above. Griffin attempts to think like a plant in that meditation, but it is clear that his own experiences and perspective color his thinking. Like

Griffin, the houseplant is struggling; it lists toward the sun, waiting to be nourished, and laments the "poor profile of my stunted growth." It is also depressed as it, like Griffin, surveys its degraded urban environment: "The view from my sill is not encouraging: colorless sky, lusterless sun, sooty field of rusted television antennas . . . the persistent dead unavoidable concrete." The plant voices its frustration with the "clumsy, irresponsible hands" of its human caregiver and the "apathy and loneliness" of its life on the windowsill. The meditation ends with an urgent cry: "Help! My stalk is starting to droop."[10] The plant's life is surprisingly similar to Griffin's own—the sense of displacement, the lack of nutrition, the loneliness and lack of companionship, even the stunted and drooping body. Indeed, we best interpret the novel's opening meditation as Griffin's imagination not of a houseplant but of *himself* as a houseplant or of a houseplant that exists only to highlight the pain of a drug-addicted Vietnam vet. The houseplant figuratively serves the human character at the center of the text in a way that positions humans as superior and plants as important only in terms of their relationships with humans.

We thus should not be surprised that other meditations in the novel depict experiences that do little to take seriously plant ontology and instead replicate basic human assumptions about plants that conveniently reflect Griffin's own figurative paralysis. Meditation 3, written in the second person, longingly imagines the pleasant emptiness of vegetal life that continues undisrupted until the appearance of a vicious intruder. "You stand in a field surrounded by family," it begins, before declaring, "The weather is kind. Nothing will ever change. . . . Centuries pass." But soon a vibration advances across this eternally pastoral idyll: "Boom pause boom pause boom pause boom pause boom boom boom pause boomboomboomboom faster now, the heavy running feet of an animal new to the forest." Helpless to defend yourself from this sonic terror, "you find yourself all at once chewed and torn."[11] The hallmarks of Wright's plant narrator are writ large here. The plant, so clearly a substitute for Griffin, is traumatized and destroyed by the violence of warlike conditions from which it cannot escape. But this third meditation also highlights some basic common assumptions about plants. The field plant is as static as the houseplant that opens the novel. Furthermore, both plants exist at the whim of a higher, more intelligent power. The houseplant must depend on broken human hands for nourishment, while the field plant is unable to counteract or respond to the destructive animals that share its space. Despite encouraging readers to imaginatively inhabit a plant's experience, the novel does little to prompt readers to shift from dominant representations of vegetal life. Instead, its narrating plant reconfirms such common ideas.

We find a wholly different scenario in Ursula Le Guin's plant-narrated short story "Direction of the Road" (1974).[12] If Wright's novel illustrates a specific human experience by reproducing basic assumptions about plant ontology, or a lack thereof, Le Guin's story is primarily concerned with illustrating the essential *differences* between human and plant experience. The narrative is above all about the ways in which a human misunderstands a plant, to that human's ultimate demise. Narrated by a roadside Oregon oak tree, Le Guin's narrative actively debunks the stories that many humans impose on the plants that surround us—stories that, in this case, disregard the agency of the oak and present readers with a model of the world that does not correspond to the oak's own experience of its environment. The story ends with the oak tree lamenting the fact that it is misperceived by a human motorcar driver whom the oak is forced to kill when, on a rainy March evening, the driver steers his car on to the wrong side of the road and disrupts what the oak calls the "Order of Things."[13] The oak explains that it does not protest having to kill the driver but does protest the fact that the driver will forever associate the oak with a static, timeless eternity. "He saw me under the aspect of eternity," the oak states. "He confused me with eternity. And because he died in the moment of a false vision, because it can never change, I am caught in it, eternally."[14]

By perceiving the oak in this way, the driver joins in a well-established Western tradition of viewing oak trees in terms of long-lasting power. As William Bryant Logan explains in *Oak*, his biography of the oak tree, throughout European history oaks have symbolized "permanence, dignity, strength, grace."[15] For Logan, this sense of fortitude and dominance stems in part from the very form of the tree, which suggests to him a "patient power." Peter Young agrees in his own study of the oak tree. He explains that, of all the trees in the forest, the oak boasts "the most enduring and solid wood."[16] The authors of yet another study of the oak assert that the species' affiliation with endurance is a result of the potential lifetimes of individual oak trees, which far exceed human life spans. Many mature oaks live for longer than three hundred years, and some are even rumored to be up to one thousand years old.[17]

For these reasons, Young suggests, oaks became associated with fertility, renewal, and permanence in cultures across Europe. Celtic cultures would each year celebrate the oak king's defeat of the holly king on the longest night of the year. Called "Yule," from a Scandinavian word for "wheel," the defeat was thought to ensure the turning of the wheel of life and signal the coming prosperity of longer days.[18] Celtic cultures also associated the oak with the doorway to the Otherworld, which they saw as a place of new beginnings and reincarnation. The ancient Greeks viewed the acorn as a symbol of reproduction and renewal given its visual similarity to the tip of the penis. Many cultures, from the Greeks to the Norse, also linked oaks to the sky gods, such as Zeus and Thor, who themselves were affiliated with fertility and power because of their ability to control rain and thunder.[19] Young explains that many early Europeans even looked to oaks to preserve their own longevity and strength; some believed that rubbing an oak would ensure good health for a year, while others drove nails into oaks after driving the same nails into their own wounds to transfer pain to the tree.[20]

In Le Guin's story, the driver's perception of the oak as eternal echoes such ancient myths and ideas. The oak stands strong, firm, and unmoving for the driver. It is an out-of-time, static thing that does not move, will not move, and will remain in place in perpetuity, both for generations to come and the everlasting moment of the driver's final vision. It is, in other words, not truly alive—not capable of experiencing the world with intelligence or agency, but a passive object that symbolizes ideas of strength and permanence. It has its story prescribed by others, and thus is not an individual capable of defining its own narrative according to its own experience of the world.

But Le Guin's oak has other ideas. For it, being imagined as immobile and timeless is tantamount to death. "I am not death. I am life. I am mortal," the oak declares shortly after it strongly rejects human presuppositions of oak life: "Eternity is none of my business. I am an oak, no more, no less. . . . Long-lived though I may be, impermanence is my right. Morality is my privilege. And it has been taken from me."[21] The tree rejects the anthropocentric notion that it is everlasting and thus somehow not alive by emphasizing its specific perception and experience of space and time. On a basic level, Le Guin draws attention to the oak's intelligence and agency by highlighting its memory via the narrative's retrospective focus. Such a representation helps readers appreciate that, while the oak's life span may greatly exceed a human time scale, this does not mean that the plant is static or that its life is unchanging. The oak recalls all of the events in the story from its past, suggesting that it is not only capable of perceiving and registering the changing circumstances that surround it but also remembering them at a later date. In light of the altercation with the driver, the plant's most significant memories at this moment in its life have to do with developments in human transportation. It begins the story reflecting on the calmer, slower days in which humans rarely moved faster than a "jigjog foot-pace," and continues to reflect upon the increasing rapidity brought to its own life by the popularization of horses and the later invention of the motorcar.[22] These memories make one thing clear to Le Guin's readers: to inhabit the oak's point of view, they must envision a life span beyond normal human conditions but one that is unquestionably aware of and concerned with change.

The oak takes such an interest in the motorcar because of its desire to uphold the order of things. Via its descriptions of its attempts to maintain the laws of physics, it tasks readers with inhabiting a wholly unfamiliar experience of space in addition to grappling with an unfamiliar scale of time. The narrator begins the story by longing for the quieter days in which humans walked: "They did not used to be so demanding . . . when one of them was on his own feet, it was a real pleasure to approach him."[23] The oak's use of the verb *approach* here is a good indication of the complexity with which it interacts with humans and its own agency in the process. Humans do not approach this plant; instead, the plant itself moves toward humans, growing and then shrinking at a rate that upholds the status quo. The tree describes how it "synchroniz[es] the rate of approach and the rate of growth perfectly" as it moves toward passers-by in those days, so that the plant "hung above . . . loomed, towered, overshadowed" the humans that walked beneath it. The oak longs for these simpler days, in which there was time to "accomplish the entire act with style."[24] As the pace of human life increases and more motorcars populate the road—oftentimes several at once—the oak is forced to increase the rate of its own growing and shrinking to maintain basic laws of relativity, such that the plant must develop new skills in approaching humans in two directions at once and in split seconds. The oak manages this "all in a hurry, without time to enjoy the action, and without rest: over and over and over."[25] Hence its frustration when the motorcar driver betrays the order of things by veering his vehicle onto the wrong side of the road. The oak has dedicated its life to upholding specific properties of space and time and takes offense when the driver betrays this perspective.

The numerous passages of spatialization in the oak's narration illustrate a fundamental difference between the ontology of humans and that of this plant. The humans, unwilling or unable to perceive the oak's mobility, fail to appreciate the plant's increases and decreases in size. This oak *does* move—it is not permanent or eternally the same—but it simply does so in a way that is unrecognized by the humans that surround it. The passages also cheekily call attention to the correlation between the laws of physics that define the oak's life and anthropocentric experiences of the world, while at the same time imaginatively upending the dominance of such human-centered experiences. The oak must grow and shrink to maintain a specific perspective of relative size—to uphold an "order of things" in which humans inside and outside of this story world so easily place themselves at the center. Yet by representing the order of things from its perspective, the narrator tasks readers with modeling a world in which they are not at the center. To comprehend the story, readers must reimagine a common roadside scene and basic laws of physics from a completely unfamiliar perspective, in which a plant lies at the heart.

Indeed, if there are any beings in the oak's story that appear unthinking or unfeeling, it is the humans who ride in the cars. For the narrator, such creatures do not appear to engage much with each other or their surroundings. Many of the humans the oak encounters do not bother to look at it or really see *anything* around them: "They seemed, indeed, not to see any more. They merely stared ahead."[26] Those humans that do engage with the narrator—such as the men who used to routinely gather to drink together while leaning on its trunk—seem more interested in dulling their perception of their surroundings than appreciating the astounding complexity of the order of things and the mechanics by which such an order works. While the oak admits a fondness for such humans, it highlights their lack of intelligence with a biting dismissal: "They have seldom lent us Grace as do the birds; but I really preferred them to squirrels."[27] The plant's brush-off of humans places them in a drastically different spot on the evolutionary chain than they tend to place themselves. In this story, humans are one step above rodents but not as advanced as birds. Humans may easily place themselves at the center of the order of things in Le Guin's narrative, but its narrating plant knows better than to conceive of them as the most intelligent and aware beings around. It is the narrating tree that reigns large in this story, and Le Guin's readers must imaginatively inhabit its consciousness and perspectives to comprehend her narrative. They must move with it and be moved by its language to interpret the text.

THE TREE'S IMAGE

Thus far we have encountered two models of plant narrators: one (Wright's) that confirms common presuppositions about plants, and one (Le Guin's) that actively resists them by foregrounding a plant ontology wholly separate from that of humans. A third plant narrator, who appears briefly in Orhan Pamuk's novel *My Name Is Red* (2001),[28] also emphasizes plant ontology but in a way that not only calls attention to an anthropocentric imagination of plant existence but also the very *material* language of plants and the meanings that language can produce. A murder mystery set in sixteenth-century Istanbul, Pamuk's novel is about many things: the triumphs and pitfalls of love, the clash between Islam and Western religions, and the philosophy and power of art, to name a few. But it is also a dazzling display of polyvocality—a novel that is narrated by a corpse in its first chapter and thereafter features sections narrated by twenty additional distinct voices, including that of a dog, a gold coin, death, a horse, Satan, the color red, and a tree. Indeed, *My Name Is Red*'s most striking feature is its unique brand of heteroglossia and Pamuk's implicit insistence that the more-than-human world plays an essential role in human affairs.

Although Pamuk's narrating plant appears only briefly in the novel, it gives voice to the text's main philosophical dilemma. The plant begins its four-page narration by declaring to an audience in an Istanbul coffee shop, "I am a tree and I am quite lonely."[29] Readers quickly discover that this vegetal loneliness does not reflect symbolically human loneliness, as it does in Wright's novel, but is an isolation of signification. Pamuk's narrating tree is not actually a speaking tree, but the speaking *image* of a tree that does not know where it belongs. The narrator was originally meant to be among the pages of a magnificent book commissioned by the nephew of a Persian shah, but was separated from its master text when the shah became jealous of the nephew, banished him from their city, and scattered the book's illustrators and illustrations to the workshops of other sultans and princes. The lonely image of a tree now lies separated from the text and other images that it believes determine its larger meaning.

This isolation causes an existential crisis—"Ah, to which story was I meant to add meaning and grace?"—and also encourages the narrator to pontificate on the meaning of art in general, and visual images in particular.[30] At the crux of Pamuk's novel lie two culturally determined styles of illustration: an Islamic style that seeks to capture things not as they actually are but their ideal images as determined by Allah, and a Frankish (European) style that privileges an anthropocentric view of the world in its attempts at mimesis. This argument over style is the motivation for the murder that opens the novel and all of the action that follows it. The image of the tree is very clear where it stands on the issue. It voices its relief that it has not been sketched in the new Frankish style, not in the least because "if I'd been thus depicted all of the dogs in Istanbul would assume I was a real tree and piss on me." The tree's image ends its chapter with a firm declaration of its allegiance to the Islamic tradition: "I don't want to be a tree, I want to be its meaning."[31]

Pamuk's narrator longs to join a larger text and symbolize the ideal tree. Its imagination of its potential fate illustrates what it believes such an ideal involves: it wonders if it is "perhaps meant to provide shade for Mejnun disguised as a shepherd as he visited Leyla in her tent," or "meant to fade into the night, representing the darkness in the soul of a wretched and hopeless man."[32] Similar to Wright's narrating plants, Pamuk's narrator conceives of a tree's life only in that plant's relationship to the humans that surround it. In this way, the narrator supports anthropocentric notions of vegetal life as inferior and existing only to serve humanity, literally or figuratively. We might pin this implicit support for an anthropocentric imagination of plant life on the fact that, later in the novel, Pamuk's readers come to realize that the image of the tree does not actually speak for itself but is given voice by a coffee-shop storyteller who is ultimately killed for his desire to impersonate other beings. The image of the tree thus acts as a sort of ventriloquist dummy, allowing the human storyteller to imagine what a tree might say if it could indeed speak. Pamuk's narrator is ultimately linked to Wright's plants, in that both are voiced by humans fantasizing about what it is like to be something other than human—what it is like to experience the world from a vegetal ontology.

But Pamuk's narrating tree departs strikingly from Wright's plants in the stories it tells. While Wright's plants are limited to recasting the experiences of a human protagonist in vegetal form, Pamuk's narrator is a skilled and creative storyteller. Indeed, the great irony of Pamuk's image is that it does not seem to realize that it is capable of producing its *own* meaning; while voicing its longing to be reunited with its master text, it weaves stories of its own to delight its audience.[33] It tells a grand origin story in narrating its autobiography that rivals any text that the shah's nephew's book might have featured. Entitled "Falling from My Story Like a Leaf Falls in Fall," the epic tale involves religion, royalty, madness, love, and journeys across great distances. No one-trick pony, the narrator displays its creativity and cross-genre capabilities when it tells two additional fictional narratives, one of a pair of Frankish painters discussing virtuosity and art while walking through a meadow, and one of a Turkish lord who swears off coffee after copulating with the devil. Both narratives, while brief, are sophisticated in their use of dialogue, setting, humor, and figurative language. In short, the narrator is a great storyteller, and its creativity betrays the human ventriloquist's anthropocentric notions of what it is like to be a plant.

Beyond this self-determination, Pamuk's novel also emphasizes plant agency in a second, more subtle way. Throughout this short chapter, Pamuk continually conflates the image of the tree with an actual tree, thereby betraying the narrator's desire to be not a tree but the meaning of a tree. The narrator's story of the Turkish lord and the devil is the most obvious moment of such slippage: the narrator describes the devil climbing "up into this branch of mine" and hiding "beneath my lush leaves."[34] At other points in the chapter, the narrator uses figurative language to illustrate its close connection to trees, such as when it describes itself as falling from its master text "like a leaf in autumn."[35] Indeed, the connection is so strong that the narrator expressly identifies itself as a tree in the chapter's opening and closing paragraphs: as I mention above, it begins by stating "I am a tree and I am quite lonely," and finishes by referring to itself as "the humble tree before you."[36] The chapter itself is even entitled "I Am a Tree." Although the narrator claims it would rather be an image of an ideal tree than the tree itself, it continually imagines and presents itself as a physical tree.

The narrator heightens this sense of slippage also by calling attention to the very vegetal material with which it is composed. Twice in the four-page chapter the narrator refers to its own materiality—it directly mentions the "nonsized, rough paper" and "coarse paper" on which it appears before its audience.[37] Such moments of materiality, coupled with the narrator's (and reader's) struggle to separate fully the tree and its image, emphasize the fact that the narrating image is made of a tree not only in its image but also in its content. In other words, the image is only made possible by tree material that has been transformed into paper. There is thus a certain amount of irony in the narrator's quip that if it had been depicted according to Frankish traditions of mimesis all of the dogs in Istanbul would assume it was a real tree and piss on it. One strong reason that the narrator cannot dissociate itself from the subject of its image is because they are composed of the same material parts.

We can make this connection even stronger if we recognize that the most common source of ink in sixteenth-century Europe was oak gall. As Logan explains, ink makers mixed crushed oak galls with vitriol, water or wine, and gum arabic.[38] This mixture produced blue ink that first became popular in Europe in the eleventh century when crusaders, returning from modern-day Turkey, brought these ingredients home with them. Oak gall ink remains throughout the traces of Western history: governments used it for official documents; architects used it for their blueprints (including Thomas Jefferson's drawings for Monticello and the University of Virginia); artists such as Leonardo da Vinci used it for their drawings; musicians, including Johann Sebastian Bach, transcribed music with it; and, importantly for Pamuk's narrator, artists including Albrecht Dürer, Rembrandt Harmenszoon van Rijn, and Vincent van Gogh drew with it.

Pamuk's narrator's comments on its very material identity remind readers that there is more than one type of vegetal language at work in this chapter. On the surface level a ventriloquist storyteller plays an imaginative game, wondering what type of language a tree would use if it could speak for itself. But on a deeper level the material language of an actual tree is at work here also, as it is the vegetal matter of that tree that permits the existence of the signifying image leading to the imaginative game. We can see this chapter as thus calling attention to the very essential and fundamental role that plants play in written narratives. After all, for the majority of human civilization it has been impossible to record and circulate a narrative *without* vegetal matter, whether this matter is the plant materials that are processed to become paper or the oak gall that is mixed to produce ink. Pamuk seems intent to foreground this material nature of narrative, not only in the image's references to its own material composition but also in the novel's closing pages. In a playfully postmodern move, *My Name Is Red*'s final paragraph informs readers that what they have been reading is in fact the transcription of Orhan, the son of one of the text's ill-fated lover protagonists. The text explains that to compose the story Orhan consulted letters written by various characters and the "rough" illustrations with the "smeared ink" of the coffee-house storyteller.[39] In addition to drawing attention to his own agency in the storytelling process, here Orhan Pamuk ends his novel by privileging the transcription of oral narratives into written ones. The written word gets the last word in Pamuk's novel, once again implicitly highlighting the importance of the material on which that word is recorded.

This emphasis on materiality is especially fitting given that Pamuk's plant narrator is a tree and not, say, a blade of grass, as the material language of many trees is one type of vegetal signification that is immediately recognizable and even familiar to humans. Tree rings are rich sources of information about plant life and have long been the wellsprings of tree biographies and broader stories of forests at large. Pamuk's narrating tree, who confuses so thoroughly the representation of a tree and the matter of a tree, recalls the same doubleness of plant material and vegetal signification that we also find in tree rings. Indeed, Pamuk's chapter and the material plant language it evokes inspire a set of startling questions: If the basic definition of a narrative is a representation of a sequence of events,[40] what simpler narrative can we identify than a series of tree rings? Are tree rings in and of themselves narratives? *Are* some plants in fact capable of narrating their own stories, as material ecocritics Iovino and Oppermann suggest they might be and Pamuk's narrating tree image implies? If so, how do these stories differ from the stories that humans commonly tell about plants?

I do not want to suggest that trees themselves are capable of producing a narrative as sophisticated or as adventurous as Wright's, Le Guin's, or Pamuk's. While trees are capable of representing sequences of events, tree rings do not contain other hallmarks of narrative, such as focalization, representations of the consciousness of characters, metalepsis, metanarration, and heteroglossia. They are also not capable of changes in chronology or temporality; tree ring narratives progress steadily, with no analepsis or prolepsis to complicate the annual recording of events. But they do suggest that the very idea of narrative is not limited to human storytellers. Although human storytellers must interpret and can elaborate upon the events that trees record in their rings, tree rings are sites of narrativity that suggest that trees are capable of producing their own meanings. They are, in other words, examples of a material language at work—a plant language that can in turn inspire human language and new human imaginations.

We find a solid example of such collaboration between plant and human language in Aldo Leopold's reading of tree rings in the "Good Oak" chapter of *A Sand County Almanac* (1949).[41] The chapter begins with what Leopold calls a "spiritual danger in not owning a farm": the idea that heat comes from a furnace.[42] This assumption is dangerous for Leopold because it ignores the very natural material on which humans depend for their survival, and thus increases the distance between human communities and their environments, the latter of which Leopold is eager to protect. To correct this assumption, Leopold dedicates his chapter to the material of a very "good oak" that warms his shins while a blizzard rages outside one cold February night. The chapter reads as a biography of the individual oak tree, revealed to Leopold and a gang of sawyers as they cut through the tree's eighty rings. Leopold presents the experience of cutting the oak as biting, "stroke by stroke, decade by decade, into the chronology of a lifetime, written in the concentric annual rings of good oak."[43] The biography at first works retrospectively, as the sawyers begin at the outer layer of oak rings and work inward, but eventually also runs the other way once the team hits the heartwood of the tree that marks its birth in 1865. As Leopold cuts he runs through an annual list of local events that affect the bank of the emigrant road on which the oak stands, from the flooding of a local river in 1913 and the first proclamation of Arbor Day in 1889 to the first stapling of factory-made barbed wire to oak trees in 1874.

Throughout the biography, Leopold is eager to stress how his experience as storyteller differs from the tree's experience as recorder of events. While the opening and closing paragraphs of the chapter are told retrospectively, Leopold writes the sections that depict the felling of the oak in the present tense and separates them by the frequent calls of the chief sawyer to "Rest!"[44] In addition to lending a sense of immediacy to these passages, Leopold's use of present tense draws attention to the human experience of the oak's demise; much as he would if he were picking up a written narrative to read, the human woodsman imaginatively relives the events recorded in the tree's biography each and every time he gazes upon the tree's rings. Leopold also places emphasis on the distinction between human and oak experience in the content of his narrative, such as when he notes that if the oak heard the stock markets fall in 1929, "its wood gives no sign."[45] Likewise, the oak did not "heed the Legislature's several protestations for the love of trees," such as the National Forest and forest-crop law of 1927, nor did it "notice the demise of the state's last marten in 1925, nor the arrival of the first starling in 1923." In these frequent acknowledgments of the oak's lack of experience, Leopold draws attention to the differing ways in which humans and the tree experience time. Indeed, Leopold's biography of the oak is odd in that it is equally as focused on what the oak does *not* register as what it does. Even events that do stand to mark the tree's wood, such as the introduction of factory-made barbed wire, appear to have no effect. Leopold "hope[s] that no such artifacts are buried in the oak now under my saw!" but never gives readers any indication that this is the case.[46]

This emphasis on immediate and past human experiences and a largely unknowable oak experience produces a biographical narrative that is best read as a delicate dance between human imagination and plant signification. The tree's rings provide the chronological bones of Leopold's narrative, allowing him to wander out and back along an anthropocentric timeline that traces changes in the local treatment of trees. He makes this relationship between human imagination and vegetal language explicit when he explains that the movement of the saw through the tree's rings provides a map of a narrative: "the saw works only across the years, which it must deal with one by one, in sequence."[47] Leopold as narrator does not attempt to impose a narrative upon the tree or incorrectly imagine the emotional texture of the tree's life. Instead he allows the oak's material language to dictate the pace and focus of his narration. The oak is not fully a narrator here, but instead provides the directions for a human narrator's story. Via his collaboration with the tree's language, Leopold encourages his readers not only to consider the history that the oak represents but also to acknowledge that history according to the oak's own timeline and language. He is aware that he cannot imagine exactly what it is like to be the tree, but by letting the tree's language prescribe the rhythm and locality of his narrative he recognizes the agency of oak that warms him.

Leopold's "Good Oak" chapter provides us with one model of narrative that is sensitive to the language of plants. The generic conventions of nonfiction limit his representation of the good oak; unlike Wright, Le Guin, and Pamuk, Leopold's interest in fact prevents him from featuring a tree narrator that uses human language to tell its story. Yet by framing human events in a way that takes into account vegetal experience and signification, we can see Leopold's chapter as engaged in a similar project to the fictional stories that I discuss above. By using a chronological structure inspired by plant ontology and acknowledging the material language of the tree, the narrative provides readers with a new way of imagining plant experience and signification. Leopold's chapter also provides us with a hint as to why three of the four texts I survey here feature tree narrators: because tree rings are such familiar forms of signification, perhaps they allow readers to imagine agentic trees more easily than intelligent mosses or legumes. But Leopold's keen sensitivity to the material nature of the oak, coupled with new research on plant signaling systems, begs a question of modern readers: How might narratives take cues from other, newly studied vegetal modes of signification? What would such narratives look like and into what new imaginative terrain might they usher readers?

Conclusion

My survey of these four texts demonstrates the various and complex ways that narratives represent plants as agentic beings capable of producing their own meanings that can construct, perpetuate, and subvert an ecosocial imaginary of vegetal life. They suggest that symbolism, chronology, spatialization, and a self-referential emphasis on the materiality of narratives themselves may be important narrative structures and techniques in the shaping and challenging of this collective imaginary. Pamuk's and Leopold's texts also suggest that in addition to rethinking our understanding of plants, we also may need to reconsider the very definition of narrative itself as a rhetorical form produced only by humans. But above all, they highlight the potential for narratives to push readers into new imaginative terrain by tasking them to experience the world as a plant. They thus demonstrate not only the power of narratives to provide us with the type of new language—a "plant" language—with which to better appreciate plant intelligence and agency but also the potential of such texts to force readers to consider the world via vegetal ontology. In other words, they show readers what plants might say if we only paused to listen.

Notes

1. Stephen Wright, *Meditations in Green* (New York: C. Scribner's Sons, 1983), 3.

2. Michael Marder, *Plant-Thinking: A Philosophy of Vegetal Life* (New York: Columbia University Press, 2013).

3. Daniel Chamovitz, *What a Plant Knows: A Field Guide to the Senses* (New York: Scientific American/Farrar, Straus and Giroux, 2012).

4. Serenella Iovino and Serpil Oppermann, "Theorizing Material Ecocriticism: A Diptych," *ISLE* 19, no. 3 (2012): 448–75 [468].

5. Richard Firn, "Plant Intelligence: An Alternative Point of View," *Annals of Botany* 93 (2004): 345–51 [346].

6. Anthony Trewavas, quoted in Michael Pollan, "The Intelligent Plant: Scientists Debate a New Way of Understanding Flora," *New Yorker*, December 23 and 30, 2013, http://www.newyorker.com/magazine/2013/12/23/the-intelligent -plant.

7. See Matthew Hall, *Plants as Persons: A Philosophical Botany* (New York: State University of New York Press, 2011). Hall makes a similar call for the study of plant-focused narratives when he states that "situated stories, songs, and poems

can be powerful aids to the recognition of autonomy and personhood in the plant kingdom." Although his focus is not on narratives narrated by plants, he does argue that "while expressing the human side of the dialogue, we also [must] allow others to 'speak' for themselves. Otherwise we risk falling back into destructive monologues" (162).

8. Michael Marder, "To Hear Plants Speak," this volume.

9. Wright, *Meditations,* 70.

10. Ibid., 3.

11. Ibid., 36.

12. Ursula K. Le Guin, "Direction of the Road," in *The Wind's Twelve Quarters: Stories* (New York: Perennial, 1987), 267–74.

13. Ibid., 274.

14. Ibid., 273–74.

15. William Bryant Logan, *Oak: The Frame of Civilization* (New York: W. W. Norton, 2006), 188.

16. Peter Young, *Oak* (London: Reaktion Books, 2013), 103.

17. For a thorough list of more than seven hundred named British oaks and their histories, see Esmond Harris, Jeanette Harris, and N. D. G. James, *Oak: A British History* (Oxford: Windgather Press, 2003), 208; on the longevity of oaks, see 146–51.

18. Young, *Oak,* 108–9.

19. For more on the worship of oak trees and oak gods, see James George Frazer, *The Golden Bough: A Study in Magic and Religion* (Mineola, N.Y.: Dover Publications, 2002), 184–87.

20. Young, *Oak,* 129.

21. Le Guin, "Direction," 274, 272.

22. Ibid., 267.

23. Ibid.

24. Ibid.

25. Ibid., 271.

26. Ibid.

27. Ibid., 268.

28. Orhan Pamuk, *My Name Is Red* (New York: Vintage International, 2002).

29. Ibid., 47.

30. Ibid., 49.

31. Ibid., 51.

32. Ibid., 49.

33. My reading of Pamuk's image as capable of producing its own narrative departs significantly from Ferma Lekesizalin's interpretation of the image as symbolizing a Lacanian lack, or a "free-floating, empty signifier." Lekesizalin

does not acknowledge the ability of the image to produce its own narratives, and instead argues that Pamuk's narrator is "a very disconcerting account of the incompleteness, imperfection of the symbolic order expressed in the confusion experienced by the missing tree as to what it represents." Ferma Lekesizalin, "Art, Desire, and Death in Orhan Pamuk's *My Name Is Red*," *English Studies in Africa* 52, no. 2 (2009): 90–103 [93].

34. Pamuk, *My Name*, 50.

35. Ibid., 47.

36. Ibid., 47, 51.

37. Ibid.

38. Logan, *Oak*, 191–92.

39. Pamuk, *My Name*, 413.

40. See Gerald Prince, *Narratology: The Form and Functioning of Narrative* (Berlin: Mouton, 1982); Prince defines "narrative" as "the representation of real or fictive events and situations in a time sequence" (1).

41. Aldo Leopold, *A Sand County Almanac, with Essays on Conservation from Round River* (New York: Oxford University Press, 1966).

42. Ibid., 6.

43. Ibid., 10.

44. Ibid.

45. Ibid., 11.

46. Ibid., 15.

47. Ibid., 18.

FOURTEEN

In the Key of Green?

The Silent Voices of Plants in Poetry

John C. Ryan

IN HIS 1826 *OBSERVATIONS ON THE GROWTH OF THE MIND,* SAMPSON Reed wrote: "Everything which is, whether animal or vegetable, is full of the expression of that use for which it is designed, as of its own existence . . . . Let [us] respect the smallest blade which grows, and permit it to speak for itself. Then may there be poetry, which may not be written perhaps, but which may be felt as a part of our being."[1] Since this plaintive appeal by Reed, allowing the "smallest blade" (or, prickliest spine or loveliest heart-shaped leaf) to speak has become a technological preoccupation for some. Let us begin with a typical example. Cactus Acoustics is a project that aims to allow saguaro cacti to vocalize.[2] We might imagine the voice of the burly saguaro as gruff and slightly imposing. Growing to considerable proportions—up to five stories high, eight tons in weight, and over a hundred years in age—*Carnegie gigantea* is endemic to the Sonoran Desert. As an adaptation to aridity, the cactus at times becomes an oversized sponge, absorbing hundreds of gallons of water a day during seasonal deluges.[3] Using an acoustic detector, researchers hope to correlate the sounds produced by the Sonoran giants to water fluctuations, temperature extremes, and ultraviolet exposure. In other words, they intend to elicit the environmental vocabulary of the saguaro: "I'm cold versus I'm really thirsty."[4] The result could be a device allowing plants to express their needs—endowing even the most hopeless plant minder with a green thumb.

The idea of plant voice takes a radical turn in this example, leading far from a poetry (but close to a technology) of being. Let us consider another instance of the "human desire for universal communicability."[5] The proprietor of a Japanese café affixed sensors to a *Hoya kerrii,* known as the "sweetheart plant" or "lucky-heart" for its cordate leaves. The equipment detects the bioelectric current of the sweetheart plant in response to its setting (including human movements in the café), and then renders the signals into Japanese words. Affectionately named Mr. Green, the plant composes a daily blog, including his observations of the weather: "Today was a sunny day. I was able to sunbathe a lot."[6] Internet users activate a lamp for his pleasure: "Being able to receive full light from the rear is delightful!"[7] Seemingly novel, the hoya project owes its lineage to digital works that have claimed to allow plants to speak, fly, and express their creativity.[8] Yet, the voices in these examples have distinctly (and perhaps eerily) human tenors. How might we permit a cactus and lucky-heart to speak for themselves without "objectifying them or, at best, speaking for them, in their defense, if not in their place"?[9] How might poetry assist us in doing so?

The (Im)Possibility of Plant Voice?

Just as the lucky-heart is delighted by the lamp's warmth, so too are humans enthralled by speaking plants, even those with artificial voice boxes. As listening to plants becomes an evermore-mediated activity, new mechanisms are purported to bridge the human–vegetal chasm while also engaging our imaginations. Such innovations are thought to promote interspecies dialogue by enabling plants, for example, to tell us when to water or feed them.[10] The problem is that, despite their endearing intents, these interventions are built upon an audiocentric *logos* privileging a narrow, monologic notion of voice. According to this *logos,* a plant voice *must* speak if it is to be heard; if it does not speak, it is mute, with neither register nor agency; and, furthermore, plant-speak requires technical prosthesis. The wild saguaro, the café sweetheart, and the household philodendron might eventually be able to communicate to us, in clumsy or sophisticated diction, but only by approximating human speaking, that is, by "ventriloquizing."[11] As Michael Marder argues, "the assumption that to have a language is to be able to speak is both erroneous and unethical . . . it ties linguistic phenomena to the voice, which only humans possess."[12]

The chortling hoya and the gruff saguaro are constructed to fulfill a desire for communion with the botanical world—one that is not normally within the scope of everyday interactions. However, these desires are met only partially in a manner that approaches plant puppetry—in which *voice* is the outcome of a kind of cyborgian vegetal subjectivity, of the humanization of the plant, of the making of the vegetal in the image of the animal. In asking "will the Saguaro give a scream when he is thirsty and make a different noise when he is too cold?"[13] the botanical imagineers of today might very well be ruling out the silent voices and embodied expressions that constitute the language of plants in all its complexity. Do these interventions disclose plant voices that have always been, or do they impose (or construct) voices that have never been—and are perhaps not meant to be, if indeed it is only humans who possess voice? Is the possibility of plant-speak valid through a scientific basis and culturally acceptable only through technologized examples like the saguaro and ventriloquized ones like Mr. Green? How might it be possible to think of voice otherwise—as nonverbal, bodily, and ecological articulation and as an ontological concern rather than an auditory phenomenon—and how might plants particularly help us in doing so? How might poetry be a medium for hearing and listening to plants—where language (rather than electric sensors and algorithms) becomes a shared and porous interface between the speaking Us and the silent plant Other?

This chapter will pursue these challenging questions in developing a concept of plant voice that resists, through language, the technological mediation of the cactus and the hoya. It will attempt to respond to Jennifer Peeples and Stephen Depoe's fundamental questions: "What is nature's voice? Does it 'speak'? If so, how? To whom? Can humans attend to the voice of nature?"[14] It will follow Reed's cue from over 150 years ago by proposing that, within critical plant studies, it is timely to reconsider the intricacies and possibilities of plant voice. The emphasis will be on language, not as human enunciation or a symbolic system but as "the house of Being,"[15] or, building upon the work of Eric King Watts, as "an original impulse of being."[16] Divided into two parts ("Theorizing Plant Voice" and "Poeticizing Plant Voice"), the chapter will develop an ecological (i.e., relational and material) concept of plant voice, first, through a critical review of ecocriticism and environmental communications studies and, second, through the contemporary poetry of Louise Glück and Elisabeth Bletsoe.

Glück and Bletsoe engage substantively with plant voice, intersecting as a result with issues of vegetal heteroglossia, embodiment, and ethics. As such, plant voice is the material elocution of the vegetal in its milieu, rather than a signifier of something outside itself. In their poetry, we witness a movement from plant voice as metaphor or literary maneuver to one intoned ecologically, as an outcome of a plant's interactions with other beings, its material environment, and human culture. Glück's *The Wild Iris* (1992) is a botanically inflected collection that received the Pulitzer Prize.[17] The work addresses themes of identity, domesticity, and plant–human communication, in which the poet, adopting the perspective of flowers, grapples with the heterogeneity of voice. Without speaking from the point of view of plants, Bletsoe's *Pharmacopœia* (2010) integrates knowledge of herbal medicines and botanical folklore as expressions of the historical plant voice articulated in the present.[18] I will argue that the poetic invocation of plant voice should not always be read pejoratively as pathetic fallacy. The examples will show that plant voice in poetry needs to be grounded in the material realities and sensory expressions of the plant in relation to its environment (including the human environment) and the creative intents of the writer. That plant voice depends on human desire (in this instance, toward poetry) affirms, rather than undermines, the extent of our interdependence with vegetal life and the need for ethical and inclusive concepts of voice.

THEORIZING PLANT VOICE

Plant Voice as Problem

First appearing in English in the late thirteenth century, the word *voice* derives from the Old French *voiz* for "speech or word" and Latin *vocem* for "sound, utterance, or cry." As a noun, *voice* can refer to the expression of feeling (as in "the voice of the people") or, as a verb, the act of expressing ("to voice an opinion").[19] However, *voice* most commonly denotes the sounds made by the human larynx, mouth, tongue, and lips to communicate in distinct tones and accents, making individuals known to other beings. To suggest that plants have voices might seem absurd or erroneous; anatomical sense tells us that plants cannot vocalize as humans do. For plant voice to be possible, then, we must think about *voice* differently, while refusing its purely metaphorical association with the vegetal. This dilemma is evident in science, where plant voice is a contentious and provocative figuration. The field of bioacoustics empirically suggests that plants have a kind of *voice*, yet the term is applied cautiously as a figure of speech and tellingly in scare quotes in relation to new scientific findings about plant communication.[20] Phytoacoustic research shows that plants emit unique sound signatures, enabling decision making and survival through communication with their habitats.[21] The ecological function of sound, by extension, implies attributes of agency and, arguably, intelligence in a form of life that has been construed as the antithesis of the animal—as mute, passive, and largely devoid of cognitive powers.[22]

Voice has also been a complex issue in philosophical and literary studies of nature. In these fields, plant voice has largely been assumed, shelved, or sidestepped as improbable—a fracture in an otherwise tenable thesis about the natural world, an imperfection in the narrative of a sentimental writer, a transgression against rational discourse, or an unfortunate flight of fancy. After all, nature cannot speak as we do. However, since the growth of the field in the 1990s, listening to nature's voice has been seen as central to the aims of ecocriticism, as Michael McDowell explicates: "Beginning with the idea that all entities in the great web of nature deserve recognition and a voice, an ecological literary criticism might explore how authors have represented the interaction of both the human and nonhuman voices in the landscape."[23] Considering these movements toward, biases against, and inconsistencies with the concept, the following discussion will stir the sleeping dragon of nature's voice as a tacit and fundamentally unresolved issue within ecological literary criticism.

Nineteenth-century British art critic John Ruskin coined *pathetic fallacy* to describe the attribution of feeling, emotion, and sentience to so-called inanimate nature. For Ruskin, this involves a "morbid state of mind, and comparatively . . . a weak one."[24] A botanical case in point from "Spring" (ca. 1850) by American writer Oliver Wendell Holmes is leveraged. The poem depicts a crocus imaginatively and sympathetically, but in androgenic terms: "The spendthrift crocus, bursting through the mould / Naked and shivering, with his cup of gold."[25] The hard facts of ecology reveal Holmes's lines to be "very beautiful, and yet very untrue. The crocus is not a spendthrift, but a hardy plant; its yellow is not gold, but saffron."[26] The distinction between human and nonhuman—and between intellect and imagination—is drawn sharply by Ruskin. (The crocus is not naked, but rather in its natural state; it does not humanly shiver, in response to cold or fear.) The protracted sentimentality of the poem would surely have caused Ruskin great anguish, from "her clustering curls the hyacinth displays" to "the robin, jerking his spasmodic throat."[27] (Retorts Ruskin: The hyacinth hasn't curls, but petals and sepals; the robin's voice is not spasmodic, but its means for attracting a mate.) Meanwhile, Holmes lamented his own powerlessness to hear nature speaking. Despite his figurative skill, he felt the "chains" of science (and ratiocination) ultimately separating him from the voices that stir and beckon him during this season: "Why dream I here within these caging walls / Deaf to her voice, while blooming Nature calls."[28] Thus we find a polarizing tension between the seeking of voice in Holmes's poetry and the refusal of voice in Ruskin's criticism.

How far has the principle of nature's voice come since Ruskin and Holmes? Not very. First, in ecocriticism, it remains overshadowed by the specter of the pathetic fallacy and the residues of modernist antisentimentality (involving an aversion toward expressions of emotion in poetry).[29] McDowell even concedes that pathetic fallacy, as the "crediting of natural objects with human qualities," is inevitable and "something to acknowledge and celebrate, not to condemn," but fails to tell us how to acknowledge and celebrate the phenomenon.[30] Second, closely associated with English Romanticism and American transcendentalism, nature's voice is usually conceptualized in sonic terms, as the speaking position of nonhuman beings. Speaking is vocalizing; if something does not have a voice, then it cannot be said to speak. Third, to attribute voice to nature is to elicit its moral consideration. If plants have voice, then they should be included within an ethical domain. Conversely, if they lack address, they are more conveniently relegated to the background.[31]

These concerns (of tradition, modality, and ethics) play out as ambivalence toward voice in mainstream ecocritical scholarship. Greg Garrard describes the "post-Romantic problem" of voice as "necessarily human and 'reflective' and yet almost naively open to the natural 'other.'"[32] The silent voice of nature, for Garrard, is an oxymoron; never lacking sonic register, voice is auditory expression. This audiocentric bias is raised in relation to Percy Shelley's poem "Mont Blanc." The mountain is endowed by the poet with voice and, by extension, agency: "Thou hast a voice, great Mountain, to repeal / Large codes of fraud and woe."[33] A Heideggerian "letting beings be" appears as a solution to the inescapable anthropocentrism of giving voice—as a means of releasing voice from sentimental human states of mind and desire.[34] How successful this decoupling of voice from reflection is we cannot be sure, as the possibility is presented in passing. However, we can be more certain that, for Garrard, following ecocritic Jonathan Bate, the viability of nature's voice as a concept is contingent on a dwelling in language, rather than constructing language as symbolic or referential. The voice of the mountain is less about *language* and more about the language of things.[35] However promising, this line of argument is only faintly sketched.

The presumption that nature's voice is out of the reach of Western consciousness is principal in Bate's *The Song of the Earth*. He posits that the "inheritors of the Enlightenment's instrumental view of nature" can only conceptualize voice as a metaphor, unable to understand dialogue with plants as a legitimate form of discourse.[36] Bate admits that this could differ for Aboriginal Australian and other indigenous peoples for whom the land sings and is impregnated with meanings and voices—where a dialogic tradition between humans and nonhumans exists in everyday awareness.[37] However, rather than confronting the stigma of pathetic fallacy and opening up possibilities for nature speaking (and, for that matter, nature writing), a series of speculations about John Clare's "The Lamentations of Round-Oak Waters" instead implies the implausibility of voice. Here, a stream vocally protests its environmental conditions. We might reflect back to the sentient saguaro ventriloquizing through an acoustic detector. But there is a difference: the stream's habitat has been fragmented by the British enclosure system of the nineteenth century. Although provocative, his queries sidestep an unsettled (and unsettling) issue in ecocriticism: "Is the voice of Round Oak Waters to be understood only as a metaphor, a traditional poetic figuration of the genius loci, or 'an extreme use of the pathetic fallacy'? Or can we conceive the possibility that a brook might really speak, a piece of land might really feel pain?"[38] To compound this ambivalence, Bate asks if attributing the ability to speak to nonhumans corresponds to their capacity for pain. Is there a potentially dangerous correlation between speaking, agency, and moral consideration? In other words, if the boundary between the animal and the plant blurs, then what will the consequences of plant voice be for societies that now callously exploit the botanical kingdom as mute material and voiceless resource? Unfortunately for plants and us, these questions are left unanswered.

Plant Voice as Potentiality

For ecocriticism, environmental communications, and critical plant studies, the voice of the vegetal should be a core theoretical concern. It is productive for scholars in these fields to reimagine the potential of nature's voice in order to understand its relationship to literary, cultural, and scientific works, rather than dismissing it as anthropocentric or questioning it open-endedly. This shift involves being aware of the preconceptions of theorists for whom voice is both figurative and contingent on human subjectivity. For instance, Christopher Manes argues that "attending to ecological knowledge means metaphorically relearning 'the language of birds'—the passions, pains and cryptic intents of the other biological communities that surround us and silently interpenetrate our existence."[39] For Manes, voice is symbolic, enigmatic, and forgotten; relearning involves deciphering, in which nature is deconstructed like a linguistic code or structurally analyzed like a text. As another example, in *The Natural Contract*, Michel Serres observes that our language allows us to "communicate with mute, passive, obscure things—things that, because of our excesses, are recovering voice, presence, activity, light."[40] Although ethically inflected, this statement reveals a tacit anthropocentrism in which "mute" things have voice only if we choose to correct our "excesses" and grant them a speaking position.

The potentiality of voice also comes up against the stigma of nostalgia—that accepting voice is an undesirable recoiling to a prediscursive idyll or an abyssal leap forward: "But where is the voice of nature calling us? *Back* to the pre-modern age? Or *forward* to a saner future?"[41] This sense of polarized temporality (past or future) is compounded by a sonic bias (the auditory dominates). As a further example, Lawrence Buell regards the voice of Walden Woods as a synecdoche rather than an actual attribute of the forest, notwithstanding Henry David Thoreau's ostensible belief otherwise.[42] Thoreau, for whom voice was something real, though inaudible, acknowledges in *Walden* (a seminal text in ecocriticism and nature writing, originally published in 1854) "the language that all things and events speak without metaphor."[43] His concept of *effluence* is illustrative. It concerns the olfactory register of fruits and flowers—an expansion through sensory awareness of nature of the bounds of voice and language. Through effluence, the perceiver and the perceived are more intimately connected; the naturalist becomes receptive to the diverse expressions (or languages) of species.[44]

Whereas ecocriticism has faltered, environmental communications theory has intensively focused on the relationship between voice and nature, offering the potential to enhance the ecocritical debate.[45] For these scholars, nature's voice is not an intractable issue to be skirted but, to the contrary, one that contributes to a critical reevaluation of the very premises of voice itself. Peeples and Depoe argue that "these questions stretch the theoretical and material dimensions of how we understand the relationship between voice and the environment."[46] The voice of the crocus or cactus necessitates rethinking language and communication, in order to postulate an affective theory of voice. As affect, voice extends to linguistic and nonlinguistic acts alike, leading to ethical implications for the rhetor and the receiver. Rather than privileging voice narrowly as the attribute of a speaking subject or an outcome of linguistic discourse, Watts characterizes voice as relational and never separable from the body.[47] Voice registers being to other beings, while also communicating identity and ideology. To develop a more inclusive voice, however, we must refuse its rigid association with speaking authority and highly individuated subjectivity.[48] Watts goes on to present an incisive yet poetic definition of voice as "the sound of affect. Voice emanates from the openings that cannot be fully closed; from the ruptures in sign systems, from the breaks in our imaginaries, from the cracks in history. It registers a powerful, some would say passionate, cluster of feelings triggered by life finding a way to announce itself."[49] It is the particular ontologies of plants (their inaudibility, lack of address, and relative fixity in place) that further stretch the "theoretical and material" limits of voice. As Marder (in response to Socrates) acknowledges, "unlike an animal, the plant has no voice (this explains its reticence), and it is incapable of spontaneously choosing its place by exercising the freedom of self-movement (which justifies its sealed character)."[50] In contrast, I argue that the plant does have a voice, but not of the animalistic kind—a voice that perfectly corresponds to its ontology and habitus, as the plant finds a way of announcing itself. Moreover, the "sound of affect" should also extend to plants in recognizing their capacity to both affect and be affected. And poetry is a means of listening to and expressing plant voice as potentiality, as "breaks in our imaginaries" and "cracks in history."

Plant Voice as Presencing

Understanding plant voice requires conceiving of possibilities beyond voice. Here is another key, already touched on: the linguisticity of plants becomes a matter of ethics. As Yi-Fu Tuan observes, when something is thought to fall silent, we tend to conceptualize it as dead (its presence is effectively lost).[51] For Reed, respect for plants is the precondition for their speaking, for their having voices in the first place and expressing something on behalf of themselves. The Genevan philosopher Jean-Jacques Rousseau (who was also a lifelong student of botany) claimed that human conscience is the voice of nature active within us.[52] For Rousseau, neglecting to listen to nature's voice leads to existential crisis; rather than incomprehensible or mute, the voice of nature is in fact accessible to those who learn to apprehend it.[53] But many of us—including myself—have tried in the most respectful and sober manner (that is, without the aid of psychedelic compounds) to hear their chatter. We have tuned our ears to green things of many shapes and sizes, and in many places in the world, only to apprehend the swoosh of the wind through leaves or the thumping of our hearts in their "caging walls" (to borrow a phrase from Holmes). Compliments spontaneously uttered to flowers—"well, aren't you a beauty," as if to prompt a conversation—are invariably met with the reverberating silence of flawed yearning, like a stone rattling inside an empty steel tank.

To understand plant voice as presencing, the voices of plants (their *internal* voices, produced by them) should be distinguished from the giving of voice to plants (their *external* voices, imposed upon or granted to them by us). However, there is also a middle ground that threads between these categories. This involves plants speaking for themselves, in which they express their voices in myriad ways as we present to them (and ourselves) the appropriate conditions for doing so (such as unfragmented habitat, pollinators, sunshine, respect). This third category—which I will call an ecology of plant voice—attempts to mediate the binaries of nature–culture, subject–object, human Us–plant Other, acting–being acted upon, and speaking–being spoken to, which limit the emergence of plant voice as dynamic interrelationality. Such a conception runs against the reduction of "the language of things to human language" and opens things to speaking "only to the extent that their linguisticity is not . . . a matter of metaphor."[54] It also attempts to refuse the alignment of voice to the speaking position of the plant. The danger there lies in reverting to an individuated, nonrelational, and aurally based conception of voice—a simply unachievable agenda for the plant, except with the aid of a prosthesis.

The first dimension (*internal* voice, which phytoacoustics points to), therefore, needs to signify something beyond audible speech or rudimentary forms of vocalization—beyond human language. An ecological plant voice encompasses plants' silent presences in space and time, their sensory articulations within an *Umwelt,* and their modes of signification— not mute but silent, not indecipherable but corporeal, not of the brain but of "mindlessness."[55] Similarly, the second dimension (*external* voice, which phytopoetics points to) must involve more than our speaking for nature (consider old-growth forest campaigns as an example) or representing plants as thinking and sensing beings. While plants cannot be said to speak in modes recognizable to us (even with acoustic detectors), they do express through manifold means that at once affirm their familiarity and strangeness. Some beings use speech, but everything, including the vegetal, speaks via presence, or "voice/presence."[56] It might, then, be possible to think of two modes of plant voice as dialogic sides of the same grape vine; that speaking for plants in poetry, as an act informed by their ecological and material realities, is in fact more ethically inflected than not writing anything from their perspectives, or, worse yet, objectifying them in language. Moving beyond a concept of voice as speech or aural utterance to one of presencing, let us now consider how these ideas of plant voice (as relational, corporeal, sensorial revealing) manifest in the poetry of Glück and Bletsoe.

Poeticizing Plant Voice

Plant Rhetorics in Glück's *The Wild Iris*

Louise Glück is an American poet (b. 1943) whose work over nearly five decades beginning with the collection *Firstborn* (1968) has been described as spare, with "subtle, psychological moments captured by the austerity of her diction."[57] In confronting states of existential unease, Glück's poetry has also been characterized as stoic and confessional in its "agitated, relentless imagery and language."[58] The Pulitzer Prize–winning volume *The Wild Iris* (1992), however, marks Glück's departure from the stable, undifferentiated writerly voice characteristic of the confessional poetics of her earlier work. Rather than a unified speaking position, *The Wild Iris* exhibits a range of subject positions in which the authorial self crosses "the border between the human and the not human, as under construction, and in a state of becoming."[59] The work has been called a "polyphonic theater," "heteroglossic text," "prayer sequence,"[60] "radically heteroglot volume,"[61] and "lieder cycle . . . written in the language of flowers."[62] Its alternation between "Matins" and "Vespers" (used as poem titles) invokes the hours, or the Roman Catholic morning and evening prayer cycles. Within this religious structure superimposed over a Vermont garden's seasonal cycle, *The Wild Iris* adopts three lyric categories: an omniscient God figure, a poet-gardener-supplicant, and fourteen flowering plant speakers.

The plant personae include wild and cultivated species such as the titular wild iris, trillium, snowdrops, Jacob's ladder, and hawthorn. Critics have described the interspecies dialogism of Glück's plant characters as figurative and inescapably mediated by the human voice. Accordingly, they stress that the plant voices are symbolic representations of God projected onto nature, metaphors of "the self when imagined as a speaking flower," and foils "for an internal conversation" in which the poet ventriloquizes plants.[63] Similarly, others observe the humanlike voices of the flowers, conceding flatly that "the human writer has no other voice to give them."[64] While it is irrefutable that the text presents flower voices in human terms, a reading of Glück's engagement with ecological plant voice will elicit nuances not possible in these analyses. The speaking flowers vocalize in English diction, of course, but also articulate spatially and sensuously as an expression of their ecological situatedness. Their speaking is plant voice as "life finding a way to announce itself"[65] with the poet and her audience becoming witnesses to their language and facilitators of its textualization (where the flowers still speak to readers like us). Glück's plant voice is underpinned by an appreciation of ecological cycles, relationality between beings, and "the shared materiality of the earth body and the personal, particularly the female, body."[66]

The first poem in the volume, "The Wild Iris," is written wholly from the plant's perspective and is redolent of the Greek myth of the vegetation goddess and ruler of the underworld, Persephone.[67] The iris flowers and leaves die but then regenerate from the bulb, offering a compelling symbol for natural cycles of life and death and signifying metaphysical reflection on nature.[68] To read the voice of the iris figuratively, however, shuns the material realities of the work, as well as the close association between voice, consciousness, and spatial articulation it develops. The iris's direct address to the poet-gardener discloses its chthonic memory, reaching into the depths of time and earth and exceeding the limited human capacity to remember: "Hear me out: that which you call death / I remember."[69] Moreover, the iris (specifically its rhizome) is self-aware, enduring seasonal interment "as consciousness / buried in the dark earth."[70] Glück's poetic assertion might not be far from the empirical reality of plant cognition. Scientific research dating back to Darwin in the nineteenth century has suggested controversially that the brain of the plant is located in its root tips.[71]

Through the iris's voice, its consciousness, spatial awareness, sound perception, and a poietic sense of time suffuse spare lines such as the following: "Overhead, noises, branches of the pine shifting."[72] Rather than an auditory phenomenon (the plant does hear noises, but it is spatially aware), plant voice here is a vinculum between the different articulations of the iris announcing its presence to other beings, including the poet-gardener-supplicant and the reader. The concluding stanzas posit plant voice as the sensuous, nonverbal expression of the flower negotiating its subterranean and aboveground milieus:

> You who do not remember
> passage from the other world
> I tell you I could speak again: whatever
> returns from oblivion returns
> to find a voice:
>
> from the center of my life came
> a great fountain, deep blue
> shadows on azure seawater.[73]

The iris's mode of address approximates Watts's description of voice as "the enunciation and acknowledgement of the obligations and anxieties of living in community with others."[74] The iris voice resonates with tactile ("great fountain"), visual ("deep blue"), and affective ("returns from oblivion") expressions, as the flower emerges from its underground dormancy not as an individual subject but in community with "the stiff earth," "birds darting in low shrubs," and other living beings and things.

Throughout *The Wild Iris,* plant speakers make themselves known spatially, while the poet observes the workings of ecological voice in her garden. A few more examples from the volume will show the polyvalence of plant voice. Empathic feelings between plants and humans are part of the text's voicing. The "small blue" flower of "The Jacob's Ladder" addresses the grieving poet, presumably lovesick, from outside her bedroom window. The self-aware flower enunciates itself gesturally in a tone of longing evident in its "naked stem / reaching the porch window."[75] The flower's language, physically pronounced, exudes empathy for the woman and shares with her a desire for transcendence from *their* earth: "Never / to leave the world! Is this / not what your tears mean?"[76] In addition to the flower-speakers, "Matins" furthers the theme of embodied empathy between plants (here, a birch tree) and humans, with the poet-speaker "actually curled in the split trunk, almost at peace / in the evening rain / almost able to feel / sap frothing and rising."[77] We find in this excerpt the birch voice as a polyphony; its inner physiology and outer presence in the garden space register in the "flesh" (to borrow from Merleau-Ponty) of contact between the human and vegetal. The poet concedes in another "Matins" that discourse with the birches—even of the silent, somatic kind—could be perceived as a flight of fancy, but nevertheless melodramatically exhorts her critics (her husband, her son, her father, her God, others) to "bury me with the Romantics / their [the birches'] pointed yellow leaves / falling and covering me."[78] Understanding birch voice, for the poet, is an empathic nonverbal act in which the tree's ecological presence engulfs her.

In these examples, plant voice is immanent rather than purely symbolic; multisensory rather than solely auditory; collective, not merely singular. The corporeal register of plant voice continues with "The White Rose," in which the flower summons the poet-gardener with its elegant gesticulations. At the same time, the rose laments its inability to conceal itself, or to turn its "voice/presence" off:

> I am not like you, I have only
> my body for a voice; I can't
> disappear into silence–

> And in the cold morning
> over the dark surface of the earth
> echoes of my voice drift,
> whiteness steadily absorbed into darkness[79]

In contrasting its linguisticity to the poet's, the rose asserts its right to expression in the mode that is particular to it. The rose voice is modulated by its earthly habitus rather than its own volitions. "Scilla" goes so far as to chide the poet-gardener for clinging to the belief that voice involves a human subject seeking a personal God. Instead, the flower speaks passionately for the collectivity of the garden through its body language: "Not I, you idiot, not self, but we, we—waves / of sky blue like / a critique of heaven."[80] For the flower-speaker, individuated voice (i.e., me speaking) is unthinkable: "why / do you treasure your voice / when to be one thing / is to be next to nothing?"[81] As a nexus of things, feelings, and memories, voice is ecological; it is neither a property of the flower, nor of the poet-supplicant, nor the garden itself, but of their interdependencies and points of contact.

The posthuman concept of plant voice that is strongly evident in Glück's *The Wild Iris* is neither metaphorical nor contingent on human vocalities. It links speaking to bodily presencing in a place (a garden) over time (the seasons). An ecology of plant voice is relational, where ecology is "thinking how all beings are interconnected, in as deep a way as possible."[82] Plant voice is not of the birch and flowers themselves (as the subjects we might wish them to be), but of the syncretism of their environmental relations (the ecological beings that they already are). This idea of plant voice reflects Richard Rogers's proposal for a "transhuman" model of communication that overcomes the privileging of human symbol construction and symbol deployment.[83] This plant voice also builds upon scholarship in the field of human–animal communication by Emily Plec, who characterizes voice as "intentional energy" exchange between humans and other forms of life. For Plec, an "anthropocentric grip on the symbolic" requires a "corporeal rhetorics of scent, sound, sight, touch, proximity, position, and so much more."[84] A plant rhetorics, beyond the symbolic, is fundamentally ecological, as plants negotiate their environments through their phenomenological gestures (see Part I of this book). Let us now turn to Elisabeth Bletsoe's *Pharmacopœia* to consider further examples of plant rhetorics and their silent, heteroglossic voices.

Plant Heteroglossia in Bletsoe's *Pharmacopœia*

Elisabeth Bletsoe is a British poet, born in 1960 and raised near the town of Wimborne in the district of Dorset, South West England. Her work has been associated with ecopoetics—a movement within contemporary poetry or poetic projects that investigates ecology.[85] Recognizing the primary importance of relationships between beings, ecopoetics questions, resists, and recasts the role of human language in the domination (or appreciation) of nature. Offering a far less idyllic image of landscape than its Romanticist antecedents, ecopoetics radicalizes the nature poetry tradition, bringing ethical and relational concerns to the fore and using the language of science to do so. Jonathan Skinner offers a four-part taxonomy of ecopoetics as *topological* (referring to a topos beyond the poem); *tropological* (hybridizing poetic and scientific languages and imparting an ecosystemic quality to a poem); *entropological* (involving or mirroring ecological materials and processes); and *ethnological* (recognizing human culture within a topos).[86] With respect for nature and in a tone of humility, ecopoetic works tend to reflect principles of ecocentrism and interdependence, while developing critiques of technology and "hyperrationality."[87] Bletsoe's poetry adheres to Skinner's criteria through its focus on the landscapes of South West England, specifically its wild (and weedy) flora. Despite its pronounced ecocentrism, her work is also deeply aware of human traditions of using plants as medicines.

Originally published in 1999, *Pharmacopœia* is a sequence of eleven poems, each titled with the common and scientific names of medicinal plants, as well as their locations: for example, "Stinking Iris *(Iris foetidissima), Kilve."[88] In addition to *Pharmacopœia*, "The Leafy Speaker"[89] from *The Regardians* (1993) demonstrates Bletsoe's particular engagement with the voices of British flora. The works' polyphonies are at once scientific, historical, and embodied. As a poetic interpretation of traditional herbal texts, *Pharmacopœia* exhibits a "double relation" of acting (humans appropriating plants as materials) and being acted upon (plants affecting human bodies by releasing therapeutic compounds or essences).[90] Bletsoe's use of the term *pharmacopoeia,* a book giving directions for the preparation of plants, animals, and minerals for medicine, positions her work within this tradition dating back to the first century C.E. text *De materia medica* of Dioscorides.[91] Central to her ecopoetics, however, is the calling into question of the instrumentalism that reduces flora to medicinal substances. Instead, her poetry widens "our vision of each plant through reference to its multiple names, its places and conditions and its mythologies."[92] Interspersed with voices of medieval herbalists, the botanical voicing in *Pharmacopœia* contrasts to Glück's tacit invocation of myths in the economic diction of *The Wild Iris.* Unlike the latter, Bletsoe's voicing refrains from speaking for plants as personae, instead engendering the poet's dialogue with historical texts and the sensuous presence of plants before her.

The plant voice of Bletsoe's ecopoetics is felt, smelled, tasted, imbibed, eaten, touched, and seen in relation to locales within South West England. The articulation of voice through plant gesture is prominent in *Pharmacopœia* and especially in "Stinking Iris." The olfactory *effluence* of the plant (to echo Thoreau) is denoted by its species name *foetidissima,* from Latin for "stinking." By the poem's end, the iris has undergone a poietic transubstantiation, "growing more grateful & aromatic / as it dries" with the closing lines drawn from the eighteenth-century herbal *Botanicum Officinale.*[93] The visual rhetorics of the iris flower are likened to a form of environmental writing—"a 'pencilling' of purple-gray / blue-gray / on tombs at Carnac"—where the gerundial "pencilling" derives from a historic source. Another poem in the sequence, "Elder *(Sambucus nigra),* Culbone," concerns the species' rich folk knowledge—"by yeast / & mùscatel"—in which its berries have been fermented for medicinal brews. The tactile act of wine making invokes the animist traditions of the elder plant that predate Christianity and the Linnaean taxonomizing of its corporeal language. The berry is transubstantiated to wine, bringing the human and the elder into an embodied double relation "where / vortices / of 5-petalled flowers /

brush lips / skin / hair." The mildly intoxicating concoction becomes a medium of co-voicing between the elder and the poet-herbalist, as evident in the poem's closing line: "we are both now / *forspoken*." This ecology of voice resists a clear boundary between internal and external plant voice—that is, the voice of the plant and the voice given to it, respectively. Through an ecopoetic practice, human speakers of the past and present come into unruly dialogue (broadly conceived and not solely verbal) with Bletsoe's elder of Culbone hamlet.

The theme of botanical folklore is strongly evident in "The Leafy Speaker," a wide-ranging reflection on the archangel Raphael as a healer and patron of writers, the Roman god Mercury, the Green Man (plant) deities, and oak-related mythologies.[94] The poem reinterprets the rich folkloric tradition of the oak (see also Chapter 13). For example, nymphs of oaks were the attendants of the Greek deity Artemis. It was Erysichthon who destroyed the fertility goddess Demeter's oak grove and was therefore cursed to endure insufferable hunger. Roman emperor Theodosius the Great's antipagan campaign involved destroying the oak at the oracle Dodona in 391 C.E.[95] In the poem, these mythological figures and moments coalesce in a presence of a roadside oak and "call to me: / a startling epiphany / though your feet seem solid enough / on the unremarkable pavement."[96] The oak's spatial and historical articulations are likened to language in the following passage:

> branches
>   in rhyming couplets
> stream from your mouth
>   leafily speaking
> the greene man the holy oake
> all that is ancient and mute
>   you give tongue to:
>   *a vision of the world*
>   *before the world*[97]

The leafy speaking of the oak is not in an aural mode of expression. The oak is not merely mouthing history. The "calling" happens as sensation and through the oak's expression of its physical being and mythological relations to "all that is ancient and mute." The oak voice, like Glück's wild iris, is prediscursive and heteroglossic.

If we cannot hear plant voice, then it *must* exist. Bletsoe's ecopoetic practice makes this plausible by giving nuance to the silent voices of plants. It shifts readers away from conventional notions of voice as the mechanical outcome of the vocal chords; as the result of the consciousness of brained organisms; and as a purely aural expression privileging those with vocal capacities. The historicized plant voice in her work resists a mechanistic paradigm in which "the voice arises in roughly the same manner for everyone. Air moving along the chords of the voice box causes vibration like the river wind against a simple reed."[98] What emerges, instead, is the location of the expressions of being—sensory, ecological, historical—that constitute vegetal situatedness. Building on Watts's notion of affect, this is voice as material enunciation. The plant voice of Bletsoe's poetry is the vegetal announcing its being—the uttering of "the [plant] body's sensory experience of its environment and of others."[99] Rather than a barrier to voice, such poetry facilitates our attentiveness to its emergence. Her ecopoetic work decouples voice from pathetic fallacy by showing that plant voice is less ethically fraught (i.e., as speaking in lieu of plants) if relationally and historically grounded; and that voice is always already a heteroglossia and, as an "original impulse of being," resists its own reduction to speaking.

CONCLUSION

The voice of plants (internal to them) and the giving of voice to plants (external to them) are intrinsically related in an interplay between human and vegetal voicing. Nonverbal, ecological voice—the central principle I have been developing in this chapter in response to pathetic fallacy—is manifested through vegetal presence and human recognition of it; through taste, smell, touch, and proprioception. As Michael Marder comments, "plants, like all living beings, articulate themselves spatially; in a body language free from gestures."[100] Contrary to the second part of Marder's assertion, the poetry of Glück and Bletsoe demonstrates that the spatial articulation of plants in languages is replete with gestures. These gestures call us neither to the past nor to the future, but to the present in which all relational possibility inheres. They announce "the [vegetal] body's presence,"[101] not by summoning through sound but by enunciating in the world's substance. As ecological presencing, plant voice is not only the susurration of speaking subjects, bearing vocal chords and enacting their discrete subjectivities. Nature speaks in the entanglement of time, space, spirit, and materiality, but not as literary conventions, communications models, and human paradigms of vocalization would have it.[102] Rather than refusing to engage for fear of ventriloquizing them, the poetry of Glück and Bletsoe reveals the diverse expressions of plant voice—in its silence and potentiality and as a part of our and their being.

Notes

1. Sampson Reed, *Observations on the Growth of the Mind*, 4th ed. (Boston: Otis Clapp, 1841), 51–52.

2. Becky Oskin, "Sound Garden: Can Plants Actually Talk and Hear?," *Livescience,* March 11, 2013, http://www.livescience.com/27802-plants-trees-talk-with-sound.html.

3. Mark A. Dimmitt, "Flowering Plants of the Sonoran Desert," in *A Natural History of the Sonoran Desert,* ed. Steven J. Phillips and Patricia Wentworth Comus (Tucson: Arizona-Sonora Desert Museum Press; Berkeley: University of California Press, 2000), 153–264.

4. Oskin, "Sound Garden," sect. "How to Listen to Plants."

5. Michael Marder, "To Hear Plants Speak," in this volume.

6. Midori-san [Mr. Green] cited in Colin Barras, "The Green Revolution: Plants Move Online," *New Scientist,* October 21, 2008, http://www.newscientist.com/blogs/shortsharpscience/2008/10/the-japanese-have-opened-up.html.

7. Mr. Green paraphrased from "Midori-san or the Slightly Potty Blog," *CORDIS,* October 24, 2008, http://cordis.europa.eu/express/20081024/finally_en.html.

8. John Charles Ryan, "Plant-Art: The Virtual and the Vegetal in Contemporary Performance and Installation Art," *Resilience: A Journal of the Environmental Humanities* 2, no. 3 (2015): 40–57.

9. Michael Marder, *Plant-Thinking: A Philosophy of Vegetal Life* (New York: Columbia University Press, 2013), 186.

10. Rob Faludi, "New York Times on Botanicalls, Again!," *Botanicalls,* April 25, 2013, http://www.botanicalls.com/2013/04/new-york-times-on-botanicalls-again/.

11. Marder, "To Hear Plants Speak," this volume.

12. Ibid.

13. Lois Wardell and Charlotte Rowe, "Cactus Acoustics," 2010, http://arapahost.com/selected_science_projects/phyto-acoustics.

14. Jennifer Peeples and Stephen Depoe, "Introduction: Voice and the Environment—Critical Perspectives," in *Voice and Environmental Communication,* ed. Jennifer Peeples and Stephen Depoe (Houndmills, Basingstoke, UK: Palgrave Macmillan, 2014), 1–17 [9].

15. Martin Heidegger, *On the Way to Language* (San Francisco: Harper & Row, 1982), 63.

16. Eric King Watts, "'Voice' and 'Voicelessness' in Rhetorical Studies," *Quarterly Journal of Speech* 87, no. 2 (2001): 179–96 [179].

17. Louise Glück, *The Wild Iris* (New York: HarperCollins, 1992).

18. Elisabeth Bletsoe, *Pharmacopœia and Early Selected Works* (Exeter: Shearsman Books, 2010).

19. Douglas Harper, "Voice," *Online Etymology Dictionary,* accessed March 11, 2016, http://www.etymonline.com/index.php?term=voice.

20. See Part I of this volume ("Science"); also Anthony Trewavas, *Plant Behaviour and Intelligence* (Oxford: Oxford University Press, 2014).

21. See Monica Gagliano, "Green Symphonies: A Call for Studies on Acoustic Communication in Plants," *Behavioral Ecology* 24, no. 4 (2013): 789–96, doi: 10.1093/beheco/ars206.

22. John Ryan, *Unbraided Lines: Essays in Environmental Thinking and Writing* (Champaign, Ill.: Common Ground Publishing, 2013), chapter 6.

23. Michael McDowell, "The Bakhtinian Road to Ecological Insight," in *The Ecocriticism Reader: Landmarks in Literary Ecology,* ed. Cheryll Glotfelty and Harold Fromm (Athens: University of Georgia Press, 1996), 371–91 [372].

24. John Ruskin, "Of the Pathetic Fallacy," in *The Genius of John Ruskin: Selections from His Writings,* ed. John D. Rosenberg (Charlottesville: University of Virginia Press, 1998), 61–72 [72].

25. Oliver Wendell Holmes, quoted in Ruskin, "Of the Pathetic Fallacy," 64.

26. Ruskin, "Of the Pathetic Fallacy," 64.

27. Oliver Wendell Holmes, *The Poems of Oliver Wendell Holmes* (Boston: Fields, Osgood, 1870), 225–27.

28. Ibid., 227.

29. Jonathan Greenberg, *Modernism, Satire, and the Novel* (Cambridge, UK: Cambridge University Press, 2011), 11–16.

30. McDowell, "The Bakhtinian Road," 373.

31. Marder, *Plant-Thinking.*

32. Greg Garrard, *Ecocriticism* (London: Routledge, 2004), 47.

33. Percy Shelley, "Mont Blanc," quoted in Garrard, *Ecocriticism,* 65.

34. Garrard, *Ecocriticism,* 47.

35. Walter Benjamin, "On Language as Such and on the Language of Man," in *Selected Writings,* vol. 1, *1913–1926,* ed. Marcus Bullock and Michael W. Jennings (Cambridge, Mass.: Belknap Press of Harvard University Press, 1996), 62–74.

36. Jonathan Bate, *The Song of the Earth* (Cambridge, Mass.: Harvard University Press, 2000), 165.

37. See Mary Graham, "Some Thoughts about the Philosophical Underpinnings of Aboriginal Worldviews," *Australian Humanities Review* 45 (2008), http://www.australianhumanitiesreview.org/archive/Issue-November-2008/graham.html; and Deborah Bird Rose, *Nourishing Terrains: Australian Aboriginal Views of Landscape and Wilderness* (Canberra: Australian Heritage Commission, 1996).

38. Bate, *Song of the Earth,* 165.

39. Christopher Manes, "Nature and Silence," in Glotfelty and Fromm, *The Ecocriticism Reader,* 15–29 [25].

40. Michel Serres, *The Natural Contract,* trans. Elizabeth MacArthur and William Paulson (Ann Arbor: University of Michigan Press, 1995), 48.

41. Bate, *Song of the Earth,* 36; italics in original.

42. Lawrence Buell, *The Environmental Imagination: Thoreau, Nature Writing, and the Formation of American Culture* (Cambridge, Mass.: Belknap Press of Harvard University Press, 1995), 152.

43. Henry David Thoreau, *Walden* (London: CRW Publishing, 2004), 119.

44. William Homestead, "The Language That All Things Speak: Thoreau and the Voice of Nature," in Peeples and Depoe, *Voice and Environmental Communication,* 183–204 [197].

45. For example, Donal Carbaugh, "Response Essay: Environmental Voices Including Dialogue with Nature, within and beyond Language," in Peeples and Depoe, *Voice and Environmental Communication,* 241–56; Watts, "'Voice' and 'Voicelessness'"; Eric King Watts, "Coda: Food, Future, Zombies," in Peeples and Depoe, *Voice and Environmental Communication,* 257–63.

46. Peeples and Depoe, "Introduction," 9.

47. Watts, "'Voice' and 'Voicelessness,'" 180.

48. Ibid., 192.

49. Watts, "Coda," 259.

50. Marder, *Plant-Thinking,* 32.

51. Yi-Fu Tuan, quoted in David Tschida, "The Ethics of Listening in the Wilderness Writings of Sigurd F. Olson," in Peeples and Depoe, *Voice and Environmental Communication,* 205–27 [220].

52. Bate, *Song of the Earth,* 35.

53. Laurence D. Cooper, *Rousseau, Nature, and the Problem of the Good Life* (University Park: Pennsylvania State University Press, 1999).

54. Peter Fenves, "The Genesis of Judgment: Spatiality, Analogy, and Metaphor in Benjamin's 'On Language as Such and on Human Language,'" in *Walter Benjamin: Theoretical Questions,* ed. David S. Ferris (Stanford, Calif.: Stanford University Press, 1996), 75–93 [89].

55. Anthony Trewavas, "Mindless Mastery," *Nature* 415 (2002): 841.

56. Yuriko Saito, "Appreciating Nature on Its Own Terms," *Environmental Ethics* 20 (1998): 135–49.

57. Joanne Feit Diehl, "Introduction," in *On Louise Glück: Change What You See,* ed. Joanne Feit Diehl (Ann Arbor: University of Michigan, 2005), 1–22 [1].

58. Elizabeth Dodd, *The Veiled Mirror and the Woman Poet: H. D., Louise Bogan, Elizabeth Bishop, and Louise Glück* (Columbia: University of Missouri Press, 1992), 151.

59. Daniel Morris, *The Poetry of Louise Glück: A Thematic Introduction* (Columbia: University of Missouri Press, 2006), 200.

60. Ibid., 191.

61. Maggie Gordon, "A Woman Writing about Nature: Louise Glück and 'the Absence of Intention,'" in *Ecopoetry: A Critical Introduction,* ed. J. Scott Bryson (Salt Lake City: University of Utah Press, 2002), 221–31 [228].

62. Helen Vendler, quoted in Glück, *The Wild Iris,* back cover.

63. Morris, *Poetry of Louise Glück,* 221, 230.

64. Linda Gregerson, "The Sower against Gardens," *Kenyon Review* 23, no. 1 (2001): 115–33 [117].

65. Watts, "Coda," 259.

66. Gordon, "A Woman," 222.

67. Glück, *The Wild Iris,* 1.

68. Morris, *Poetry of Louise Glück,* 200.

69. Glück, *The Wild Iris,* p. 1, lines 3–4.

70. Ibid., lines 9–10.

71. Peter Barlow, "Charles Darwin and the Plant Root Apex: Closing a Gap in Living Systems Theory as Applied to Plants," in *Communication in Plants: Neuronal Aspects of Plant Life,* ed. František Baluška, Stefano Mancuso and Dieter Volkmann (Berlin: Springer-Verlag, 2006), 37–61.

72. Glück, *The Wild Iris,* p. 1, line 5.

73. Ibid., lines 16–23.

74. Watts, "'Voice' and 'Voicelessness'," 180.

75. Glück, *The Wild Iris,* p. 24, lines 10–11.

76. Ibid., lines 13–15.

77. Ibid., p. 2, lines 10–13.

78. Ibid., p. 13, lines 15–17.

79. Ibid., p. 47, lines 12–18.

80. Ibid., p. 14, lines 1–3.

81. Ibid., lines 3–6.

82. Tim Morton, "The Dark Ecology of Elegy," in *The Oxford Handbook of the Elegy,* ed. Karen Weisman (Oxford: Oxford University Press, 2010), 251–71 [255].

83. Richard Rogers, "Overcoming the Objectification of Nature in Constitutive Theories: Toward a Transhuman, Materialist Theory of Communication," *Western Journal of Communication* 62, no. 3 (1998): 244–72.

84. Emily Plec, "Perspectives on Human–Animal Communication," in *Perspectives on Human–Animal Communication: Internatural Communication,* ed. Emily Plec (London: Routledge, 2013), 1–13 [7].

85. Tom Bristow, "Ecopoetics," in *The Facts on File Companion to World Poetry, 1900 to the Present,* ed. R. Victoria Arana (New York: Infobase Publishing, 2008), 156–59.

86. Nerys Williams, *Contemporary Poetry* (Edinburgh: Edinburgh University Press, 2011), 158.

87. J. Scott Bryson, "Introduction," in Bryson, *Ecopoetry,* 5–6.

88. Bletsoe, *Pharmacopœia,* 101.

89. Ibid., 75–78.

90. David Nowell Smith, "Scaping the Land: The New British Pastoral," *Chicago Review* 57 (2012): 182–93 [189].

91. Dioscorides Pedanius (of Anazarbos.), *De materia medica,* trans. Lily Y. Beck (Hildesheim: Olms-Weidmann, 2005).

92. Harriet Tarlo, "Introduction," in *The Ground Aslant: An Anthology of Radical Landscape Poetry,* ed. Harriet Tarlo (Exeter: Shearsman Books, 2011), 7–18 [14].

93. Joseph Miller, *Botanicum Officinale; or, A Compendious Herbal* (London: Printed for E. Bell, J. Senex, W. Taylor, and J. Osborn, 1722), 13.

94. Bletsoe, *Pharmacopœia,* 97.

95. Peter Young, *Oak* (London: Reaktion Books, 2013), 106.

96. Bletsoe, *Pharmacopœia,* 75.

97. Ibid., 77; italics in original.

98. David Appelbaum, *Voice* (Albany: State University of New York, 1990).

99. Watts, "'Voice' and 'Voicelessness,'" 180.

100. Marder, *Plant-Thinking,* 75.

101. Watts, "'Voice' and 'Voicelessness,'" 180.

102. Tschida, "Ethics of Listening," 206.

Acknowledgments

THIS BOOK WOULD NOT HAVE BEEN POSSIBLE WITHOUT THE GENEROUS support of several academic institutions. We are grateful to the Department of Spanish and Portuguese, the Comparative Literature Program, and the Graduate School of Georgetown University; the Australian Research Council Discovery Early Career Researcher Award and the Early Career Fellowship Support Programs at the University of Western Australia; and the CREATEC Postdoctoral Research Fellowship Program at Edith Cowan University.

We thank Douglas Armato, director of the University of Minnesota Press, for his enthusiastic support of the project from the very first steps of the publication process onward, and Erin Warholm-Wohlenhaus, editorial assistant, for her continued engagement and help in the various stages of preparation of the manuscript for publication.

Contributors

JONI ADAMSON is professor of environmental humanities in the Department of English and senior sustainability scholar at the Julie Ann Wrigley Global Institute of Sustainability at Arizona State University. She is the author or coeditor of seven books and critical editions that helped to establish and expand the environmental humanities and environmental studies. She is a principal investigator for "Humanities for the Environment" (HfE), an international networking project for the environmental humanities funded by Andrew W. Mellon, and leads the 2.0 development team for the HfE international website, http://hfe-observatories.org.

NANCY E. BAKER recently retired from forty years of teaching philosophy at Sarah Lawrence College. She taught Wittgenstein's *Philosophical Investigations* and a course on language and religious experience. She has contributed to *Feminist Interpretations of Wittgenstein* and *Language, Ethics, and Animal Life: Wittgenstein and Beyond*, and is working on a collection of essays about Wittgenstein and nonduality.

MONICA GAGLIANO is a former research fellow of the Australian Research Council and currently a research associate professor of evolutionary ecology and adjunct senior research fellow at the University of Western Australia. She is author of numerous scientific articles in the fields of animal and plant behavioral and evolutionary ecology and is coeditor of *The Green Thread: Dialogues with the Vegetal World*. She has pioneered the new research field of plant bioacoustics and recently extended the concept of cognition to plants, reigniting the discourse on plant subjectivity and ethical standing. For more information, visit http://www.monicagagliano.com.

KAREN L. F. HOULE is professor of philosophy at the University of Guelph in Canada. Her studies in posthumanist thought are informed by her background in science, her work with the Canadian Community of Practice in Ecosystem Approaches to Health, and her capacity as a poet (*During,* 2008; *Ballast,* 2000). She is author of *Responsibility, Complexity, and Abortion: Toward a New Image of Ethical Thought* and coeditor, with Jim Vernon, of *Hegel and Deleuze: Together Again for the First Time.*

LUCE IRIGARAY is director of research in philosophy at the Centre National de la Recherche Scientifique, Paris. A doctor of philosophy, she is also trained in linguistics, philology, psychology, and psychoanalysis. Through more than thirty books, translated into various languages, her thought develops in a range of literary forms, from the philosophical to the scientific, the political, and the poetic.

ERIN JAMES is assistant professor of English at the University of Idaho, where she directs the MA program in English. She is author of *The Storyworld Accord: Econarratology and Postcolonial Narratives.*

RICHARD KARBAN is professor of entomology at the University of California at Davis. He is interested in how plants communicate and how they defend themselves against herbivores. He has authored more than 140 scientific articles and two key books, *Induced Plant Responses to Herbivory* (with Ian Baldwin) and *Plant Sensing and Communication.*

ANDRÉ KESSLER is professor of chemical ecology at Cornell University, where he studies the chemical and molecular mechanisms, ecological consequences, and evolutionary aspects of plant-induced responses to herbivory. As a specific focus his research tries to understand how plant chemistry functions as a vehicle of information transfer between organisms and so affects ecological and evolutionary processes.

ISABEL KRANZ is Fritz Thyssen Research Fellow at the Institute for German Literature at the University of Vienna, working on a project titled "Literary Botany: Plants as Figures of Knowledge, 1700 to 2000." Her research focuses on the interrelations between literature and the history of knowledge, media, and historiography, as well as on the work of Walter Benjamin. Her publications include *Sprechende Blumen: Ein ABC der Pflanzensprache* and *Floriographie: Die Sprachen der Blumen,* edited by Isabel Kranz, Alexander Schwan, and Eike Wittrock.

MICHAEL MARDER is Ikerbasque Research Professor of Philosophy at the University of the Basque Country (UPV-EHU), Spain, and Professor-at-Large at the Humanities Institute of Diego Portales University (UDP), Chile. He is author of numerous articles and books in the fields of environmental philosophy, phenomenology, and political thought. His most recent monographs include *Pyropolitics: When the World Is Ablaze; Dust; Through Vegetal Being,* coauthored with Luce Irigaray; and *Energy Dreams: Of Actuality.*

TIMOTHY MORTON is Rita Shea Guffey Chair in English at Rice University. He is author of numerous books, including *Hyperobjects: Philosophy and Ecology after the End of the World; Dark Ecology: For a Logic of Future Coexistence; Nothing: Three Inquiries in Buddhism; Realist Magic: Objects, Ontology, Causality; The Ecological Thought;* and *Ecology without Nature,* as well as more than 150 essays on philosophy, ecology, literature, music, art, architecture, design, and food. He blogs regularly at http://www.ecologywithoutnature.blogspot.com.

CHRISTIAN NANSEN is assistant professor in insect ecology and remote sensing in the Department of Entomology and Nematology at the University of California at Davis. Much of his research is devoted to teaching and researching reflectance profiling of crop plants in response to biotic and abiotic stressors, with the goal of developing ways to use remotely sensed reflectance profiling of crop plants to improve agricultural management practices.

ROBERT A. RAGUSO is professor in the Department of Neurobiology and Behavior at Cornell University. He studies chemical communication in a plant–pollinator context, with emphases on context-dependent interactions and synergism with other sensory modalities. He has helped pioneer the interdisciplinary study of floral scent by establishing a Gordon Research Conference on that theme, by authoring more than one hundred scientific papers, including several influential reviews on the subject, and by training numerous students and colleagues on gas chromatography-mass spectrometry (GC-MS) in his laboratory.

JOHN C. RYAN is postdoctoral research fellow in the School of Arts at the University of New England in Australia and honorary research fellow in the Department of English and Cultural Studies at the University of Western Australia. He is the author of the books *Green Sense*; *Two with Nature* (with Ellen Hickman); *Unbraided Lines*; *Digital Arts* (with Cat Hope); *Being With*; and *Posthuman Plants*.

CATRIONA SANDILANDS is professor in the Faculty of Environmental Studies at York University in Toronto, where she teaches and writes at the intersections of environmental literature and philosophy, queer and feminist studies, and ecological politics. The author of more than sixty published works on topics ranging from national parks to lesbian communities, Walter Benjamin to honeybees, melancholic natures to monstrous plants, she has written *The Good-Natured Feminist: Ecofeminism and the Quest for Democracy* (Minnesota, 1999) and coedited (with Bruce Erickson) *Queer Ecologies: Sex, Nature, Politics, Desire* and the forthcoming *Green Words / Green Worlds: Writing, Environment, and Politics*.

PATRÍCIA VIEIRA is associate professor of Spanish and Portuguese, comparative literature, and film and media studies at Georgetown University and associate research professor at the Center for Social Studies (CES) of the University of Coimbra. She is the author of *Seeing Politics Otherwise: Vision in Latin American and Iberian Fiction* and *Portuguese Film, 1930–1960: The Staging of the New State Regime*, and coeditor of *Existential Utopia: New Perspectives on Utopian Thought*; *Found Images: Documentary, Politics, and Social Change in Portugal*; and *The Green Thread: Dialogues with the Vegetal World*. For more information, visit www.patriciavieira.net.

Index

abstraction, 91, 95, 119

Acuña, Cristóbal de, 226, 232n27

adaptation, process of, viii, 45, 49–50, 92, 273

aesthetics, ix, xiv, xvi, 158, 178–81, 186, 202, 223, 225

agency, xiii, xvi, xxvi, 28, 95, 174, 188, 218, 231n13, 235, 254–56, 258–60, 264, 266, 269–70, 274, 277, 279

agriculture: crops, 47–48, 66, 74, 76–80, 187, 242, 246–47, 268; fertilizer, 70, 76–79; pesticides, 76–78, 155; politics, 244; "push–pull" technology, 47–48

algorithm, xxv, 103, 177, 275

Algorithmic Beauty of Plants, The, 177

Allee effect, 249

allegory, xviii, xxvi, 109, 212n37, 212n40, 235

allelopathy, 248

Allston, Washington, x–xi, xvii

Amazon (rainforest), xxvi, 159, 166, 218, 225–30

Amazonian literature, 226–30

anthropocentrism, xi, 87, 89, 104, 113–14, 123, 139, 145, 148, 152n3, 174–76, 179–80, 183, 188, 245, 259, 261–64, 269, 279–80, 288

anthropomorphism, xviii, 112, 116, 136, 218, 232n22, 238, 240, 256

arche-writing, xxvi, 223–25

Aristotle, ix, 110, 120, 168, 169n4

articulation, xvii–xix, xxiv, xxvii, 104–5, 119–23, 225, 256, 275, 283, 285, 289–91

Attis, 107

Augustine, St., 104, 106–7, 109, 121, 123n6, 123n8, 123n9, 125n44; *The City of God against the Pagans,* 106–7; *Confessions,* 106

Austin, J. L., 130, 160; *How to Do Things with Words,* 130, 160

autoimmunity, 185

awareness, xv, 96, 176–78, 183, 279, 281, 285; ecological, 176, 248; self-, 11, 185

Backster, Cleve, 215

bacteria, 6–7, 15–16, 28, 31, 88, 139, 172n19, 180

Bataille, Georges, 110–11, 206

Bates, Henry Walter, 226

Bateson, Gregory, 241

Baudelaire, Charles, 181

behavior, xi–xiii, xv–xvi, xxiii, xxiv, xxxn15, 3, 10–11, 16–17, 29, 33–36, 38, 44, 46, 48, 50, 84, 86–87, 89–95, 113, 117, 127–28, 134, 136–39, 141–52, 152n3, 218, 241, 245, 248, 252n40, 254

Benjamin, Walter: "On Language as such and on the Language of Man," xviii, xxvi, 104, 120, 122, 220–23, 232nn19–22

Bergson, Henri, 144, 185

12, 28, 42, 45, 47, 117; processing and use of, xii, 7, 11, 36–37, 39, 63, 89, 137; representation of, 68–69, 72, 201; transmission and sharing of, xv, 9, 12, 14–17, 29, 37–40, 45–48, 93, 104, 116–18, 195, 198, 242

inscription, xxvi, 115, 119, 217–30, 231n13

instrumentalization, 279; of plants, xiv–xv, 114, 235, 237, 239, 243–44, 289

intelligence, viii, x, xvi, xxiv–xxv, 11, 95, 114, 118, 150; artificial, 184; collective, 237, 239, 241, 247, 249; distributed, 235, 247; foreign, 178–79, 181; human, 104, 188, 262; plant, 95, 117–18, 136–52, 177–78, 235, 239–40, 244, 249, 254–55, 257, 259–60, 262, 269–70, 277; of triffids, 237

intentionality, ix, xvi, 121, 130, 136, 141, 147, 162, 179, 217, 238; and language, 288

interaction, xiii, xviii, xxiii, 11, 15, 27, 29, 32–33, 39–49, 89, 95, 103, 225, 277; antagonistic and competitive, 27–28, 30, 39, 93; context-dependent, 29, 92, 94; ecological, xviii, 31, 96; human–plant, xvi, 215, 218, 260; mutualistic and cooperative, 27–28, 31–32, 39; outcome of, 28, 276; plant, xxii, 15, 27, 36, 38–39, 45, 47–48, 90, 109, 115, 218, 274; plant–herbivore, 28–30; plant–pollinator, 29, 32–34, 42–43, 46, 91; of processes and systems, 74, 84, 92; sender–receiver, xxiii, 27

interconnectedness, xi, xiii, xx, 122, 241, 287

interdependence, xviii, xx, xxvii, 276, 287–88

interdisciplinarity, xv, xxiii, xxviiin2, 118. *See also* transdisciplinarity

Jeffers, Robinson, xx

Jonas, Hans, 144

Jörmungandr, 173, 186

Joyce, James, 104, 110–11; *Ulysses,* 111

justice, xxv, 123, 169nn1–3, 170n5; distributive, 156–58, 159, 167; and injustice, 160–61; linguistic, 158, 165; procedural, 158

Kant, Immanuel: *The Critique of Pure Reason,* 183; Kantian, 109, 173, 178–83, 187

Keats, John, 186

kin recognition, 37

kinship, 11, 93, 95, 142, 145–46, 149, 238–39, 280

kitsch, 186, 190n32

knowledge, xvii, xxii, 94, 136, 138, 216; botanical, 194, 205–6, 276; folk and popular, xv, 207, 276, 289; foundation of, xxi, 153; human, 64, 104, 109, 115, 232; plant, 93, 112, 195; production, acquisition, and use of, 64, 93, 194; scientific, 63, 67, 74, 205, 217

Krampen, Martin, xix. *See also* phytosemiotics

Lacan, Jacques, 177, 271–72n33

language, ix, xii, xv, 168; of emotion, 203; extrinsic and intrinsic, xvii–xviii; inadequacy of, xii; and justice, 158, 160–61; materiality of, 266–67; as mode of action, 160; as performativity, 165; of things,